SCHAUM'S OUTLINE OF

THEORY AND PROBLEMS

OF

COMBINATORICS

•

V. K. BALAKRISHNAN, Ph.D.
Professor of Mathematics
University of Maine
Orono, Maine

SCHAUM'S OUTLINE SERIES
McGRAW-HILL, INC.

New York St. Louis San Francisco Auckland Bogotá Caracas
Lisbon London Madrid Mexico City Milan Montreal
New Delhi San Juan Singapore
Sydney Tokyo Toronto

V.K. BALAKRISHNAN is a professor of mathematics at the University of Maine, where he coordinates an interdepartmental program on operations research. He has an honors degree in mathematics from the University of Madras, a master's degree in pure mathematics from the University of Wisconsin at Madison, and a doctorate degree in applied mathematics from the State University of New York at Stony Brook. He is a Fellow of the Institute of Combinatorics and its Applications and a member of the American Mathematical Society, Mathematical Association of America, and the Society for Industrial and Applied Mathematics. He is the author of *Introductory Discrete Mathematics* (1991) and *Network Optimization* (1995).

This book is printed on recycled paper containing a minimum of 50% total recycled fiber with 10% postconsumer de-inked fiber. Soybean based inks are used on the cover and text.

Schaum's Outline of Theory and Problems of
COMBINATORICS

1 2 3 4 5 6 7 8 9 10 BAW BAW 9 8 7 6 5 4

ISBN 0-07-003575-X

Sponsoring Editor: Arthur Biderman
Production Supervisor: Paula Keller
Editing Supervisor: Patty Andrews

Library of Congress Cataloging-in-Publication Data

Balakrishnan, V. K., date–
Schaum's outline of theory and problems of combinatorics / V.K. Balakrishnan.
p. cm.--(Schaum's outline series)
Includes index.
ISBN 0-07-003575-X
1. Combinatorial analysis--Outlines, syllabi, etc. I. Title.
QA164.B35 1995
511′.6′076--dc20 94-26728
CIP

Dedicated to

Paul Erdös

who as the founder of modern combinatorics has been posing problems, coining conjectures and tackling theorems in number theory, graph theory and combinatorics besides showing the world the way to count the number of ways in more than one way for more than half a century.

Say mathematician, how many are the combinations in one composition with ingredients of six different tastes—sweet, pungent, astringent, sour, salt and bitter—taking them by ones, twos, or threes, etc.?

[From *Lilavathi* of Bhaskara (the great twelfth century mathematician of India) as quoted in N.L. Biggs: The Roots of Combinatorics, Historia Mathematica 6 (1979), 109–116.]

Preface

At an introductory level, combinatorics is usually considered as a branch of discrete mathematics in which the main problem is that of counting the number of ways of arranging or choosing objects from a finite set according to some simple specified rules. Thus the crux of the problem, at the beginning stage at least, is mainly that of enumeration. But if the prescribed rules and constraints become complicated the question to ask naturally is whether an arrangement satisfying the given requirements exists in the first place; if so, in the subsequent analysis one investigates the methods of constructing such arrangements. In some cases these arrangements also have to meet certain optimality criteria, in which case we seek an optimal solution of the problem. A typical statement in some of these optimal situations will assert that the minimum for one kind of a selection will correspond to the maximum for another kind, yielding a "max-min theorem." Thus in a wider sense, combinatorics deals with the enumeration, existence, analysis, and optimization of discrete structures.

Combinatorial mathematics has a variety of applications. It is utilized in several physical and social sciences, for example, chemistry, computer science, operations research, and statistics. Consequently, there has recently been a rapid growth in the depth and breadth of the field of combinatorics. The subject is becoming an increasingly important component of the curriculum both at the graduate and undergraduate levels at universities and colleges in the United States and abroad.

In this book I have attempted to present the important concepts of contemporary combinatorics in a sequence of four chapters. I hope that students will find this book useful for a course in combinatorics or discrete mathematics either as the main text or as a supplementary text. It is designed for students with a wide range of maturity and can also serve as a useful and convenient reference book for many professionals in industry, research, and academe.

In each chapter the basic ideas are developed in the first few pages by giving definitions and statements of theorems to familiarize the reader with concepts that will be fully exploited in the selection of solved problems that follow the text. These problems are grouped by topic and are presented in increasing order of maturity and sophistication. A beginning student may therefore stop at any point and proceed to the next chapter without losing the continuity of the development of the material. The collection of solved problems is the unifying feature of the book.

Unlike other branches of mathematics, in combinatorics the solutions of problems play a special role because in many instances a problem may need an *ad hoc* argument based on some kind of special insight; that is, it may not be possible to solve it by applying results of known theorems alone. I present a variety of problems covering various branches of the subject. Students are encouraged to try to solve a problem without looking at the solution. The thrill is in solving the problem independently, and the reward is invariably heightened if the student can solve the problem by a different (and possibly more elegant) method. I have used these problems as assignments and projects in my courses on combinatorics and discrete mathematics during the past few years, and the contributions and encouragements of my students—too numerous to mention individually—are gratefully acknowledged.

In writing a book like this, I have benefitted enormously from the contributions of other mathematicians and scientists in the field. Since this book is meant to provide basic theory and solved problems, I have provided a short list of books for further study for the discriminating student, instead of an exhaustive bibliography. Let me take this opportunity to express my deep sense of gratitude to some outstanding mathematicians who have made significant contributions in modern combinatorics. I have come into contact with them over the past two decades either at national and international conferences or by private correspondence. They include K. Bogart, R. Brualdi, V. Chvátal, J. Conway, R.P. Dilworth, P. Erdös, M. Gardner, R. Graham, M. Hall, Jr., F. Harary, P. Hilton, A.J. Hoffman, V. Klee, D. Kleitman, D. Knuth, E. Lawler, G. Polya, A. Ralston, F. Roberts, G.C. Rota, H. Ryser, E. Snapper, R. Stanley, R. Stanton, W. Trotter, A. Tucker, and H. Wilf.

I wish to thank my colleagues and friends at the University of Maine for giving me the facilities and encouragement for writing this book. In particular I am grateful to Ali Ozluk and Frank Curtis for invaluable suggestions and critical review of the manuscript. If I have not given proper credit in this book to anyone who deserves recognition for specific results, I apologize for the omission, and I will make every effort to include such acknowledgment in subsequent editions. Likewise, it is possible that there are errors and misprints. I accept complete and total responsibility for them. If they are brought to my attention they too will be rectified. Any feedback in this regard from the reader is welcome.

In conclusion I would like to express my immense gratitude to the editorial and production staff at McGraw-Hill, particularly to my sponsoring editor Arthur Biderman for his unfailing cooperation and encouragement and to my editing supervisor Patricia Andrews for the indefatigable assistance she gave me during the final stage of the production of this book. The manuscript was completely and thoroughly edited by Dr. David Beckwith, who was my initial sponsoring editor. After retiring from McGraw-Hill he undertook to complete the editing of the manuscript. He showed his special love of combinatorics by combing every single detail in the manuscript. For this admirable and enviable professionalism I am truly beholden to him.

V.K. Balakrishnan
University of Maine

Contents

Chapter 1

Basic Tools

1.1 THE SUM RULE AND THE PRODUCT RULE

How many arrangements of a specified kind can be undergone by a given set of objects? Or, in how many ways can a prescribed event occur? *Combinatorics* is the branch of mathematics that seeks to answer such questions without enumerating all possible cases; it depends on two elementary rules.

DISJUNCTIVE OR SUM RULE If an event can occur in m ways and another event can occur in n ways, and if these two events cannot occur simultaneously, then one of the two events can occur in $m + n$ ways. More generally, if E_i $(i = 1, 2, \ldots, k)$ are k events such that no two of them can occur at the same time, and if E_i can occur in n_i ways, then one of the k events can occur in $n_1 + n_2 + \cdots + n_k$ ways.

Example 1. If there are 18 boys and 12 girls in a class, there are $18 + 12 = 30$ ways of selecting 1 student (either a boy or a girl) as class representative.

Example 2. Suppose E is the event of selecting a prime number less than 10 and F is the event of selecting an even number less than 10. Then E can happen in 4 ways, and F can happen in 4 ways. But, because 2 is an even prime, E or F can happen in only $4 + 4 - 1 = 7$ ways.

SEQUENTIAL OR PRODUCT RULE If an event can occur in m ways and a second event can occur in n ways, and if the number of ways the second event occurs does not depend upon how the first event occurs, then the two events can occur simultaneously in mn ways. More generally, if E_i $(i = 1, 2, \ldots, k)$ are k events and if E_1 can occur in n_1 ways, E_2 can occur in n_2 ways (no matter how E_1 occurs), E_3 can occur in n_3 ways (no matter how E_1 and E_2 occur), $\ldots$, E_k can occur in n_k ways (no matter how the previous $k - 1$ events occur), then the k events can occur simultaneously in $n_1 n_2 n_3 \cdots n_k$ ways.

Example 3. A bookshelf holds 6 different English books, 8 different French books, and 10 different German books. There are (*i*) $(6)(8)(10) = 480$ ways of selecting 3 books, 1 in each language; (*ii*) $6 + 8 + 10 = 24$ ways of selecting 1 book in any one of the languages.

Example 4. The scenario is as in Example 3. An English book and a French book can be selected in $(6)(8) = 48$ ways; an English book and a German book, in $(6)(10) = 60$ ways; a French book and a German book, in $(8)(10) = 80$ ways. Thus there are $48 + 60 + 80 = 188$ ways of selecting 2 books in 2 languages.

Example 5. If each of the 8 questions in a multiple-choice examination has 3 answers (1 correct and 2 wrong), the number of ways of answering all questions is $3^8 = 6561$.

1.2 PERMUTATIONS AND COMBINATIONS

Suppose X is a collection of n distinct objects and r is a nonnegative integer less than or equal to n. An **r-permutation of X** is a selection of r out of the n objects. Selections are *ordered*; thus, for example, 2 5 4 and 2 4 5 are different 3-permutations of $X = \{1, 2, 3, 4, 5\}$. An n-permutation of X is called simply a **permutation of X**.

The number of r-permutations of a collection of n distinct objects is denoted by $\boldsymbol{P(n, r)}$; this number is evaluated as follows. A member of X can be chosen to occupy the first of the r positions in n ways. After that, an object from the remaining collection of $n - 1$ objects can be chosen to occupy the second position in $n - 1$ ways. Notice that the number of ways of placing the second object does not depend upon how the first object was chosen or placed. Thus, by the product rule, the first two positions can be filled in $n(n-1)$ ways ... and all r positions can be filled in

$$P(n, r) = n(n-1)\cdots(n-r+1) = \frac{n!}{(n-r)!}$$

ways. Here we have introduced the **factorial function**, $m! \equiv (1)(2)\cdots(m)$ and $0! \equiv 1$. In particular, the number of permutations of n objects is $P(n, n) = n!$.

Example 6. There are $P(6, 6) = 6! = 720$ 6-letter "words" that can be made from the letters of the word NUMBER, and there are $P(6, 4) = 6!/2! = 360$ 4-letter "words."

An *unordered* selection of r out of the n elements of X is called an ***r*-combination of *X***. In other words, any subset of X with r elements is an r-combination of X. The number of r-combinations or r-subsets of a set of n distinct objects is denoted by $\mathbf{C(n, r)}$ ("*n choose r*"). For each r-subset of X there is a unique complementary $(n-r)$-subset, whence the important relation $C(n, r) = C(n, n-r)$.

To evaluate $C(n, r)$, note that an *r-permutation* of an n-set X is necessarily a *permutation* of some r-subset of X. Moreover, distinct r-subsets generate distinct r-permutations. Hence, by the sum rule,

$$P(n, r) = P(r, r) + P(r, r) + \cdots + P(r, r)$$

The number of terms on the right is the number of r-subsets of X; i.e. $C(n, r)$. Thus $P(n, r) = C(n, r)P(r, r) = C(n, r)\, r!$. The following theorem summarizes our results.

Theorem 1.1.

(*i*) $P(n, r) = \dfrac{n!}{(n-r)!}$

(*ii*) $C(n, r) = \dfrac{P(n, r)}{r!} = \dfrac{n!}{r!\,(n-r)!} = C(n, n-r)$

Example 7. From a class consisting of 12 computer science majors, 10 mathematics majors, and 9 statistics majors, a committee of 4 computer science majors, 4 mathematics majors, and 3 statistics majors is to be formed. There are

$$C(12, 4) = \frac{12!}{4!8!} = \frac{12\cdot 11\cdot 10\cdot 9}{4\cdot 3\cdot 2\cdot 1} = 11\cdot 5\cdot 9 = 495$$

ways of choosing 4 computer science majors, $C(10, 4) = 210$ ways of choosing 4 mathematics majors, and $C(9, 3) = 84$ ways of choosing 3 statistics majors. By the product rule, the number of ways of forming a committee is thus $(495)(210)(84) = 8{,}731{,}800$.

Example 8. Refer to Example 7. In how many ways can a committee consisting of 6 or 9 members be formed such that all 3 majors are equally represented?

A committee of 6 (with 2 from each group) can be formed in $C(12, 2)\cdot C(10, 2)\cdot C(9, 2) = 106{,}920$ ways. The number of ways of forming a committee of 9 (with 3 from each group) is $C(12, 3)\cdot C(10, 3)\cdot C(9, 3) = 2{,}217{,}600$. Then, by the sum rule, the number of ways of forming a committee is $106{,}920 + 2{,}217{,}600 = 2{,}324{,}520$.

1.3 THE PIGEONHOLE PRINCIPLE

Some of the most profound and complicated results in modern combinatorial theory flow from a very simple proposition: **If *n* pigeonholes shelter *n* + 1 or more pigeons, at least 1 pigeonhole shelters at least 2 pigeons.**

Example 9. To ensure that a class includes at least 2 students whose last names begin with the same letter of the (English) alphabet, the class should have at least 27 students. (Here the letters are the pigeonholes.)

Example 10. Suppose there are many red socks, many white socks, and many blue socks in a box. What is the least number of socks that one should grab from the box (without looking at the contents) to be sure of getting a matching pair?

If each color is considered as a pigeonhole, then $n = 3$. Therefore, if one grabs $n + 1 = 4$ pigeons (socks), at least 2 of them will share a color.

A straightforward generalization of the pigeonhole principle is as follows: **If n pigeonholes shelter $kn + 1$ pigeons, where k is a positive integer, at least 1 pigeonhole shelters at least $k + 1$ pigeons.**

Example 11. Rework Example 10 if 3 pairs, all of one color, are desired.

There are still $n = 3$ pigeonholes, and we want to ensure that one (or more) of them shall contain $k + 1 = 6$ (or more) pigeons. Thus we grab $kn + 1 = (5)(3) + 1 = 16$ pigeons.

Example 12. A chest contains 20 shirts, of which 4 are tan, 7 are white, and 9 are blue. At the least, how many shirts must one remove (blindfolded) to get $r = 4, 5, 6, 7, 8, 9$ shirts of the same color?

Case 1. $r = 4 = k + 1$. So $k = 3$, and since there are 3 colors, $n = 3$. Thus, at least $kn + 1 = 10$ shirts must be removed.

Case 2. $r = 5 = k + 1$. Here the analysis is simplest if we imagine that shirts are drawn from the chest sequentially. In a longest chain—which is what we are looking for—the first 4 draws are "wasted" in removing the $4 < r$ tan shirts, and the remainder of the sequence consists of as many draws of white and blue shirts ($n = 2$) as are required to ensure $r = 5$ shirts of the same color. But this latter number is given by the pigeonhole principle as $kn + 1 = 4(2) + 1 = 9$. Thus, $4 + 9 = 13$ shirts must be removed.

Case 3. $r = 6 = k + 1$. The situation is like that of Case 2; so $4 + (kn + 1) = 4 + [5(2) + 1] = 15$ shirts must be removed.

Case 4. $r = 7 = k + 1$. As in Cases 2 and 3, $4 + (kn + 1) = 4 + [6(2) + 1] = 17$ shirts.

Case 5. $r = 8 = k + 1$. Now both the tan and the white shirts are worthless, so that $4 + 7 + (kn + 1) = 4 + 7 + [7(1) + 1] = 19$ shirts must be removed.

Case 6. $r = 9 = k + 1$. Like Case 5; $4 + 7 + (kn + 1) = 4 + 7 + [8(1) + 1] = 20$ shirts.

The conclusions of Example 12 can be summarized as a theorem.

Theorem 1.2. Let S be a set composed of x_1 objects of type 1, $x_2 \geq x_1$ objects of type 2, $x_3 \geq x_2$ objects of type 3, ..., $x_n \geq x_{n-1}$ objects of type n. Let ν_r denote the smallest integer with the property that *every* subset of S of size ν_r contains r or more objects of the same type. Then,

$$\nu_r = \begin{cases} n(r-1)+1 & r \leq x_1 \\ (n-1)(r-1)+1+x_1 & x_1 < r \leq x_2 \\ (n-2)(r-1)+1+x_1+x_2 & x_2 < r \leq x_3 \\ \cdots\cdots\cdots\cdots\cdots\cdots & \\ (1)(r-1)+1+x_1+x_2+\cdots+x_{n-1} & x_{n-1} < r \leq x_n \end{cases}$$

The following theorem is an extension of the pigeonhole principle.

Theorem 1.3. If there are $p_1 + p_2 + \cdots + p_n - n + 1$ or more pigeons among hole 1, hole 2, ..., and hole n, then, for some j, hole j has p_j or more pigeons.

Proof. Suppose the conclusion of the theorem is false. Then, hole 1 has at most $p_1 - 1$ pigeons and hole 2 has at most $p_2 - 1$ pigeons and ..., and hole n has at most $p_n - 1$ pigeons. So the total number of pigeons is at most $p_1 + p_2 + \cdots + p_n - n$, resulting in a contradiction.

Example 13. If an undergraduate club is to include at least 6 freshmen or at least 5 sophomores or at least 4 juniors or at least 3 seniors, no more than $6 + 5 + 4 + 3 - 4 + 1 = 15$ students need be invited to join.

Definition: If x is a real variable, the **floor of x**, denoted $\lfloor x \rfloor$, is the greatest integer less than or equal to x.

Theorem 1.4. Given m pigeons within n pigeonholes, at least 1 hole contains at least $p + 1$ pigeons, where $p = \lfloor (m - 1)/n \rfloor$.

Proof. If, on the contrary, every hole contained at most p pigeons,

$$\text{No. of pigeons} \leq np \leq n\left(\frac{m-1}{n}\right) = m - 1 < m$$

Example 14. Suppose there are 26 students ($m = 26$) and 7 cars ($n = 7$) to transport them. Then $p = \lfloor 25/7 \rfloor = 3$, so that at least 1 car must have 4 or more passengers.

Solved Problems

THE SUM AND PRODUCT RULES

1.1 There are 15 married couples in a party. Find the number of ways of choosing a woman and a man from the party such that the two are (*a*) married to each other, (*b*) not married to each other.

(*a*) A woman can be chosen in 15 ways. Once a woman is chosen, her husband is automatically chosen. So the number of ways of choosing a married couple is 15. (*b*) A woman can be chosen in 15 ways. Among the 15 men in the party, one is her husband. Out of the 14 other men, one can be chosen in 14 ways. The product rule then gives $(15)(14) = 210$ ways.

1.2 Find the number of (*a*) 2-digit even numbers, (*b*) 2-digit odd numbers, (*c*) 2-digit odd numbers with distinct digits, and (*d*) 2-digit even numbers with distinct digits.

Let E be the event of choosing a digit for the units' position, and F be the event choosing a digit for the tens' position.

(*a*) E can be done in 5 ways; F can be done in 9 ways. The number of ways of doing F does not depend upon how E is done; hence, the sequence $\{E, F\}$ can be done in $(5)(9) = 45$ ways. Likewise, $\{F, E\}$ can be done in $(9)(5) = 45$ ways.

(*b*) The argument is as in (*a*): there are 45 2-digit odd numbers.

(*c*) If F is done first, the number of ways of doing E depends upon how F was done; so we cannot apply the product rule to the sequence $\{F, E\}$. But we can apply the product rule to the sequence $\{E, F\}$. There are 5 choices for the units' digit, and for each of these there are 8 choices for the tens' digit. So the sequence $\{E, F\}$ can be done in 40 ways; i.e., there are 40 2-digit odd numbers with distinct digits.

(*d*) We distinguish two cases. If the units' digit is 0—which can be accomplished in 1 way—the tens' digit can be chosen in 9 ways. If 2, 4, 6, or 8 is chosen as the units' digit, the tens' digit can be chosen in 8 ways. Thus the sum and product rules give a total of $(1)(9) + (4)(8) = 41$ ways.

1.3 A computer password consists of a letter of the alphabet followed by 3 or 4 digits. Find (*a*) the total number of passwords that can be created, and (*b*) the number of passwords in which no digit repeats.

(*a*) The number of 4-character passwords is $(26)(10)(10)(10)$, and the number of 5-character passwords is $(26)(10)(10)(10)(10)$, by the product rule. So the total number of passwords is $26{,}000 + 260{,}000 = 286{,}000$, by the sum rule.

(*b*) The number of 4-character passwords is (26)(10)(9)(8) = 18,720, and the number of 5-character passwords is (26)(10)(9)(8)(7) = 131,040, for a total of 149,760.

1.4 How many among the first 100,000 positive integers contain exactly one 3, one 4, and one 5 in their decimal representation?

It is clear that we may consider instead the 5-place numbers 00000 through 99999. The digit 3 can be in any one of the 5 places. Subsequently the digit 4 can be in any one of the 4 remaining places. Then the digit 5 can be in one of 3 places. There are 2 places left, either of which may be filled by 7 digits. Thus there are (5)(4)(3)(7)(7) = 2940 integers in the desired category

1.5 Find the number of 3-digit even numbers with no repeated digits.

By Problem 1.2(*d*), The hundreds' and units' positions can be simultaneously filled in 41 ways. For each of these ways, the tens' position can be filled in 8 ways. hence the desired number is (41)(8) = 328 ways.

1.6 A *palindrome* is a finite sequence of characters that reads the same forwards and backwards [GNU DUNG]. Find the numbers of 7-digit and 8-digit palindromes, under the restriction that no digit may appear more than twice.

By the mirror-symmetry of a palindrome (of length n), only the first $\lfloor (n+1)/2 \rfloor$ positions need be considered. In our case this number is 4 for both lengths. Since the first digit may not be 0, there are 9 ways to fill the first position. There are then $10-1=9$ ways to fill the second position; $10-2=8$ ways for the third; $10-3=7$ ways for the fourth. Thus there are (9)(9)(8)(7) = 4536 palindromic numbers of either length.

1.7 Prove that a palindromic (decimal) number of even length is divisible by 11.

The inductive proof exploits the fact that when the first and last characters are stripped from a palindrome, a palindrome remains. Thus, let N be a palindromic number of length $2k$. If $k=1$, the theorem obviously holds. If $k \geq 2$, we have

$$\begin{aligned} N &= a_{2k-1}10^{2k-1} + a_{2k-2}10^{2k-2} + \cdots + a_k 10^k + a_k 10^{k-1} + \cdots + a_{2k-2}10^1 + a_{2k-1}10^0 \\ &= a_{2k-1}(10^{2k-1} + 10^0) + (a_{2k-2}10^{2k-2} + \cdots + a_{2k-2}10^1) \\ &\equiv a_{2k-1}P + Q \end{aligned}$$

Here,
$$P = \underbrace{100\cdots001}_{\text{length } 2k} = 11 \times \underbrace{9090\cdots9091}_{\text{length } 2k-2}$$

and either $Q=0$ (divisible by 11) or, for some $1 \leq r \leq k-1$,

$$Q = 10^r\ \{\text{palindrome of length } 2(k-r)\} = 10^r\{11R\}$$

where the last step follows from the induction hypothesis. Therefore, N is divisible by 11, and the proof is complete.

1.8 In a **binary palindrome** the first digit is 1 and each succeeding digit may be 0 or 1. Count the binary palindromes of length n.

See Problem 1.6. Here we have $\lfloor (n+1)/2 \rfloor - 1 = \lfloor (n-1)/2 \rfloor$ free positions, so the desired number is

$$2^{\lfloor (n-1)/2 \rfloor}$$

1.9 Find the number of proper divisors of 441,000. (A **proper divisor** of a positive integer n is any divisor other than 1 and n.)

Any integer can be uniquely expressed as the product of powers of prime numbers; thus, $441{,}000 = (2^3)(3^2)(5^3)(7^2)$. Any divisor, proper or improper, of the given number must be of the form $(2^a)(3^b)(5^c)(7^d)$, where $0 \le a \le 3$, $0 \le b \le 2$, $0 \le c \le 3$, and $0 \le d \le 2$. In this paradigm the exponent a can be chosen in 4 ways; b in 3 ways; c in 4 ways; d in 3 ways. So, by the product rule, the total number of proper divisors will be $(4)(3)(4)(3) - 2 = 142$.

1.10 Count the proper divisors of an integer N whose prime factorization is

$$N = p_1^{n_1} p_2^{n_2} \cdots p_k^{n_k}$$

By Problem 1.9, the number of proper divisors of N is $(n_1 + 1)(n_2 + 1) \cdots (n_k + 1) - 2$.

1.11 Refer to Problem 1.9. Find the number of ways of factoring 441,000 into 2 factors, m and n, such that $m > 1$, $n > 1$, and the only common divisor of m and n is 1. (In other words, m and n are **relatively prime.**)

Consider the set $X = \{2^3, 3^2, 5^3, 7^2\}$ associated with the prime factorization of 441,000. It is clear that each element of X must appear in the prime factorization of m or in the prime factorization of n, but not in both. Moreover, the 2 prime factorizations must be composed exclusively of elements of X. It follows that the number of relatively prime pairs m, n is equal to the number of ways of partitioning X into 2 unordered nonempty subsets (*unordered* because mn and nm represent the same factorization).

The possible unordered partitions are the following:

$$X = \{2^3\} + \{3^2, 5^3, 7^2\} = \{3^2\} + \{2^3, 5^3, 7^2\} = \{5^3\} + \{2^3, 3^2, 7^2\} = \{7\} + \{2^3, 3^2, 5^3\}$$

and

$$X = \{2^3, 3^2\} + \{5^3, 7^2\} = \{2^3, 5^3\} + \{3^2, 7^2\} = \{2^3, 7^2\} + \{3^2, 5^3\}$$

Our answer is therefore $4 + 3 = 7 = 2^{4-1} - 1$.

1.12 Generalize Problem 1.11 by showing that the integer of Problem 1.10 has $2^{k-1} - 1$ factorizations into relatively prime pairs m, n $(m > 1,\ n > 1)$.

Make an induction on k. For $k = 1$, the result holds trivially. For $k \ge 2$, we must show that a set of k distinct elements,

$$Z = \{a_1, a_2, \ldots, a_{k-1}, a_k\}$$

has $2^{k-1} - 1$ unordered partitions into 2 nonempty parts, provided that the corresponding result holds for all $(k-1)$-sets. Now, one partition of Z is

$$Z = \{a_l\} \cup \{a_1, a_2, \ldots, a_{k-1}\} \equiv \{a_k\} \cup W$$

All the remaining partitions may be obtained by first partitioning W into two parts—which, by the induction hypothesis, can be done in $2^{k-2} - 1$ ways—and then including a_k in one part or the other—which can be done in 2 ways. By the product rule, the number of partitions of Z is therefore $1 + (2^{k-2} - 1)(2) = 2^{k-1} - 1$. QED

1.13 In a **binary sequence** every element is 0 or 1. Let X be the set of all binary sequences of length n. A **switching function (Boolean function) of *n* variables** is a function from X to the set $Y = \{0, 1\}$. Find the number of distinct switching functions of n variables.

The cardinality of X is $r = 2^n$. So the number of switching functions is 2^r.

1.14 A switching function f is **self-dual** if the value of f remains unchanged when, in each element of the domain of f, the digits 0 and 1 are interchanged. For example, when $n = 6$, $f(101101) = f(010010)$ if f is a self-dual switching function. List all self-dual switching functions of 2 variables.

There are 4 self-dual switching functions from the set $X = \{00, 01, 10, 11\}$ to the set $Y = \{0, 1\}$:

(*a*) $f_1(00) = f_1(11) = f_1(01) = f_1(10) = 0$.

(*b*) $f_2(00) = f_2(11) = f_2(01) = f_2(10) = 1$.

(*c*) $f_3(00) = f_3(11) = 0$; $f_3(01) = f_3(10) = 1$.

(*d*) $f_4(00) = f_4(11) = 1$; $f_4(01) = f_4(10) = 0$.

1.15 Find the number of self-dual switching functions of n variables.

In the notation of Problem 1.13, X can be partitioned into $r/2 = 2^{n-1}$ pairs $(\xi, \bar{\xi})$, where sequence $\bar{\xi}$ is sequence ξ after a 0-1 interchange. A self-dual switching function maps each *pair* into 0 or into 1; hence there are $2^{r/2}$ such functions. (This is the square root of the total number of switching functions.)

1.16 A collection consists of n_i identical objects of type i, where $i = 1, 2, \ldots, k$. Find the number of ways of selecting at least 1 object from the collection.

Suppose the "objects" of type i are all p_i, the ith prime factor of an integer N. Then we are asked for the number of divisors of N, with only the divisor 1 (no object selected) excluded. From Problem 1.10, this number is $(n_1 + 1)(n_2 + 1) \cdots (n_k + 1) - 1$.

1.17 How many 3-digit numbers can be formed by using the 6 digits 2, 3, 4, 5, 6, and 8, if (*a*) repetitions of digits are allowed? (*b*) repetitions are not allowed? (*c*) the number is to be odd and repetitions are not allowed? (*d*) the number is to be even and repetitions are not allowed? (*e*) the number is to be a multiple of 5 and repetitions are not allowed? (*f*) the number must contain the digit 5 and repetitions are not allowed? (*g*) the number must contain the digit 5 and repetitions are allowed?

(*a*) (6)(6)(6).

(*b*) (6)(5)(4).

(*c*) The last digit can be chosen in 2 ways; so the answer is (2)(5)(4).

(*d*) From (*b*) and (*c*), $(6)(5)(4) - (2)(5)(4) = (4)(5)(4)$.

(*e*) The last digit can be chosen in only 1 way; so the answer is (1)(5)(4).

(*f*) The digit 5 can be located in 3 ways; the answer is (3)(5)(4).

(*g*) An indirect approach, as in (*d*), is simplest. With repetitions allowed, there are $(6)(6)(6) = 216$ 3-digit numbers. Of these, $(5)(5)(5) = 125$ do not contain the digit 5. So the answer is $216 - 125 = 91$.

1.18 A **bit** is either 0 or 1; a **byte** is a sequence of 8 bits. Find (*a*) the number of bytes, (*b*) the number of bytes that begin with 11 and end with 11, (*c*) the number of bytes that begin with 11 and do not end with 11, and (*d*) the number of bytes that begin with 11 or end with 11.

(*a*) $2^8 = 256$. (*b*) The 4 open positions can be filled in $2^4 = 16$ ways. (*c*) Use (*b*): $2^6 = 64$ bytes begin with 11; so the answer is $64 - 16 = 48$. (*d*) Again use (*b*): 64 bytes begin with 11; likewise, 64 bytes end with 11. In the sum of these numbers, $64 + 64 = 128$, and each byte that both begins and ends with 11 is counted twice. Hence our answer is $128 - 16 = 112$ bytes.

1.19 There are 10 members—A, B, C, D, E, F, G, H, I, and J—in the executive council of a club. The council has to choose a chair, a secretary, and a treasurer among themselves. It is understood that nobody can hold more than one office at a time. Find the number of ways of selecting these office-bearers, if (*a*) any member can hold any office; (*b*) A has to be the chair; (*c*) B declines the chair; (*d*) C has to be either the secretary or the treasurer; (*e*) either D or E has to be the treasurer; and (*f*) I and J decline all offices.

(*a*) (10)(9)(8). (*b*) (1)(9)(8). (*c*) There are 9 ways of choosing a chair, 9 ways of choosing a secretary, and 8 ways of choosing a treasurer; altogether, (9)(9)(8) ways. (*d*) (9)(8) + (9)(8). (*e*) (9)(8) + (9)(8). (*f*) (8)(7)(6).

1.20 Find the number of 5-digit integers that contain the digit 6 exactly once.

Case 1. The first digit in the number is 6. There are $p = (1)(9)(9)(9)(9)$ ways of accomplishing this.

Case 2. The first digit is not 6. As the first digit cannot be 0 either, it can be supplied in 8 ways. The digit 6 must occupy one of the remaining 4 places; this can be done in 4 ways. The 3 remaining positions can be filled in (9)(9)(9) ways. So in this case there are $q = (8)(4)(9)(9)(9)$ ways.

By the sum rule, the number of integers in the desired category is $p + q$.

1.21 Find the total number of positive integers with distinct digits.

We have to consider numbers with at least 1 digit and at most 10 digits. The number of positive integers in the desired category will be

$$[9] + [(9)(9)] + [(9)(9)(8)] + [(9)(9)(8)(7)] + [(9)(9)(8)(7)(6)] + \cdots + [(9)(9)(8)(7)(6)(5)(4)(3)(2)(1)]$$

1.22 The class of all subsets of a set X is called the *power set* of X and is denoted by 2^X. If X has n elements, how many elements has 2^X?

Any subset A of X may be defined by making an "include in A/exclude from A" choice for each of the n elements of X. Thus there are 2^n subsets A (including the null set and X itself); i.e., 2^X has 2^n elements.

1.23 If a set X has $2n + 1$ elements, find the number of subsets of X with at most n elements.

Partition the power set of X (Problem 1.22) into a set P_1 consisting of all subsets of X with at most n elements, and a set P_2 consisting of all subsets of X with at least $n + 1$ elements. It is obvious that if $\sigma \in P_1$, then σ', the complement of σ in X, belongs to P_2; and vice versa. Thus P_1 and P_2 are in one-to-one correspondence and so have the same number of elements. This number—which is the answer to our problem—is one-half the number of elements of 2^X: $(1/2)2^{2n+1} = 2^{2n}$.

1.24 Find the number of common divisors of 2 positive integers, m and n, having prime factorizations $m = p^a q^b r^c$ and $n = p^\alpha q^\beta s^\delta$.

Any common divisor must have in its prime factorization only those primes that are common to m and n, each to a power than exceeds neither the power in m nor the power in n. Thus, with $A \equiv \min(a, \alpha)$ and $B \equiv \min(b, \beta)$, *a typical common divisor is* $p^i q^j$, where i is in the set $\{0, 1, 2, \ldots, A\}$ and j is in the set $\{0, 1, 2, \ldots, B\}$ So the number of common divisors will be $(A + 1)(B + 1)$.

1.25 (*a*) Find the number of positive integers with n digits in which no 2 adjacent digits are the same (call this property *redness*). (*b*) Count the even red numbers.

(*a*) Let $X(n)$ be the number of n-digit red integers. The first place can be filled in 9 ways. The second place can be filled in 9 ways. In fact, each place can be filled in 9 ways; thus $X(n) = 9^n$.

(*b*) Let $X(n) = Y(n) + Z(n)$, where $Y(n)$ and $Z(n)$, respectively, count the even and odd red numbers. Since 9^n is odd, $Y(n)$ cannot be equal to $Z(n)$; but is seems likely that one will be equal to $(9^n - 1)/2$ and the other to $(9^n + 1)/2$. This is indeed the case: we shall prove by induction that $Y(n) = [9^n + (-1)^n]/2$ and $Z(n) = [9^n - (-1)^n]/2$.

The result is true for $n = 1$, since $Y(1) = 4$ and $Z(1) = 5$. Now, any even red number with $n + 1$ digits can be constructed by appending a final even digit to a red number with n digits. If the n-digit number is even, this can be done in 4 ways; otherwise, in 5 ways. Thus we have established that

$$Y(n+1) = 4Y(n) + 5Z(n) = 4Y(n) + 5[X(n) - Y(n)]$$

$$= 5X(n) - Y(n) = \frac{9^{n+1} + (-1)^{n+1}}{2}$$

Also

$$Z(n+1) = X(n+1) - Y(n+1) = \frac{9^{n+1} - (-1)^{n+1}}{2}$$

As the result is true for $n+1$ whenever it is true for n, the induction is complete.

1.26 Find the number of binary sequences of length n that contain an even number of 1s.

If a binary sequence of length $n-1$ has an even number of 1s, we append the digit 0 to it, to obtain a binary sequence of length n. If a binary sequence of length $n-1$ has an odd number of 1s, we append the digit 1 to it. In other words, the number of binary sequences of length n with an even number (or, for that matter, with an odd number) of 1s is equal to the number of binary sequences of length $n-1$, which is 2^{n-1}.

1.27 Find the number of n-character strings that can be formed using the letters A, B, C, D, and E such that each string has an even number of A's.

The total number of strings is 5^n, out of which there are 3^n strings which contain zero (an even number) A's and zero B's. The set X of the residual $5^n - 3^n$ strings can be partitioned into a class $\{X_i\}$ of subsets as follows: two strings s and s' belong to the same subset X_i if and only if the locations of C's, D's, and E's in s are exactly the same as in s'. This means that there are exactly as many strings in X_i as there are sequences of A's (1s) and B's (0s) of some fixed length $n_i \leq n$. Half of these strings will have an even number of A's, by Problem 1.26. Thus the total number of strings in X with an even number of A's is $(5^n - 3^n)/2$; and the solution to our problem is:

$$3^n + \frac{5^n - 3^n}{2} = \frac{5^n + 3^n}{2}$$

PERMUTATIONS AND COMBINATIONS

1.28 A function f from a set A to a set B is **one-one** if distinct elements x and y of A have distinct images $f(x)$ and $f(y)$ in B. Find the number of one-one functions from A to B if A has m elements and B has $n \geq m$ elements (a necessary condition).

There are $P(n, m)$ choices for the range of the function; thus there are $P(n, m)$ distinct functions.

1.29 Find the probability p_n that a randomly assembled group of n people includes at least 2 people with the same birthday (day of the year).

Here we deal not with a sample of *people*, but with a sample of *birthdays*—i.e., integers from 1 to 365 inclusive. Our notion of probability is:

$$\text{Probability} = \frac{\text{number of ``favorable'' samples}}{\text{total number of samples}}$$

In this problem it is simplest to consider the complementary event: all n birthdays are distinct. This event is realized in $P(365, n)$ samples; and the total number of samples is 365^n. Hence, $1 - p_n = P(365, n)/365^n$, or

$$p_n = 1 - \frac{P(365, n)}{365^n} = 1 - \frac{(365)(365-1)(365-2)\cdots[365-(n-1)]}{365^n}$$

$$= 1 - \left(1 - \frac{1}{365}\right)\left(1 - \frac{2}{365}\right)\cdots\left(1 - \frac{n-1}{365}\right)$$

It may be verified that $p_n > 1/2$ when $n > 25$.

1.30 There are n married couples at a party. Each person shakes hands with every person other than her or his spouse. Find the total number of handshakes.

The $2n$ people fall into $C(2n, 2) = (2n)(2n-1)/2$ unordered pairs, out of which n pairs will be the married couples. Thus the number of pairs who are not married couples is $n(2n-1) - n = 2n(n-1)$, and this is also the number of handshakes.

1.31 In how many ways may n girls and n boys be seated in a row of $2n$ chairs, if the two sexes must alternate?

The chairs are marked $1, 2, 3, \ldots, 2n$. There are $P(n, n) = n!$ ways of seating the boys in the odd-numbered chairs. After the boys are seated, the girls can be seated in $n!$ ways. So this pattern will give rise to $(n!)^2$ seating arrangements. There are the same number of arrangements with the girls in the odd-numbered chairs. Thus the total number of arrangements $2(n!)^2$.

1.32 Ten different paintings are to be allocated to n office rooms so that no room gets more than 1 painting. Find the number of ways of accomplishing this, if *(a)* $n = 14$ and *(b)* $n = 6$.

(a) $P(14, 10)$, by Problem 1.28. *(b)* Now the supply exceeds the demand; so we map from rooms to paintings, obtaining $P(10, 6)$ from Problem 1.28.

1.33 Solve Problem 1.32 if there are 10 identical posters instead of 10 distinct paintings.

(a) Choose any 10 of the 14 rooms; this can be done in $C(14, 10)$ ways. Then hang 1 poster in each room. Since any poster is like any other, there is only one way of doing so. Hence the answer is $C(14, 10)$. *(b)* Choose 6 posters from the collection and hang 1 poster in each room; there is $(1)(1) = 1$ way of doing this.

1.34 A **circular permutation** is an arrangement of distinct objects around a circle (or other simple closed curve). Find the number of circular permutations of n distinct objects.

If the n sites where the objects are to be placed are distinct and marked $1, 2, \ldots, n$, there will be $n!$ arrangements, as usual. If that is not the case, the number will be smaller, owing to the fact that 2 distinct linear permutations can define the same circular permutation. (For example, the permutations *ABCD* and *BCDA* represent the same circular permutation.) This smaller number is, in fact, $(n-1)!$ The proof is simple: A distinguished object may be placed somewhere on the circle in 1 way. On the "punctured" circle defined by this first placement, the remaining $n-1$ objects may be ranged in clockwise order (say) in $P(n-1, n-1) = (n-1)!$ ways.

1.35 Two linear permutations of n objects, p and q, are called **reflections** of each other if the first object in p is the last object in q, the second object in p is the $(n-1)$st object in $q, \ldots,$ the last object in p is the first object in q. A circular permutation of n objects is a **ring permutation** if the defining linear permutation of $n-1$ objects (see Problem 1.34) and its reflection are not considered distinct. Find the number of ring permutations of n distinct objects.

Each ring permutation defines 2 circular permutations; so the number of ring permutations is $(n-1)!/2$.

1.36 A **combinatorial proof** is one that uses combinatorial reasoning instead of calculation; e.g., our proof that $C(n, r) = C(n, n-r)$. Give a combinatorial proof of the identity

$$C(m+n, 2) - C(m, 2) - C(n, 2) = mn$$

Consider a collection of m mathematics majors and n nursing majors. By the product rule, the number of ways of choosing a mathematics major and a nursing major is mn. A different route must lead to the same number: from the $C(m+n, 2)$ possible pairs of students we eliminate the $C(m, 2)$ pairs of 2 mathematics majors and the $C(n, 2)$ pairs of 2 nursing majors.

1.37 Give a combinatorial proof of **Pascal's identity**,

$$C(n, r) = C(n-1, r) + C(n-1, r-1)$$

Consider a set X with n elements. Let Y be any subset of X with $n-1$ elements. Every subset of X with r elements is either a subset of Y with r elements or the union of a subset of Y with $r-1$ elements and the singleton set consisting of the unique element of X which is not in Y. There are $C(n-1, r)$ subsets in the former category and $C(n-1, r-1)$ subsets in the latter category. The sum of these 2 numbers is necessarily $C(n, r)$.

Pascal's identity is perhaps the most important single formula in combinatorics.

1.38 Prove the binomial theorem for a positive integer n:

$$(x+y)^n = x^n + C(n, 1)x^{n-1}y + \cdots + C(n, r)x^{n-r}y^r + \cdots + y^n \equiv \sum_{r=0}^{n} C(n, r)x^{n-r}y^r$$

A typical term in the expansion of $(x+y)^n$ is $x^{n-r}y^r$ multiplied by an integral coefficient. If we write out $(x+y)^n$ as $(x+y)_1(x+y)_2 \cdots (x+y)_n$, we see that this integral coefficient is just the number of ways of selecting the r out of n parentheses that shall furnish the r y's going into $x^{n-r}y^r$. The integral coefficient therefore equals $C(n, r)$.

Because of their occurrence in the binomial theorem, the integers $C(n, r)$ are known as **binomial coefficients.**

1.39 Prove:

(*a*) $\sum_{r=0}^{n} C(n, r) = 2^n$ (*b*) $\sum_{r=0}^{n} (-1)^r C(n, r) = 0$

(*c*) $\sum_{r \text{ even}}^{n} C(n, r) = \sum_{r \text{ odd}}^{n} C(n, r) = 2^{n-1}$

(*a*) Set $x = y = 1$ in the binomial theorem. (*b*) Set $x = -y = 1$ in the binomial theorem. (*c*) Add and subtract the results of (*a*) and (*b*).

1.40 Retrieve the results of Problem 1.39 by combinatorial arguments.

(a) Consider a set X with n elements. Then (Problem 1.22) X has 2^n subsets, giving the right-hand side of the identity. But also the set X has $C(n, r)$ subsets of cardinality r, where r runs from 0 to n. So the total number of subsets is $C(n, 0) + C(n, 1) + \cdots + C(n, r)$, as on the left-hand side of the identity.

(*b*), (*c*) It is only necessary to establish that X has just as many subsets with an even number of elements as subsets with an odd number. But this is obviously true when n is odd, because each even (odd) subset can be paired off with its odd (even) complement. Suppose, then, that n is even (and positive). With θ a designated element of X, make the decomposition $X = X' \cup \{\theta\}$. Because X' is odd (has odd cardinality), it has equal numbers of even and odd subsets. (Remember that the null set is included among the even subsets.) Now, all subsets of X may be obtained by either including or not including θ in a subset of X'. Since inclusion of θ changes the parity of the subsets, it is evident that X must have twice as many even subsets and twice as many odd subsets as does X'—that is, X must have equally many even and odd subsets.

1.41 Establish the identity $C(2n, 2) = 2C(n, 2) + n^2$ by a combinatorial argument.

Consider a set X with $2n$ elements which is partitioned into 2 subsets, Y and Z, each of cardinality of n. Every subset of X with 2 elements belongs to one of the following three mutually exclusive classes of subsets: (*i*) the class of all 2-element subsets of Y; (*ii*) the class of all 2-element subsets of Z; (*iii*) the class of all subsets $\{y, z\}$ with $y \in Y$ and $z \in Z$. There are $C(n, 2)$ subsets in each of the first two classes and n^2 subsets in the third class.

1.42 In buying a ticket in a state-sponsored lottery, one chooses a subset T consisting of 6 distinct numbers from the set of the first 48 positive integers. After the sales are closed, a computer selects at random 6 numbers out of these same 48; these 6 numbers constitute the winning set, W. If $T = W$, the ticket holder wins the jackpot. If T and W have 5 numbers in common, the ticket holder wins the second prize. If T and W have 4 numbers in common, the ticket holder wins the third prize. Find (*a*) the number of *different* tickets that can be sold; (*b*) the number of distinct second-prize tickets; (*c*) the number of distinct third-prize tickets.

(*a*) $C(48, 6)$. (*b*) To win a second prize, one has to choose 5 numbers out of the set W and 1 number out of the complement of W, which consists of 42 numbers. So the number of distinct second-prize tickets will be $C(6, 5)C(42, 1) = 252$. (*c*) $C(6, 4)C(42, 2) = 12{,}915$.

1.43 (*a*) Establish the identity (**convolution rule** or **Vandermonde identity**)

$$C(p+q, r) = \sum_{j=0}^{r} C(p, j)C(q, r-j)$$

(*b*) Connect the convolution rule with Problem 1.42.

(*a*) By the binomial theorem, the left-hand side of the convolution rule is the coefficient of x^r in $(1+x)^{p+q}$; the right-hand side is the coefficient of x^r in $(1+x)^p(1+x)^q$. The two coefficients must, of course, be equal.

(*b*) In the convolution rule, let $p = 42$ and $q = r = 6$:

$$C(48, 6) = C(42, 0)C(6, 6) + C(42, 1)C(6, 5) + C(42, 2)C(6,4) + \cdots + C(42, 6)C(6, 0)$$

which represents the same classification (by overlap with the winning set) of distinct tickets.

1.44 Establish **Newton's identity**,

$$C(n, r)C(r, k) = C(n, k)C(n-k, r-k)$$

Suppose there are n faculty members in a university. The faculty assembly consists of r members, and the faculty senate consists of k members; every member of the senate is also a member of the assembly. (Hence, $n \geq r \geq k$.) The number of ways of forming a senate can be computed by two methods: (*i*) First choose r people from the set of n faculty members to constitute the assembly; this can be done in $C(n, r)$ ways. After that, choose k senators from the r assembly people; this can be done in $C(r, k)$ ways. So the total number of ways is $C(n, r)C(r, k)$, as in the left-hand side of Newton's identity. (*ii*) First choose k faculty members to constitute the senate; this can be done in $C(n, k)$ ways. These k individuals are ipso facto members of the assembly; the additional $r-k$ assembly people may be chosen from the $n-k$ remaining faculty in $C(n-k, r-k)$ ways. Hence the total number of ways is $C(n, k)C(n-k, r-k)$, as in the right-hand side.

1.45 Prove that if n is a prime number, $C(n, r)$ is divisible by n for $r = 1, 2, \ldots, n-1$.

Set $k = 1$ in Newton's identity to obtain $rC(n, r) = nC(n-1, r-1)$. Since n divides the right-hand side of this equation, it divides the left-hand side. Because n is a prime, it must then divide either r or $C(n, r)$. But n cannot divide r, for it is greater than r.

1.46 Use a combinatorial argument to prove

$$C(n+r+1, r) = \sum_{j=0}^{r} C(n+j, j)$$

The left-hand side is the number of ways of choosing r elements from a set $X = \{x_1, x_2, x_3, \ldots, x_{n+r+1}\}$. As for the right-hand side, consider the construction of an r-element subset A of X. If x_1 is not to be in A, the r elements of A have to be chosen from the remaining $n+r+1-1$ elements of X; this can be done in $C(n+r, r)$ ways. If x_1 is to be in A, then we need $r-1$ additional elements, which may or may not include x_2. If x_2 is not included, these $r-1$ elements have to come from the remaining $n+r+1-2$ elements of X; they can be chosen in $C(n+r-1, r-1)$ ways. If x_1 and x_2 are included in A, then.... The process terminates when $x_1, x_2, \ldots, x_r$ are included in A and we must pick 0 elements from the remaining $n+1$, which can be done in $C(n+1, 0) = C(n, 0) = 1$ way. Thus we have obtained, in reverse order, the summands in the right-hand side of our equation and have shown that the sum also counts the r-subsets of X.

1.47 Prove:

$$C(m+n, n) = C(m, 0)C(n, 0) + C(m, 1)C(n, 1) + \cdots + C(m, n)C(n, n)\,.$$

In the convolution formula [Problem 1.43(*a*)] set $q = r$ and note that $C(q, q-j) = C(q, j)$.

1.48 Find the number of 5-digit positive integers such that in each of them every digit is greater than the digit to its right.

There are $C(10, 5)$ ways of selecting 5 distinct (necessary) digits. Once these digits are chosen, there is only 1 way of arranging them in a decreasing order from left to right. So the number of these integers is $C(10, 5)$.

1.49 Find the number of ways of arranging the letters which appear in (*a*) ELASTIC and (*b*) ASBESTOS.

(*a*) As the 7 letters are distinct, there are 7! ways. (*b*) The 5 letters which do not repeat are *A*, *B*, *E*, *O*, and *T*. The 5 positions to be occupied by these letters may be chosen from among 8 positions in $P(8, 5)$ ways. There is then only 1 way of assigning the 3 *S*'s to the 3 remaining positions. So the answer is $P(8, 5)$.

1.50 A and B are 2 students in a group of n. Find the number of ways of assigning the n students to a line of n single rooms such that (*a*) A and B are in adjacent rooms, and (*b*) A and B are not in adjacent rooms.

(*a*) There are $n-1$ ways of choosing a pair of adjacent rooms, and 2 ways of installing A and B in the chosen pair. So there are $2(n-1)$ ways of accommodating A and B. The other $n-2$ students can be assigned to the other $n-2$ rooms in $(n-2)!$ ways. So the total number of ways is

$$2(n-1)(n-2)! = 2(n-1)!$$

(*b*) $n! - 2(n-1)! = (n-2)(n-1)!$.

1.51 There are 18 chairs (marked $1, 2, \ldots, 18$) in a row to seat 5 chemistry majors, 6 mathematics majors, and 7 physics majors. Find the number of seating arrangements, if (*a*) the chemistry majors occupy the first 5 seats; (*b*) the chemistry majors are barred from the first 5 seats; (*c*) the chemistry majors occupy the first 5 seats and the mathematics majors occupy the next 6 seats; and (*d*) students with the same major sit in a block.

(*a*) (5!)(13!). (*b*) The 5 chemistry majors can occupy the 13 seats marked $6, \ldots, 18$ in $P(13, 5)$ ways. Once the chemistry majors are seated, the other students can occupy the remaining chairs in 13! ways. So the answer is

$P(13, 5)(13!) = (13!)^2/8!$. (*c*) $(5!)(6!)(7!)$. (*d*) First permute the 3 groups and then permute the students within each group: $(3!)(5!)(6!)(7!)$.

1.52 Find the number of ways of seating n women and n men at a round table so that between every 2 women there is a man.

By Problem 1.34, there are $(n-1)!$ ways of seating the women. Now put a chair between every 2 women; these chairs can be occupied by the men in $n!$ ways. So the number of seating arrangements is $n!\,(n-1)!$.

1.53 The n members of the board of directors include the president and 2 vice presidents. Find the number of ways of seating the board at a round table so that the vice presidents are on either side of the president.

The president occupies a chair. The 2 chairs on either side of the president can be occupied by the 2 vice presidents in 2 ways. The other $n-3$ members can be seated in $(n-3)!$ ways. Thus the answer is $2(n-3)!$.

1.54 Find the number of ways of seating r out of n people around a circular table, and the others around another circular table.

First choose the r individuals for the first table—this can be done in $C(n, r)$ ways. These r individuals can be seated in $(r-1)!$ ways (Problem 1.34). The remaining $n-r$ individuals can be seated in $(n-r-1)!$ ways. So the answer is $C(n, r)(r-1)!\,(n-r-1)!$.

1.55 There are 3 apartments—A, B, and C— for rent in a building. Each unit will accept either 3 or 4 occupants. Find the number of ways of renting the apartments to 10 students.

This is clearly a case of ''a four and two threes.'' Now A can be given 4 occupants, and B and C 3 each, in $C(10, 4)C(6, 3)C(3, 3)$ ways. The total number will be 3 times this, or $3C(10, 4)C(6, 3)$.

1.56 Let $1, 2, 3, \ldots, n$ label n fixed points on the circumference of a circle. Each of these points is joined to every one of the remaining $n-1$ points by a straight line, and the points are so positioned on the circumference that at most 2 straight lines meet in any interior point of the circle. Find the number of such interior intersection points.

Any interior intersection point corresponds to 4 of the labeled points—namely, the 4 endpoints of the intersecting line segments. Conversely, any 4 labeled points determine a quadrilateral, the diagonals of which intersect once within the circle. Thus the answer is $C(n, 4)$.

1.57 Find the number of ways of seating m women and n men $(m < n)$ at a round table so that no 2 women sit side by side. (Compare Problem 1.52.)

Place n chairs around the table, in which the man may be seated in $(n-1)!$ ways. Then place a chair between every 2 men, creating n distinct vacant places. By Problem 1.28, the m women can be assigned to these n places in $P(n, m)$ ways. Once the women are seated, the extra chairs are removed (which can be done in 1 way). So the answer is $(n-1)!\,P(n, m)$.

1.58 A point in the cartesian plane whose coordinates are integers is called a **lattice point**. Consider a path from the origin to the lattice point $A(m, n)$, where m and n are nonnegative, that (*i*) starts from the origin; (*ii*) is always parallel to the x-axis or the y-axis; (*iii*) makes turns only at a lattice point, either along the positive x-axis or along the positive y-axis; and (*iv*) terminates at A. Determine the number of such paths.

A typical path is a sequence of $m+n$ unit steps, m of them horizontal and n of them vertical. Hence the number of paths is $C(m+n, m) = C(m+n, n)$, the number of ways of reserving positions in the sequence for one or the other kind of step.

1.59 Establish the identity

$$C(n+1, r+1) = C(n, r) + C(n-1, r) + C(n-2, r) + \cdots + C(r, r)$$

The identity is readily verified for $n = 1$. For $n > 1$, use Pascal's identity (Problem 1.37) to replace the left-hand member by $C(n, r+1) + C(n, r)$. Obviously the induction will succeed.

1.60 Prove that $1 + 2 + 3 + \cdots + n = n(n+1)/2$.

Set $r = 1$ in the identity of Problem 1.59.

1.61 A woman has 11 colleagues in her office, of whom 8 are men. She would like to have some of her colleagues to dinner. Find the number of her choices if she decides to invite (*a*) at least 9 of them, and (*b*) all her women colleagues and sufficient men colleagues to make the numbers of women and men equal.

(*a*) $C(11, 9) + C(11, 10) + C(11, 11) = 67$.

(*b*) She has to invite 4 men, since there will be 4 women dining, including herself. So the answer is $C(8, 4)$.

1.62 Prove:

$$1^2 + 2^2 + 3^2 + \cdots + n^2 = \frac{n(n+1)(2n+1)}{6}$$

Sum the easily derived identity $k^2 = C(k, 1) + 2C(k, 2)$ on k:

$$\sum_{k=1}^{n} k^2 = \sum_{k=1}^{n} C(k, 1) + 2\sum_{k=2}^{n} C(k, 2)$$

By Problem 1.59 the two sums on the right equal $C(n+1, 2) = n(n+1)/2$ and $C(n+1, 3) = (n+1)(n)(n-1)/6$. Hence,

$$\sum_{k=1}^{n} k^2 = \frac{n(n+1)}{2}\left[1 + \frac{2(n-1)}{3}\right] = \frac{n(n+1)(2n+1)}{6}$$

1.63 Evaluate the sum $S = (1)(2) + (2)(3) + (3)(4) + \cdots + (n)(n+1)$.

Because $k(k+1)$ can be written as $2C(k+1, 2)$,

$$S = 2[C(2, 2) + C(3, 2) + \cdots + C(n+1, 2)]$$

But then, by Problem 1.59 with $r = 2$ and n replaced by $n+1$, $S = 2C(n+2, 3)$.

1.64 Show that

$$\sum_{k=0}^{n-r} P(r+k, r) = \frac{1}{r+1} P(n+1, r+1)$$

Multiply both sides of the identity of Problem 1.59 by $r!$ to obtain the desired identity. Note that the left-hand side may be expanded as

$$[(1)(2)(3)\cdots(r)] + [(2)(3)(4)\cdots(r+1)] + \cdots + [(n-r+1)\cdots(n-1)n]$$

We therefore have a generalization of Problem 1.63.

1.65 According to Problem 1.28, a permutation of $X = \{1, 2, 3, \ldots, n\}$ is a one-one mapping of X *onto* itself. If P and Q are 2 permutations of X, their **product**, $P \circ Q$, is the permutation of X obtained by following the mapping Q with the mapping P. Moreover, the **inverse**, P^{-1}, of P is the permutation of X represented by the mapping inverse to the mapping P. Letting $n = 5$, $Q = 2\,3\,4\,1\,5$—i.e.,

$$\langle 1, 2, 3, 4, 5 \rangle \xrightarrow[Q]{} \langle 2, 3, 4, 1, 5 \rangle$$

—and $P = 1\,2\,5\,3\,4$, find (*a*) $P \circ Q$, (*b*) $Q \circ P$, (*c*) Q^{-1}, and (*d*) P^{-1}.

(*a*)
$$\langle 1, 2, 3, 4, 5 \rangle \xrightarrow[Q]{} \langle 2, 3, 4, 1, 5 \rangle \xrightarrow[P]{} \langle 2, 5, 3, 1, 4 \rangle$$

so that $P \circ Q = 2\,5\,3\,1\,4$. (*b*) $Q \circ P = 2\,3\,5\,4\,1$. (*c*) $Q^{-1} = 4\,1\,2\,3\,5$. (*d*) $P^{-1} = 1\,2\,4\,5\,3$.

1.66 Show that if P and Q are 2 permutations on X, then $(P \circ Q)^{-1} = Q^{-1} \circ P^{-1}$.

Refer to Problem 1.65. An equivalent definition of P^{-1} is $P \circ P^{-1} = P^{-1} \circ P = I$, where I is the identity mapping on X. Bearing in mind that multiplication of permutations is associative, we have

$$(P \circ Q) \circ (Q^{-1} \circ P^{-1}) = P \circ I \circ P^{-1} = P \circ P^{-1} = I$$

and

$$(Q^{-1} \circ P^{-1}) \circ (P \circ Q) = Q^{-1} \circ I \circ Q = Q^{-1} \circ Q = I$$

1.67 A permutation P of $x = \{x_1, x_2, \ldots, x_n\}$ is a **derangement** if $P(x_i) \neq x_i$ for $i = 1, 2, \ldots, n$. Prove that the inverse of a derangement is a derangement.

If $P(x_i) = x_j$ $(j \neq i)$, then $P^{-1}(x_j) = x_i$ $(i \neq j)$.

1.68 Is the product of 2 derangements necessarily a derangement?

No; for example, the product of derangements P and P^{-1} (Problem 1.67) is I, which is certainly not a derangement. Thus, while the permutations compose a group under multiplication (the **symmetric group**), the derangements constitute merely a subset, not a subgroup.

1.69 A particle starts from a fixed point O which is taken as the origin of x coordinates. At every unit of time, starting with $t = 0$, the particle either remains at its present position or moves 1 unit in the positive x direction. The probability that the particle stays in its present position is q and the probability that it moves is p; thus $p + q = 1$. Let $P_n(r)$ be the probability that the particle has moved r units when $t = n$. Show that:

(*a*) $P_n(r) = pP_{n-1}(r-1) + qP_{n-1}(r)$.

(*b*) $P_n(r) = C(n, r)p^r q^{n-r}$; i.e., $P_n(r)$ is the coefficient of x^r in $(px + q)^n$.

(The motion of the particle is known as a **one-dimensional binomial random walk**.)

(*a*) At $t = n$, either the particle has just arrived from the point $x = r - 1$, the probability of which is $P_{n-1}(r-1)p$, or it had already reached $x = r$ at $t = n - 1$ and stayed there, the probability of which is $P_{n-1}(r)q$.

(*b*) This is proved by induction on n, as follows. $P_0(0) = 1 = C(0, 0)p^0 q^0$, and $P_0(r) = 0$ when $r > 0$; so the theorem is true when $n = 0$. Suppose the theorem is true for $n - 1$. Then,

$$P_{n-1}(r) = C(n-1, r)p^r q^{n-r-1} \qquad \text{and} \qquad P_{n-1}(r-1) = C(n-1, r-1)p^{r-1}q^{n-r}$$

whence $P_n(r) = qP_{n-1}(r) + pP_{n-1}(r-1) = [C(n-1, r-1) + C(n-1), r)]p^r q^{n-r} = C(n, r)p^r q^{n-r}$

So the theorem is true for n as well.

1.70 In a **two-dimensional binomial random walk** a particle starts ($t = 0$) from the origin $O(0, 0)$ of a cartesian coordinate system and moves in one unit of time one unit of distance either parallel to the $+X$-axis (with probability p) or parallel to the $+Y$-axis (with probability q), where $p + q = 1$. Determine the probability $\Pi_n(r, s)$ that the coordinates of the particle at $t = n$ are (r, s).

In each unit of time the particle must move one way or the other. Hence, at $t = n$, $s = n - r$. We can therefore view the two-dimensional walk as the walk of Problem 1.69 with the "remains at its present position" option replaced by "moves 1 unit in the positive y direction." This gives at once:

$$\Pi_n(r, s) = \begin{cases} 0 & s \neq n - r \\ P_n(r) & s = n - r \end{cases}$$

1.71 In Problem 1.70 let $p = 1/3$, and $q = 2/3$. Compute the probabilities of the following events: (*a*) the particle passes through the point $(5, 2)$; (*b*) the particle passes through $(5, 2)$ and $(7, 1)$; and (*c*) the particle passes through $(5, 2)$ and $(6, 3)$.

(*a*) This event can occur only at $t = 7$, with probability $P_7(5) = C(7, 5)(1/3)^5(2/3)^2$.

(*b*) First to hit $(5, 2)$ and then $(7, 1)$ would require a decrease in the y coordinate; the other order, a decrease in the x coordinate. But either coordinate can only increase, so the probability here is zero.

(*c*) The probability of the path $(5, 2)$-$(6, 2)$-$(6, 3)$ is $P_7(5)(1/3)(2/3)$ and the probability of the path $(5, 2)$-$(5, 3)$-$(6, 3)$ is $P_7(5)(2/3)(1/3)$. The desired probability is the sum, or $(4/9)P_7(5)$.

1.72 Consider a bidirectional random walk on the X axis. The particle starts (at time $t = 0$) from the origin and can make steps of $+1$ (with fixed probability p) or -1 (with fixed probability $q = 1 - p$). Show that $P_n(r)$—the probability that the particle is at $x = r$ after n steps—is the coefficient of x^r in the binomial expansion of $(px + q/x)^n$.

Obviously, $P_n(r)$ obeys

$$P_n(r) = pP_{n-1}(r-1) + qP_{n-1}(r+1) \tag{i}$$

which is a recurrence relation in 2 integer variables, n and r. First, get rid of r by introducing the **generating function**

$$F_n(x) \equiv \sum_{r=-\infty}^{\infty} P_n(r)x^r$$

[Observe that $P_n(r) = 0$ for $|r| > n$.] When (*i*) is multiplied by x^r and summed over all r, the result is

$$F_n(x) = \left(px + \frac{q}{x}\right) F_{n-1}(x) \tag{ii}$$

The solution to (*ii*)—a difference equation in n alone—is evidently

$$F_n(x) = \left(px + \frac{q}{x}\right)^n F_0(x) = \left(px + \frac{q}{x}\right)^n \tag{iii}$$

since $F_0(x) = P_0(0) = 1$. Thus $P_n(r)$ is the coefficient of x^r in $(px + q/x)^n$, as asserted.

1.73 In Problem 1.72 let $p = 3/4$ and $q = 1/4$. Compute the probability that the particle has $x \geq 1$ during $1 \leq t \leq 4$.

There are precisely three sequences of four steps (starting from the origin) that satisfy the prescribed condition:

Sequence	Probability
$\langle +1, +1, -1, +1 \rangle$	$p^3 q$
$\langle +1, +1, +1, -1 \rangle$	$p^3 q$
$\langle +1, +1, +1, +1 \rangle$	p^4

The required probability is therefore

$$2p^3 q + p^4 = p^3(q + 1) = \frac{135}{256}$$

The reader should note that the first two sequences land the particle at $x = 2$; and the third sequence, at $x = 4$. However, the answer *is not* $P_4(2) + P_4(4)$, because the forbidden sequences $\langle +1, -1, +1, +1 \rangle$ and $\langle -1, +1, +1, +1 \rangle$ contribute to $P_4(2)$.

1.74 A finite sequence $\langle a_1, a_2, \ldots, a_n \rangle$ of real numbers is **unimodal** if there exists a positive integer $1 < j < n$ such that $(a_1 < a_2 < \cdots < a_{j-1} \leq a_j > a_{j+1} > \cdots > a_n$. Show that the sequence $\langle C(n, 0), C(n, 1), \ldots, C(n, n) \rangle$ is unimodal for any $n > 1$, and find the largest number(s) in the sequence.

By Theorem 1.1(*ii*), $C(n, r + 1)/C(n, r) = (n - r)/(r + 1)$. Thus the sequence is strictly increasing for $n - r > r + 1$, or $r < (n - 1)/2$, and strictly decreasing for $r > (n - 1)/2$. Explicitly, for $n = 2k$ $(k = 1, 2, \ldots)$,

$$C(2k, 0) < \cdots < C(2k, k) > \cdots > C(2k, 2k)$$

and, for $n = 2k + 1$,

$$C(2k+1, 0) < \cdots < C(2k+1, k) = C(2k+1, k+1) > \cdots > C(2k+1, 2k+1)$$

1.75 Show that the sequence $\langle a_0, a_1, a_2, \ldots, a_n \rangle$, where $a_r \equiv C(n, r)x^{n-r}y^r$ and x and y are positive, is unimodal.

As in Problem 1.74 we determine that the sequence is strictly increasing for

$$r < t \equiv \frac{ny - x}{x + y}$$

and strictly decreasing for $r > t$. There are four possibilities:

(*a*) $t < 0$. Then $a_0 > a_1 > a_2 > \cdots > a_n$.

(*b*) $t = 0$. Then $a_0 = a_1 > a_2 > a_3 > \cdots > a_n$.

(*c*) $t > 0$ and t is not an integer. Let $k = \lfloor t \rfloor + 1$; then

$$a_0 < a_1 < \cdots < a_k > a_{k+1} > \cdots > a_n$$

(*d*) $t > 0$ and t is an integer. Then $a_0 < a_1 < \cdots < a_t = a_{t+1} > a_{t+2} > \cdots > a_n$.

THE PIGEONHOLE PRINCIPLE

1.76 Let $X = \{0, 1, 2, 3, 4, 5, 6, 7, 8, 9, 10\}$. Show that if S is any subset of X with 7 elements, then there are 2 elements of S whose sum is 10.

The subsets $H_1 = \{0, 10\}$, $H_2 = \{1, 9\}$, $H_3 = \{2, 8\}$, $H_4 = \{3, 7\}$, $H_5 = \{4, 6\}$, and $H_6 = \{5\}$ may be considered as 6 pigeonholes; and the elements of S, as 7 pigeons.

1.77 Show that in any group of people there will be at least 2 people who know the same number of people in the group.

Suppose that in the group $X = \{1, 2, \ldots, n\}$ there are k people who do not know anybody in the group.

(*a*) If $k > 1$, there are at least 2 people who know nobody in the group.

(*b*) If $k = 0$ let x_i be the number of people known to i, where $i = 1, 2, \ldots, n$. Since $1 \le x_i \le n - 1$ for each i, the n numbers x_i cannot all be distinct. So there are at least 2 integers i and j such that $x_i = x_j$.

(*c*) If $k = 1$, we ignore the person who does not know anyone in the group. We are then back in situation (*b*), with n replaced by $n - 1$.

1.78 Consider a tournament in which each of n players plays against every other player and each player wins at least once. Show that there are at least 2 players having the same number of wins.

The number of wins for a player is at least 1 and at most $n - 1$. These $n - 1$ numbers correspond to $n - 1$ pigeonholes to accommodate n player-pigeons.

1.79 Show that any set of n integers has a subset such that the sum of the integers in the subset is divisible by n.

Let $X = \{x_1, x_2, \ldots, x_n\}$ and $s_i = x_1 + x_2 + \cdots + x_i$, where $i = 1, 2, 3, \ldots, n$. If any s_i is divisible by n, we are done. Suppose this is not the case. Then the remainder r_i obtained when s_i is divided by n is at least 1 and at most $n - 1$; so that, by the pigeonhole principle, we must have $r_p = r_q$ for some $p < q$. But then

$$s_q - s_p = x_{p+1} + x_{p+2} + \cdots + x_q$$

leaves a remainder 0, i.e., is divisible by n.

1.80 Let X denote a set of 9 positive integers, and, for any subset E of X, let $s(E)$ represent the sum of the elements of E. Find the range of values of the largest element, n, of X for which there must exist two subsets A and B such that $s(A) = s(B)$.

For any subset E,

$$1 \le s(E) \le n + (n-1) + \cdots + (n-8) = 9n - 36$$

So the number of distinct values of $s(E)$ is at most $9n - 36$. As there are $2^9 - 1 = 511$ nonempty subsets E, the pigeonhole argument yields

$$511 > 9n - 36 \qquad (i)$$

as a sufficient condition for the existence of two equal-sum subsets. Clearly, (*i*) is satisfied for values of n from 9 (the smallest possible) through 60.

1.81 If 5 points are chosen at random in the interior of an equilateral triangle each side of which is 2 units long, show that at least 1 pair of points has a separation of less than 1 unit.

The equilateral triangle can be partitioned into 4 equilateral triangles, and each side is 1 unit. We have 5 points and 4 triangles; the conclusion is obvious.

1.82 If 10 points are chosen at random in the interior of an equilateral triangle each side of which is 3 units long, show that some pair of points are within 1 unit of each other.

Subdivide the original triangle into 9 equilateral triangles (each side is 1 unit) by trisecting each side and drawing parallel lines through the points of subdivision. There are 9 triangles and 10 points.

1.83 If 5 points are chosen at random in the interior of a square and each side is 2 units, show that the distance between some pair of points is smaller than $\sqrt{2}$ units.

Divide the square into 4 congruent squares by joining the midpoints of opposite edges. The diagonal of each of the small squares is $\sqrt{2}$. We have 4 squares and 5 points.

1.84 Show that any set of 7 distinct integers includes 2 integers x and y such that either $x+y$ or $x-y$ is divisible by 10.

Let $X=\{x_1, x_2, \ldots, x_7\}$ be a set of 7 distinct integers and let r_i be the remainder when x_i is divided by 10. Consider the following partition of X:

$$H_1=\{x_i : r_i=0\} \qquad H_2=\{x_i : r_i=5\}$$
$$H_3=\{x_i : r_i=1 \text{ or } 9\} \qquad H_4=\{x_i : r_i=2 \text{ or } 8\}$$
$$H_5=\{x_i : r_i=3 \text{ or } 7\} \qquad H_6=\{x_i : r_i=4 \text{ or } 6\}$$

There are 6 pigeonholes for 7 pigeons. If x and y are in H_1 or in H_2, then both $x+y$ and $x-y$ are divisible by 10. If x and y are in one of the other 4 subsets, then either $x-y$ or $x+y$ is divisible by 10, but not both.

1.85 The total number of games played by a team in a 15-day season was 20. The rules required the team to play at least 1 game daily. Show that there was a period of consecutive days during which exactly 9 games were played.

Let x_i be the number of games played by the team up to and including the ith day. The 15 numbers $x_1, x_2, \ldots, x_{15}$ (set A) are all distinct and increasing; hence the 15 numbers $x_1+9, x_2+9, \ldots, x_{15}+9$ (set B) are also distinct and increasing. Thus we have a set $(A \cup B)$ of 30 positive integers (pigeons) with at most $x_{15}+9=29$ distinct values (pigeonholes). No 2 elements of A, nor of B, can be equal. Therefore, for some i and j, $x_j=x_i+9$, or

$$9=x_j-x_i=\text{number of games played in days } i+1, i+2, \ldots, j$$

1.86 Show that in any assignment of n objects to r places there will be at least 2 places with the same number of recipients, if $n<r(r-1)/2$.

Let x_i be the number of objects assigned to place i, where $i=1, 2, \ldots, r$. No 2 places get the same number of recipients if and only if the r integers x_i are all distinct. If this condition is fulfilled, we can relabel the places so as to make $x_1<x_2<\cdots<x_i<\cdots<x_r$. Then $x_i \geq i-1$ for all i, whence, by addition

$$\Sigma x_i \geq \Sigma (i-1) \qquad \text{or} \qquad n \geq \frac{r(r-1)}{2}$$

So, if $n<r(r-1)/2$, the x_i cannot all be distinct.

1.87 There are 12 microcomputers and 8 laser printers in an office. Find the minimum number of connections to be made which will guarantee that if 8 or fewer computers want to print at the same time, each of them will be able to use a different printer.

We shall show that 40 connections will do the job, leaving it to the reader to prove that this number is minimal. Suppose the printers are denoted by P_j $(j = 1, 2, \ldots, 8)$ and the computers by C_i $(i = 1, 2, \ldots, 12)$. Connect the first printer to the first 5 computers. Then connect the second printer to the 5 consecutive printers starting with C_2. Then connect the third printer to the 5 consecutive printers starting with C_3. Continue like this, generating the the connection matrix of Fig. 1-1.

	P_1	P_2	P_3	P_4	P_5	P_6	P_7	P_8
C_1	1	0	0	0	0	0	0	0
C_2	1	1	0	0	0	0	0	0
C_3	1	1	1	0	0	0	0	0
C_4	1	1	1	1	0	0	0	0
C_5	1	1	1	1	1	0	0	0
C_6	0	1	1	1	1	1	0	0
C_7	0	0	1	1	1	1	1	0
C_8	0	0	0	1	1	1	1	1
C_9	0	0	0	0	1	1	1	1
C_{10}	0	0	0	0	0	1	1	1
C_{11}	0	0	0	0	0	0	1	1
C_{12}	0	0	0	0	0	0	0	1

Fig. 1-1

Let the 8 computers requiring a printer be $C_{i_1}, C_{i_2}, \ldots, C_{i_8}$, where $i_1 < i_2 < \cdots < i_8$. (Obviously, if any 8 computers can be accommodated, any smaller number can be accommodated.) The crucial observation is that

$$s \le i_s \le s + 4 \qquad (s = 1, 2, \ldots, 8) \qquad (i)$$

Indeed, if $i_s < s$ there would be s positive integers smaller than s; and if $i_s \ge s + 5$, at most $12 - (s + 6) + 1 = 7 - s$ values would be available to the $8 - s$ remaining indices. It follows from *(i)* and Fig. 1-1 that P_1 can be reserved for C_{i_1}; P_2 for C_{i_2}; ...; P_8 for C_{i_8}.

1.88 If each row and each column of an $n \times n$ matrix is a permutation of the first n positive integers, the matrix is known as a **latin square of order *n***. Two latin squares of order n, $A = [a_{ij}]$ and $B = [b_{ij}]$, are **orthogonal** if the n^2 ordered pairs (a_{ij}, b_{ij}) are all distinct. Suppose $A_1, A_2, \ldots, A_t$ are pairwise orthogonal latin squares of order n. Show that t cannot exceed $n - 1$.

Obviously $t = 0$ for $n = 1$ and $n = 2$; thus we assume $n > 2$. The pairwise orthogonality of $A_1, A_2, \ldots, A_t$ is not disturbed if the first row of each of these matrices is transformed into $[1\ \ 2\ \ 3\ \cdots\ n]$ through suitable permutations of the columns. Now the first element in the second row of any one of these matrices must come from the set $\{2, 3, \ldots, n\}$. Moreover, if i and j are the first elements in the second rows of 2 of these matrices, then i and j are distinct. (Otherwise, the orthogonality condition is violated, because all first rows are $[1\ \ 2\ \cdots\ n]$. Hence the set $\{2, 3, \ldots, n\}$ has to supply t distinct values for the (2, 1)-element; perforce $t \le n - 1$.

1.89 Prove that any set of 3 distinct integers includes 2 integers x and y such that $F(x, y) \equiv x^3y - xy^3$ is divisible by 10.

The result is true if the set includes $x = 0$ or $y = 0$. Also,

$$F(-x, y) = F(x, -y) = -F(x, y) \qquad \text{and} \qquad F(-x, -y) = F(x, y)$$

so that we may assume without loss of generality that the 3 distinct integers are all positive. Now, for any x and y, $F(x, y)$ is even. So it is enough to show that $F(x, y)$ is divisible by 5, which will certainly be the case if either x or y is divisible by 5. Since $F(x, y) = xy(x - y)(x + y)$, what we have to prove is this: given any 3 positive integers none of which is divisible by 5, the sum or difference of 2 of them is divisible by 5.

Now, the last digit of any number not divisible by 5 belongs to the set $A = \{1, 2, 3, 4, 6, 7, 8, 9\}$. Let $B = \{1, 4, 6, 9\}$ and $C = \{2, 3, 7, 8\}$—two pigeonholes. Of the 3 integers (pigeons) in our set, at least 2 belong to B or at least 2 belong to C. In either case, either their sum or their difference is divisible by 5. This completes the proof.

1.90 Show that any sequence of $n^2 + 1$ distinct real numbers contains a subsequence of at least $n + 1$ terms that is either an increasing sequence or a decreasing sequence. In particular, every sequence of n distinct numbers has a monotone subsequence of length at least $\sqrt{n}$.

For the sequence $\langle a_i : i = 1, 2, \ldots, n^2 + 1\rangle$, let p_i be the number of terms in the longest increasing subsequence that starts with a_i. If $p_i \geq n + 1$ for some i, we are done. Suppose, on the contrary, $p_i \leq n$ for every i. Let $H_j = \{a_i : p_i = j\}$, where $j = 1, 2, \ldots, n$. The $n^2 + 1$ elements of the sequence are thus partitioned into n sets. By Theorem 1.3 (choose $p_1 = \cdots p_n = n + 1$), at least $n + 1$ of these elements belong to one of the n sets, say, H_r. Let a_i and a_j be two numbers in H_r, where $i < j$. If $a_i < a_j$, there is a subsequence with at least $r + 1$ terms starting from a_i, which is a contradiction. Thus $a_i > a_j$ whenever $i < j$. So take any $n + 1$ elements from H_r and arrange them in increasing order of their subscripts, to obtain a decreasing sequence of $n + 1$ elements.

In particular, every sequence of k distinct numbers has a monotone subsequence (decreasing or increasing) of at least $\sqrt{k - 1} + 1 > \sqrt{k}$ numbers.

1.91 Suppose X is the set of the first $2n$ positive integers and S is any subset of X with $n + 1$ elements. Show that S contains 2 integers such that 1 is divisible by the other.

Any element r of S can be written as $r = 2^t s$, where t is a nonnegative integer and s is an odd number from the set X. There are at most n choices for s. So there are at least 2 numbers x and y in S such that $x = 2^p s$ and $y = 2^q s$, with $p \neq q$. Hence, either x divides y or vice versa.

1.92 (*a*) Suppose $P_i(x, y_i)$, where $i = 1, 2, \ldots, 5$, are 5 lattice points (Problem 1.58) in the plane. Show that at least 1 of the line segments determined by pairs of these lattice points has a lattice point as its midpoint. (*b*) Generalize the result of (*a*) to n-dimensional Euclidean space.

(*a*) The set A of all lattice points in Euclidean 2-space can be partitioned into 4 subsets: A_1 is the subset where both the coordinates are odd; A_2 is the subset where both the coordinates are even; A_3 is the subset in which the first coordinate is odd and the second coordinate is even; and A_4 is the subset in which the first coordinate is even and the second coordinate is odd; Out of the 5 given lattice points, at least 2 must belong to 1 of these 4 subsets. The midpoint of the segment joining this pair is a lattice point.

(*b*) Given $2^n + 1$ lattice points in Euclidean n-space, the midpoint of at least one of the line segments determined by these points is a lattice point.

RAMSEY NUMBERS

1.93 Show that in any group of 6 people there will always be a subgroup of 3 people who are pairwise acquainted or a subgroup of 3 people who are pairwise strangers.

Let {A, B, C, D, E, F} be a group of 6 people. Suppose that the people known to A are seated in room Y and the people not known to A are seated in room Z; A is not in either room. Then there are necessarily at least 3 people in either room Y or in room Z. (*a*) Suppose B, C, and D to be in room Y. Either these 3 people are mutual strangers (and the theorem is true) or at least 2 of them (say, B and C) know each other. In the latter case, A, B, and C form a group of 3 mutual acquaintances—and the theorem is true. (*b*) In (*a*), replace room Y by room Z and interchange the notions of "acquaintance" and "strangers."

1.94 Show that in any group of 10 people there is always (*a*) a subgroup of 3 mutual strangers or a subgroup of 4 mutual acquaintances, and (*b*) a subgroup of 3 mutual acquaintances or a subgroup of 4 mutual strangers.

(*a*) Let A be 1 of the 10 people; the remaining 9 people can be assigned to 2 rooms: those who are known to A are in room Y and those who are not known to A are in room Z. Either room Y has at least 6 people or room Z has at least 4 people. (*i*) Suppose room Y has at least 6 people. Then, by Problem 1.93, there is either a subgroup of 3 mutual acquaintances or a subgroup of 3 mutual strangers (validating the theorem) in this room. In the former case, A and these 3 people constitute 4 mutual acquaintances. (*ii*) Suppose room Z has at least 4 people. Either these 4 people know one another or at least 2 of them, B and C, do not know each other. In the former case we have a subgroup of 4 mutual acquaintances. In the latter case A, B, and C constitute 3 mutual strangers.

(*b*) In the previous scenario, let people who are strangers become acquaintances, and let people who are acquaintances pretend they are strangers. The situation is symmetric.

1.95 Show that in any group of 20 people there will always be either a subgroup of 4 mutual acquaintances or a subgroup of 4 mutual strangers.

Suppose A is one of these 20 people. People known to A are in room Y and people not known to A are in room Z. Either room Y has at least 10 people or room Z has at least 10 people. (*i*) If Y has at least 10 people, then by Problem 1.94(*b*), there is either a subgroup of 3 mutual acquaintances or a subgroup of 4 mutual strangers—as asserted—in this room. In the former case A and these mutual acquaintances will form a subgroup of 4 mutual acquaintances. (*ii*) Interchange "acquaintances" and "strangers" in (*i*).

1.96 Let p and q be 2 positive integers. A positive integer r is said to have the **(p, q)-Ramsey property** if in any group of r people either there is a subgroup of p people known to one another or there is a subgroup of q people not known to one another. [By **Ramsey's theorem** all sufficiently large integers r have the (p, q)-Ramsey property.] The smallest r with the (p, q)-Ramsey property is called the **Ramsey number**, $R(p, q)$. Show that (*a*) $R(p, q) = R(q, p)$, (*b*) $R(p, 1) = 1$, and (*c*) $R(p, 2) = p$.

(*a*) See Problems 1.93(*b*), 1.94(*b*), and 1.95(*b*). (*b*) This is obvious. (*c*) In any group of p people, if all of them are not known to one another, there will be at least 2 people who do not know each other.

1.97 Show that $R(3, 3) = 6$.

Problem 1.93 implies that $R(3, 3) \leq 6$. To show that $R(3, 3) > 5$, it is enough to consider a seating arrangement of 5 people about a round table in which each person knows only the 2 people on either side. In such a situation there is no set of 3 mutual acquaintances and no set of 3 people not known to one another.

1.98 Show that if m and n are integers both greater than 2, then

$$R(m, n) \leq R(m-1, n) + R(m, n-1)$$

[This recursive inequality gives an (unsharp) upper bound for $R(m, n)$.]

Let $p \equiv R(m-1, n)$, $q \equiv R(m, n-1)$, and $r \equiv p + q$. Consider a group $\{1, 2, \ldots, r\}$ of r people. Let L be the set of people known to person 1 and M be the set of people not known to person 1. The 2 sets together have $r - 1$ people; so either L has at least p people or M has at least q people. (*a*) If L has $p = R(m-1, n)$ people, then, by definition, it contains a subset of $m - 1$ people known to one another or it contains a subset of n people unknown to one another. In the former case the $m - 1$ people and person 1 constitute m people known to one another.

Thus, in their case, a group of $R(m-1, n) + R(m, n-1)$ people necessarily includes m mutual acquaintances or n mutual strangers; i.e.,

$$R(m, n) \le R(m-1, n) + R(m, n-1)$$

(*b*) By the usual symmetry argument the same conclusion follows when M contains q people.

1.99 Show that if m and n are integers greater than 1, then

$$R(m, n) \le C(m+n-2, m-1) \tag{i}$$

(a nonrecursive upper bound).

When $m = 2$ or $n = 2$, (*i*) holds with equality (Problem 1.96). The proof is by induction on $k = m + n$. As we have just seen, the result is true when $k = 4$. Assume the result true for $k - 1$. Then

$$R(m-1, n) \le C(m+n-3, m-2) \qquad \text{and} \qquad R(m, n-1) \le C(m+n-3, m-1)$$

Now, Pascal's identity gives $C(m+n-3, m-2) + C(m+n-3, m-1) = C(m+n-2, m-1)$; so that

$$R(m-1, n) + R(m, n-1) \le C(m+n-2, m-1)$$

But (Problem 1.98) $R(m, n) \le R(m-1, n) + R(m, n-1)$.

1.100 If $R(m-1, n)$ and $R(m, n-1)$ are both even and greater than 2, prove that

$$R(m, n) \le R(m-1, n) + R(m, n-1) - 1$$

As in Problem 1.98, let $p \equiv R(m-1, n)$, $q \equiv R(m, n-1)$, and $r \equiv p + q$. It suffices to establish that in any group $X = \{1, 2, \ldots, r-1\}$ of $r - 1$ people there is either a subgroup of m people who know one another or a subgroup of n people who do not know one another. Let d_i be the number of people known to person i, for $i = 1, 2, \ldots, r-1$. Since knowing is mutual, $d_1 + d_2 + \cdots + d_{r-1}$ is necessarily even. But $r - 1$ is odd; so d_i is even for at least 1 i, which we may take to be $i = 1$. Let L be the set of people known to person 1 and let M be the set of people not known to person 1. Since there are an even number of people in L, there must be an even number of people in M as well. Now either L has at least $p - 1$ people or M has at least q people. But $p - 1$ is odd. So either L has at least p people or M has at least q people. (*a*) Suppose L has at least p people. Because $p = R(m-1, n)$, L must contain either $m - 1$ people known to one another or n people not known to one another (in which case the theorem holds). In the former case these $m - 1$ people and person 1 will constitute m people known to one another (and the theorem holds). (*b*) The case of q or more people in M is handled by symmetry.

1.101 Show that $R(4, 3) = 9$.

By Problems 1.97, 1.96(*c*), and 1.100,

$$R(4, 3) \le R(3, 3) + R(4, 2) - 1 = 9$$

To prove that $R(4, 3) = R(3, 4) > 8$, we exhibit a group of 8 people which has no subgroup of 3 people known to one another and no subgroup of 4 people not known to one another. Here is a scenario: 8 people sit about a round table. Each person knows exactly 3 people: the 2 people sitting on either side of him and the person sitting farthest from him.

1.102 Show that $R(5, 3) = 14$.

$R(5, 3) \le R(4, 3) + R(5, 2) = 9 + 5 = 14$. To see that $R(5, 3) = R(3, 5) > 13$, consider a group of 13 people sitting at a round table such that each person knows only the fifth person on his right and the fifth person on his left. In such a situation there is no subgroup of 3 mutual acquaintances and no subgroup of 5 mutual strangers.

1.103 Show that $R(4,4)=18$.

$R(4,4) \le R(3,4)+R(4,3)=9+9=18$. To show that $R(4,4)>17$, consider an arrangement of 17 people about a round table such that each person knows exactly 6 people: the first, second, and fourth persons on one's right and the first, second, and fourth persons on one's left. It can be verified that in this arrangement there is no subgroup of 4 mutual acquaintances or of 4 mutual strangers.

[Ramsey numbers $R(p,q)$ with $p,q>2$ are called **nontrivial**. In Problems 1.97–1.103, 4 of the 7 known nontrivial Ramsey numbers have been computed.]

1.104 Let k_i $(i=1,2,\ldots,t)$ and m be positive integers, with each $k_i \ge m$ and $t \ge 2$. Let $\langle C_1, C_2, \ldots, C_t \rangle$ be an ordered partition of the class C of all m-element subsets of an n-element set X. [There are thus $C(n,m)$ elements in C.] Then the positive integer n has the **generalized** $(\boldsymbol{k_1, k_2, \ldots, k_t; m})$**-Ramsey property** if, for some value of i in the range 1 to t, X possesses a k_i-element subset B such that all m-element subsets of B belong to C_i. The smallest such n is the **generalized Ramsey number**, $R(k_1, k_2, \ldots, k_t; m)$. Show that $R(p,q)=R(p,q;2)$.

Let $n=R(p,q)$ and suppose that $X=\{1,2,\ldots,n\}$ is a group of n people. The class of all 2-element subsets of X is $C=\{\{i,j\}: i \ne j\}$. Let $C=C_1 \cup C_2$ be any partition of C; this partition defines and is defined by a relation of "knowing" whereby i and j know each other if and only if $\{i,j\}$ belongs to C_1. Now, since $n=R(p,q)$, either X has a subgroup of p people who know one another—i.e., a p-element subset B all 2-element subsets of which belong to C_1—or X has a subgroup of q mutual strangers—i.e., a q-element subset B' all 2-element subsets of which belong to C_2. Hence $n \ge R(p,q;2)$; and it is easy to see that the inequality cannot hold.

1.105 Show that the pigeonhole principle is equivalent to the proposition that

$$R(k_1, k_2, \ldots, k_t; 1) = k_1 + k_2 + \cdots + k_t - t + 1$$

Let $R(k_1, k_2, \ldots, k_t; 1)=n$. Thus n is the smallest positive integer such that when any n-element set X is arbitrarily partitioned as $X=C_1 \cup C_2 \cup \cdots \cup C_t$, then C_1 contains at least k_1 elements, or C_2 contains at least k_2 elements, or ..., or C_t contains at least k_t elements. The proof of Theorem 1.3 demonstrates that this minimal n has the value $k_1+k_2+\cdots+k_t-t+1$.

1.106 Let A be any $n \times n$ matrix. Matrix P is an $\boldsymbol{m \times m}$ **principal submatrix** of A if P is obtained from A by removing any $n-m$ rows and the same $n-m$ columns. Show that for every positive integer m, there exists a positive integer n such that every $n \times n$ *binary* matrix A has an $m \times m$ principal submatrix P in one of the following four categories:

(*i*) P is diagonal.
(*ii*) Every nondiagonal entry of P is 1.
(*iii*) P is lower triangular and every element in the lower triangle is 1.
(*iv*) P is upper triangular and every element in the upper triangle is 1.

Let n be any positive integer greater than $R(m,m,m,m;2)$ and let $A=[a_{ij}]$ be any $n \times n$ binary matrix, the rows of which constitute the set $X=\{r_1, r_2, \ldots, r_n\}$.

The class C of all 2-element subsets of X is partitioned into 4 classes, as follows:

$$C_1=\{\{r_i,r_j\}: a_{ji}=0, a_{ij}=0\} \qquad C_3=\{\{r_i,r_j\}: a_{ji}=0, a_{ij}=1\}$$
$$C_2=\{\{r_i,r_j\}: a_{ji}=1, a_{ij}=1\} \qquad C_4=\{\{r_i,r_j\}: a_{ji}=1, a_{ij}=0\}$$

Since $n \ge R(m,m,m,m;2)$, there exists a subset X' of X with m elements (rows) such that all 2-element subsets of X' are contained in one of these 4 classes. This implies the existence of an $m \times m$ principal submatrix in one of the categories (i) through (iv).

1.107 A collection of points in the plane are **in general position** if no 3 of the points are collinear. A polygon with n sides, or ***n*-gon**, is **convex** if the line segment joining any 2 interior points is also within the n-gon. Show that if 5 points in the plane are in general position, then 4 of them are the vertices of a convex quadrilateral.

Let the smallest convex polygon that contains the 5 points be a convex m-gon; obviously, all the vertices of this m-gon belong to the given set of points. If $m = 5$ or $m = 4$, there is nothing to prove. If $m = 3$ (the only other possibility), there is a triangle formed by 3 of the 5 points (say, A, B, and C), and the other 2 points, D and E, are inside the triangle. Then the line determined by D and E will divide the triangle into 2 parts such that 1 of these 2 parts contains 2 vertices of the triangle (say, A and B); $ABDE$ is the sought convex quadrilateral.

1.108 If n points are located in general position in the plane, and if every quadrilateral formed from these n points is convex, then the n points are the vertices of a convex n-gon.

Suppose the n points do not form a convex n-gon. Consider the smallest convex polygon that contains the n points. At least one of the n points (say, the point P) is in the interior of this polygon. Let Q be one of the vertices of the polygon. Divide the polygon into triangles by drawing line segments joining Q to every vertex of the polygon. The point P then will be in the interior of one of these triangles, which contradicts the convexity hypothesis.

1.109 Show that for any integer $m \geq 3$ there exists an integer n such that whenever n points in the plane are in general position, some m of these points are the vertices of a convex m-gon.

Let $n \geq R(5, m; 4)$ and let X be any set of n points in general position. The class C of all 4-element subsets of X is partitioned into 2 subclasses, C_1 and C_2, the former being the subclass of quartets of points which determine convex quadrilaterals. Now, according to Ramsey's theorem, there exists an m-element subset, B, of X such that every 4-element subset of B belongs to C_1, or there exists a 5-element subset, B', of X such that every 4-element subset of B' belongs to C_2. The latter alternative is impossible, by Problem 1.107. The former alternative must then hold; and Problem 1.108 at once gives the proof.

1.110 An **arithmetic progression** of length n is a sequence of the form $\langle a, a+d, a+2d, \ldots, a+(n-1)d \rangle$. Show that in any partition of $X = \{1, 2, \ldots, 9\}$ into 2 subsets, at least 1 of the sets contains an arithmetic progression of length 3.

Suppose that the theorem is false. Let X be partitioned into P and Q, and let 5 be an element of P. Obviously both 1 and 9 $[d = 4]$ cannot be in P; so that there are 3 cases to consider.

Case 1. 1 is in P and 9 is in Q. Since 1 and 5 are in P, 3 is in Q. Since 3 and 9 are in Q, 6 is in P. Since 5 and 6 are in P, 4 is in Q. Since 3 and 4 are in Q, 2 is in P. Since 5 and 6 are in P, 7 is in Q. Since 7 and 9 are in Q, 8 is in P. But then P contains the arithmetic progression 2, 5, 8—a contradiction.

Case 2. 9 is in P and 1 is in Q. Set X is invariant when each element is replaced by its tens-complement. Under this transformation the present case becomes Case 1, which has already been disposed of.

Case 3. 1 and 9 are in Q. The number 7 is either in P or in Q; suppose it is in P. Since 5 and 7 are in P, both 3 and 6 are in Q. That means Q has the arithmetic progression 3, 6, 9. On the other hand, if 7 is in Q, then 8 is in P. Since 1 and 7 are in Q, 4 is in P. Since 4 and 5 are in P, 3 is in Q. Since 1 and 3 are in Q, 2 is in P. Then P has the arithmetic progression 2, 5, 8.

1.111 A **geometric progression** of length n is a sequence of the form $\langle a, ad, ad^2, ad^3, \ldots, ad^{n-1} \rangle$. Show that in any partition of $X = \{1, 2, 3, \ldots, 2^8\}$ into 2 sets, at least 1 of the sets contains a geometric progression of length 3.

Given the partition $X = X_1 \cup X_2$, let $P \equiv \{1, 2, \ldots, 9\}$, $P_1 \equiv \{k \in P : 2^{k-1} \in X_1\}$, and $P_2 \equiv P - P_1$. If P_1 is empty, then X_2 must contain the geometric progression 1, 2, 4. If P_2 is empty, then X_1 contains 1, 2, 4. Finally, if $P = P_1 \cup P_2$ is a partition, Problem 1.110 ensures that one of the subsets—say, P_1—contains on arithmetic progression $k, k+l, k+2l$. In consequence, X_1 must contain the geometric progression $2^{k-1}, 2^{k-1}2^l, 2^{k-1}(2^l)^2$.

CATALAN NUMBERS

1.112 A **path from P_0 to P_m** in the cartesian plane is a sequence $\langle P_0, P_1, \ldots, P_m \rangle$ of lattice points (Problem 1.58), $P_i(x_i, y_i)$, such that, for each $i = 0, 1, \ldots, m-1$, $x_{i+1} = x_i + 1$, $y_{i+1} = y_i$ or $x_{i+1} = x_i$, $y_{i+1} = y_i + 1$. This path is **good** if $y_i < x_i$ $(i = 0, 1, \ldots, m)$; otherwise it is **bad**. (*a*) Find the number of paths from P_0 to P_m. (*b*) Obtain a necessary and sufficient condition for given endpoints P_0 and P_m to be linked by paths of both categories.

(*a*) In Problem 1.58 set $m = x_m - x_0$ and $n = y_m - y_0$ to obtain the required number as $C(x_m - x_0 + y_m - y_0, x_m - x_0)$. (*b*) See Fig. 1-2: a good path is one that lies entirely below the 45° line. Thus the conditions $y_0 < x_0$ and $y_m < x_m$ are necessary for a good path, to which may be adjoined $x_0 \le x_m$ and $y_0 \le y_m$ (the x and y coordinates can never decrease along the path). Under these 4 conditions *all* paths will be good, unless it is possible for a path to intersect the 45° line at some ordinate less than or equal to y_m; i.e., unless $x_0 \le y_m$. Thus the desired criterion is

$$y_0 < x_0 \le y_m < x_m$$

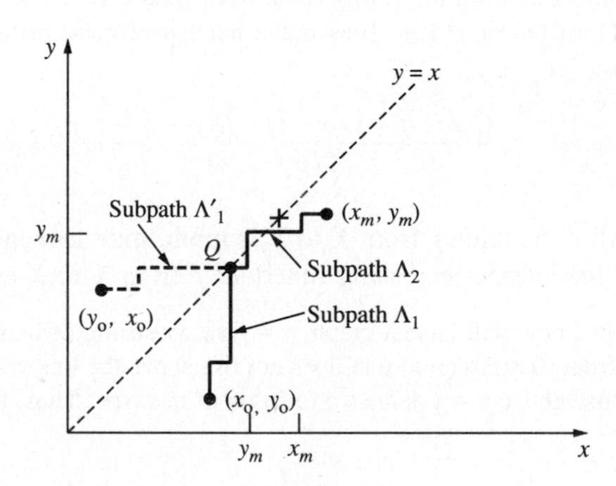

Fig. 1-2

1.113 Count the good paths from (x_0, y_0) to (x_m, y_m).

Figure 1-2 shows a bad path from (x_0, y_0) to (x_m, y_m); it *first* intersects the line $y = x$ in the lattice point Q. If subpath Λ_1, from (x_0, y_0) to Q, is reflected in the 45° line, then $\Lambda_1' + \Lambda_2$ is a *path* from (y_0, x_0) to (x_m, y_m). [All paths from (y_0, x_0) are bad, but that is of no importance here.] Conversely, any path from (y_0, x_0) to (x_m, y_m) defines by partial reflection a bad path from (x_0, y_0) to (x_m, y_m). By Problem 1.58 there are $C(x_m - y_0 + y_m - x_0, x_m - y_0)$ bad paths—and hence

$$C(x_m - x_0 + y_m - y_0, x_m - x_0) - C(x_m - x_0 + y_m - y_0, x_m - y_0)$$

good paths—from (x_0, y_0) to (x_m, y_m).

1.114. The ***n*th Catalan number**, C_n, is defined as the number of good paths from $(1, 0)$ to $(n, n-1)$. Show that

$$C_n = \frac{1}{n} C(2n-2, n-1)$$

Making the appropriate substitutions in Problem 1.113, we have

$$C_n = C(2n-2, n-1) - C(2n-2, n) = C(2n-2, n-1)\left[1 - \frac{n-1}{n}\right] = \frac{1}{n}C(2n-2, n-1)$$

1.115 Find the number of paths from $(0, 0)$ to (n, n) such that (*a*) either $x > y$ at all *interior* lattice points or $y > x$ at all interior lattice points; and (*b*) $y \le x$ at every lattice point on the path; and (*c*) the path never crosses the line $y = x$.

(*a*) The number of paths of this type will be twice the number of good paths from $(1, 0)$ to $(n, n-1)$, or $2C_n$.

(*b*) Let A be the point (n, n). Suppose the origin $O(0, 0)$ is transferred to $O'(-1, 0)$. The new coordinates are $O'(0, 0)$, $O(1, 0)$, and $A(n+1, n)$. The number of good paths (in the new system) from O to A—namely, C_{n+1}—is equal to the number of paths (in the old system) from O *to* A in which $y \le x$ at every lattice point.

(*c*) By reflectional symmetry, the required number is twice the number found in (*b*), or $2C_{n+1}$.

1.116 (*The Ballot Problem*) Suppose P and Q are 2 candidates for a public office who secured p votes and q votes, respectively. If $p > q$, find the probability that P stayed ahead of Q throughout the counting of votes.

In the cartesian plane let x and y, respectively, denote the votes accumulated by P and Q at any stage. Every path (Problem 1.112) from $(0, 0)$ to (p, q) represents a possible history of the voting, and conversely. Thus (Problem 1.58) the number of ways the voting could have gone is $C(p+q, p)$, out of which P leads continually in $C(p+q-1, p-1) - C(p+q-1, p)$ [this is the number of good paths from $(1, 0)$ to (p, q)]. The desired probability is therefore

$$\frac{C(p+q-1, p-1) - C(p+q-1, p)}{C(p+q, p)} = \frac{p-q}{p+q}$$

1.117 Let $X = \{1, 2, \ldots, n\}$. A function f from X to X is **monotonic increasing** if $f(i) \le f(j)$ whenever $i < j$. Find the number of monotonic increasing functions f from X to X such that $f(i) \le i$ for every i in X.

Any function of this type will have a graph, $y = f(x)$, consisting of lattice points that can be embedded in a unique way in a path from $(0, 0)$ to (n, n) that does not rise above the line $y = x$. Conversely, any path from $(0, 0)$ to (n, n) that does not rise above $y = x$ defines a function of this type. Thus, from Problem 1.115(*b*), the answer is C_{n+1}.

1.118 Find the number of sequences of the form $\langle u_1\ u_2, \ldots, u_{2n}\rangle$ such that

(*i*) u_i is either -1 or $+1$, for every i;

(*ii*) $u_1 + u_2 + \cdots + u_k \ge 0$, for $1 \le k \le 2n-1$; and

(*iii*) $u_1 + u_2 + \cdots + u_{2n} = 0$.

Consider a path from $(0, 0)$ to (n, n) as traced by a particle which makes unit steps in the x and y directions. Let the particle's location after i steps be (x_i, y_i) and define

$$u_i \equiv (x_i - x_{i-1}) - (y_i - y_{i-1})$$

Then, if the particle never rises above the line $y = x$, the integers u_i $(i = 1, 2, \ldots, 2n)$ satisfy (*i*), (*ii*), and (*iii*) above.

Conversely, every sequence $\langle u_i\rangle$ that obeys (*i*), (*ii*), and (*iii*) defines a path from $(0, 0)$ to (n, n) that never rises above $y = x$. Hence [Problem 1.115(*b*)], the number of such sequences is C_{n+1}.

1.119 Find the number of sequences of the form $\langle a_1, a_2, \ldots, a_{2n+1}\rangle$, where

(*i*) each a_i is a nonnegative integer;

(*ii*) $a_1 = a_{2n+1} = 0$; and

(*iii*) $a_{i+1} - a_i$ is either -1 or 1, for every i.

Define $u_i = a_{i+1} - a_i$ $(i = 1, 2, \ldots, 2n)$, or, inversely,

$$a_{k+1} = \sum_{i=1}^{k} u_i \qquad (k = 0, 1, \ldots, 2n)$$

Then the a_i obey (*i*), (*ii*), and (*iii*) above if and only if the u_i obey (*i*), (*ii*), and (*iii*) of Problem 1.118. Thus the number of a sequences equals the number of u sequences, which is C_{n+1}.

1.120 To obtain the product of n numbers, $n - 1$ successive multiplications are to be performed, involving 2 factors at a time. A pair of parentheses is used to indicate the 2 components of a multiplication whenever the lack of these would cause ambiguity. Find the number of ways of expressing the product of n numbers in this manner, if the n numbers can be arranged in any order.

The product of a and b can be expressed as ab or ba. The product of a, b, c, and d can be expressed as $(ab)(cd)$, $(cd)(ab)$, $a((bc)d)$, and so on. Let ω_n be the number of ways of expressing the product of $x_1, x_2, \ldots, x_n$ by inserting parentheses. From each product of these n numbers, we can obtain a product of $x_1, x_2, \ldots, x_n, x_{n+1}$ by one of the following methods:

(*a*) Suppose the product of the n numbers is y. Then we can form either $x_{n+1}y$ or yx_{n+1} (2 possibilities).

(*b*) Suppose u is the product of the first r numbers out of the n numbers, and v is the product of the remaining $n - r$ numbers. Then we have the 4 possibilities $(x_{n+1}u)v$, $(ux_{n+1})v$, $u(x_{n+1}v)$, and $u(vx_{n+1})$.

Now r can vary from 1 to $n - 1$. Thus each representation involving n numbers defines $2 + 4(n - 1) = 4n - 2$ representations involving $n + 1$ numbers; or

$$\omega_{n+1} = (4n-2)\omega_n = (4n-2)(4n-6)\omega_{n-1} = \cdots = (4n-2)(4n-6)\cdots(3)(1) = \frac{2^n(2n)!}{2^n n!} = \frac{(2n)!}{n!}$$

It then follows from Problem 1.114 that $\omega_n = n!C_n$.

We note that exactly $n - 2$ pairs of parentheses—one **opening** and one **closing**—are required in any **parenthesisation** of an n-factor product.

1.121 Rework Problem 1.120 if the order of the n numbers is fixed.

It is clear that any parenthesisation of n numbers in fixed order gives rise to $n!$ homologous parenthesisations when the numbers are permuted among themselves. Moreover, every parenthesisation counted in Problem 1.120 can be derived by permutation from the homologous fixed-order parenthesisation. Thus the required number of ways is $\omega_n/n! = C_n$—an alternate definition of the nth Catalan number.

1.122. A *diagonal triangulation* of a convex polygon is a division of the polygon into triangles by diagonals which do not intersect except at vertices of the polygon. Show that the number of triangles and the number of diagonals in any diagonal triangulation of a convex polygon with n vertices are $n - 2$ and $n - 3$, respectively.

Suppose that the numbers of triangles in 2 different triangulations are p and q. Then

$$p\pi = q\pi = \text{sum of the interior angles of the polygon}$$

whence $p = q$. With the n vertices of the polygon marked $\mathbf{1, 2, \ldots, n}$, the diagonals $\mathbf{13, 14, \ldots, 1n-1}$ yield a particular triangulation. This—and therefore every—triangulation involves $n - 2$ triangles.

In counting the edges of the $n - 2$ triangles in any triangulation, each diagonal involved is counted twice, since it is an edge of exactly 2 triangles. Each side of the polygon is counted once, since it is an edge of exactly 1 triangle. Therefore, with $x \equiv$ number of diagonals, we have $3(n - 2) = 2x + n$, or $x = n - 3$.

1.123 Consider a triangulation of a convex n-gon ($n \geq y$). Let us call a constituent triangle *type 0*, *type 1*, or *type 2*, according as the triangle has 0, 1, or 2 sides in common with the polygon. Prove that there must be at least two type 2 triangles in the triangulation.

Let f_0, f_1, and f_2 be the respective numbers of the 3 types. From Problem 1.122,

$$f_0 + f_1 + f_2 = n - 2 \qquad (i)$$

and, by a double counting of the sides of the polygon,

$$f_1 + 2f_2 = n$$

Subtract (i) from (ii): $f_2 - f_0 = 2$, or $f_2 = 2 + f_0 \geq 2$.

1.124 A particular diagonal triangulation of a convex hexagon is shown in Fig. 1-3(a); 5 of the 6 sides are labeled in clockwise order $x_1, x_2, \ldots, x_5$. Show that the triangulation induces a unique parenthesisation of the "product" $x_1 x_2 \cdots x_5$.

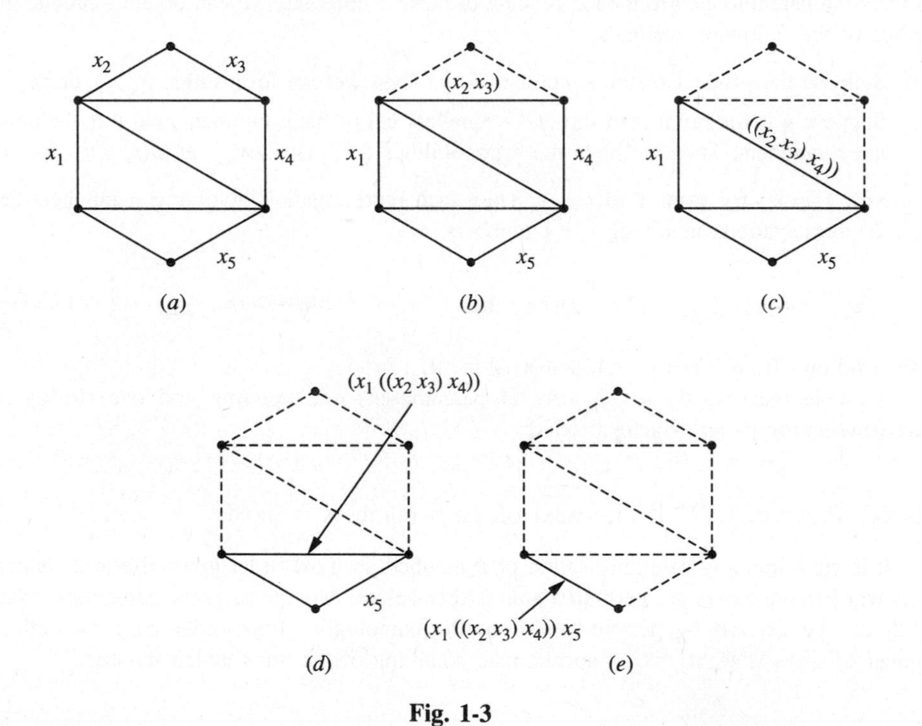

Fig. 1-3

Problem 1.123 guarantees that the triangulation must involve a type 2 triangle that does not contain the unlabeled side. Take any such triangle (in the present case there is just 1) and collapse it into that side which is a diagonal of the hexagon; label the surviving side with the parenthesized product of the labels of the 2 sides that vanish. The resulting labeled pentagon is shown in Fig. 1-3(b). Now iterate the process—see Figs. 1-3(c), (d), (e)—until the initially unlabeled side carries the sought parenthesisation of $x_1 x_2 \cdots x_5$.

1.125 Show that the number of diagonal triangulations of a convex polygon with $n + 1$ vertices is C_n (whereby still another definition is given for the Catalan numbers).

The general "collapse procedure" of Problem 1.124, together with the reverse "explosion procedure," establishes a one-to-one correspondence between triangulations of a convex $(n + 1)$-gon and parenthesisations of an ordered n-factor product. An appeal to Problem 1.121 gives the required result.

1.126 Find the number of triangulations of a convex polygon with $n \geq 4$ vertices such that every triangle is type 1 or type 2 (Problem 1.123).

The triangulation must involve $n-3$ diagonals (Problem 1.122), and, by Problem 1.123, there must be precisely two type 2 triangles. Number the vertices consecutively in such fashion that diagonal $d_1 = \mathbf{2n}$ cuts off one of these type 2 triangles (see Fig. 1-4). Now our triangulation must induce a triangulation of the convex $(n-1)$-gon $\mathbf{2\,3} \cdots \mathbf{n\,2}$ (interior shaded in Fig. 1-4) that also has the property $f_0 = 0$. This implies that side d_1 must be covered by a type 2 triangle (else the n-gon would have three type 2 triangles). Hence there are 2 choices for d_2: $d_2 = \mathbf{3n}$ and $d_2 = \mathbf{2n-1}$. Repeating the argument for the residual $(n-2)$-gon, etc., we obtain $(1)(2)(2)\cdots(2) = 2^{n-4}$ triangulations of the n-gon, in each of which the chosen d_{n-3} cuts off the second type 2 triangle. Because vertex **1** can be any vertex of the n-gon, we would seem to have altogether $n2^{n-4}$ triangulations. But in this grand total each triangulation is counted twice—once according to its beginning type 2 triangle and once according to its ending type 2 triangle. Our answer is therefore $n2^{n-5}$.

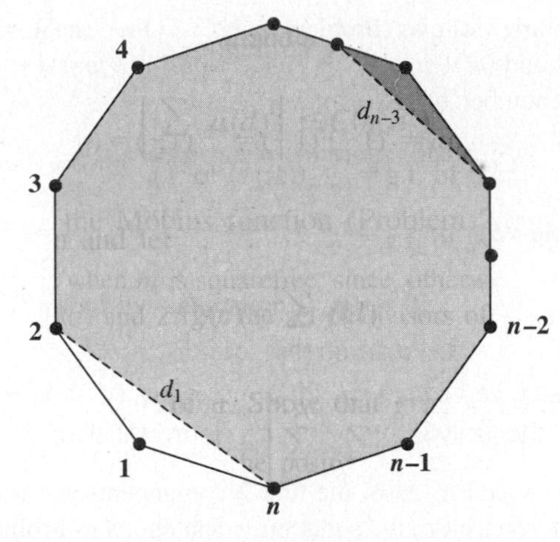

Fig. 1-4

1.127 To see a talent show, one has to buy an entrance ticket worth $5.00. A customer is allowed to buy only 1 ticket. Some people come with exactly one $10 bill, and some people come with exactly one $5 bill; suppose there are m customers in the former category and n customers in the latter category. The box office has no money initially. The customers stand in front of the box office in a line. Find the number of ways a line can be formed so that each customer gets a ticket and every customer who presents a $10 bill gets a $5 bill along with the ticket.

Obviously m cannot exceed n. Let the symbol **T** denote a $10 customer and let **F** denote a $5 customer. Each line in front of the box office can then be considered as a vector with $m+n$ components, such that (*i*) m components are the symbol **T** and n components are the symbol **F**; (*ii*) the ith component represents the ith customer away from the box office. Conversely, every such vector is a line of customers. The number of vectors is $C(m+n, m)$. A vector is **feasible** if every **F** gets a ticket and if every **T** gets a ticket and a $5 bill in change. Any other vector is **infeasible**. For example, a vector whose first component is **T** is infeasible.

We now count the infeasible $(m+n)$ vectors by showing that each such vector corresponds to a unique $(m+n+1)$ vector having first component **T** and having m **T**'s in all. Suppose v is an infeasible vector; i.e., at some stage a customer with a $10 bill comes to the box office and (for the first time) the box office has no $5 bill to give back. If this customer is represented by a **T** in the ith position, then among the previous $i-1$ components there occur equal numbers of **T**'s and **F**'s. In other words, among the first i components the number of **F**'s is p and the number of **T**'s is $p+1$, for some nonnegative integer p. Let an additional **F** be introduced as the first component. Then, up to and including the new $(i+1)$st component, the number of **T**'s and the number of **F**'s will be $p+1$. Suppose that each **T** is changed into a **F**, and each **F** is changed into a **T**, among these first $2p+2$ components. The unique result is a vector with $m+n+1$ components, the first component being **T** and with m **T**'s in all. Conversely, consider any vector with $m+n+1$ components, starting with **T** and having a total of m **T**'s. Since $m \leq n$, there must be subvector (starting from the first component) with equal numbers of **T**'s and **F**'s.

In this subvector, change **T**'s into **F**'s and vice versa, discard the first component, and adjoin the remaining portion of the vector to this subvector at its end. Then we have an infeasible vector for the given problem.

By virtue of the one-one correspondence just demonstrated, there are $C(m+n, m-1)$ infeasible vectors. Thus, by difference, the number of feasible vectors is

$$C(m+n, m) - C(m+n, m-1) = \frac{n-m+1}{n+1} C(m+n, m) \qquad (i)$$

By Problem 1.114, this number is the Catalan number C_{n+1} if $m = n$.

1.128 Repeat Problem 1.127 if it is given that the first q customers are **F**'s.

The number of possible line-vectors is equal to the number of sequences of m **T**'s and $n-q$ **F**'s, or $C(m+n-q, m)$. If $q \geq m$, all these vectors are feasible (the box office has accumulated at least m \$5 bills by the time the first \$10 customer arrives).

If $q < m$, the reasoning follows Problem 1.127: One establishes a one-one correspondence between infeasible $(m+n)$ vectors and $(m+n+1)$ vectors *of which the first* $q+1$ *components are* **T**'s, there being m **T**'s in all. Consequently, the number of infeasible vectors is

$$C(m+n+1-q-1, m-q-1) = C(m+n-q, n+1)$$

and the number of feasible vectors is

$$C(m+n-q, m) - C(m+n-q, n+1)$$

1.129 Find the number of binary vectors in which the number of 1s is m and the number of 0s is n, such that every component after the first is preceded by more 0s than 1s.

Obviously m cannot exceed n. Also, the first 2 components are both 0. We distinguish between two cases: $m < n$ and $m = n$. In the former case, the situation is analogous to Problem 1.127, where **F** corresponds to 0 and **T** corresponds to 1. Of the two 0s at the beginning, we do not take the first into consideration. We then have a feasible vector (with $m+n-1$ components) in which the number of 1s is m. By (i) of Problem 1.127 there are

$$\frac{n-m}{n} C(m+n-1, m)$$

such vectors.

If $m = n$, let a 0 represent a unit step in the x direction and a 1 represent a unit step in the y direction. Then there are just as many binary vectors of the given type as there are paths from $(0, 0)$ to (n, n) with the property that $x > y$ at each interior lattice point. By Problem 1.115(a), the number of these paths is C_n.

1.130 Find the number of ways of arranging $2n$ distinct real numbers as two n-vectors, $u = [u_1 \ u_2 \ \cdots \ u_n]$ and $v = [v_1 \ v_2 \ \cdots \ v_n]$, such that (i) in each vector the components are in strict decreasing order and (ii) $u_i > v_i$ for all i.

Imagine that the $2n$ numbers correspond to $2n$ customers (of distinct heights) waiting in front of a box office to buy a \$5 ticket, as in Problem 1.127, such that there are n customers who have exactly one \$5 bill and n who have exactly one \$10 bill. Let each u_i correspond to a person with a \$5 bill (an **F**) and each v_i to a **T**. Now assume that in the waiting line the $2n$ customers stand in order of decreasing heights (first in line is tallest). Then we have a vector of **F**'s and **T**'s with $2n$ components out of which the number of **F**'s is n. We assert that this vector is feasible in the sense of Problem 1.127. In fact, consider the **T** customer who corresponds to v_i in the vector v. The number of **T**'s ahead of this customer in the line is $i-1$, whereas—by condition (ii)—the number of **F**'s ahead of him is at least i. So this vector is indeed feasible.

Conversely, every feasible vector w can be decomposed into 2 vectors, u and v, as above: the vector v is the subline of \$5 customers and v is the subline of \$10 customers, each arranged in order of decreasing heights. In the feasible vector w each **T** is preceded by more **F**'s than **T**'s; hence u and v must obey condition (ii). This one-to-one correspondence shows that the required number of ways is C_{n+1}.

1.131 Prove combinatorially that the Catalan numbers satisfy the nonlinear recursion relation

$$C_{n+1} = \sum_{i=1}^{n} C_i C_{n-i+1} \qquad (n \geq 1)$$

The proof is obvious from Fig. 1-5. By Problem 1.115(*b*) there are C_{n+1} paths from $(0, 0)$ to (n, n) with $y \leq x$ at each lattice point. Let X_i $(i = 1, 2, \ldots, n)$ denote the subset of these paths that have (i, i) as their *second* lattice point on the line $y = x$. By Problem 1.115(*a*) there are C_i ways of going from $(0, 0)$ to (i, i); and by Problem 1.115(*b*) [with the origin of coordinates translated to (i, i)] there are C_{n-i+1} ways of going from (i, i) to (n, n). Hence, by the product rule, X_i contains $C_i C_{n-i+1}$ paths.

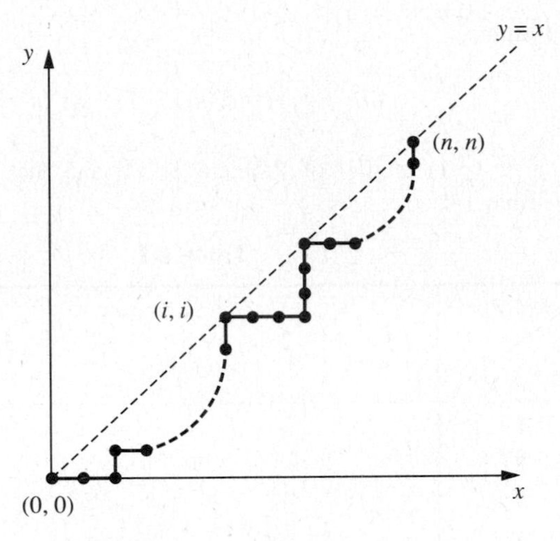

Fig. 1-5

STIRLING NUMBERS

1.132 Define the **falling factorial polynomials** by $[x]_0 \equiv 1$ and

$$[x]_n \equiv x(x-1)(x-2)\cdots(x-n+1) \qquad (n = 1, 2, 3, \ldots) \qquad (i)$$

The coefficient of x^r in $[x]_n$ is known as the **Stirling number of the first kind**, $s(n, r)$; thus

$$[x]_n \equiv \sum_{r=0}^{\infty} s(n, r)x^r \qquad \text{where} \qquad s(n, r) = 0 \text{ for } r > n \qquad (ii)$$

Prove the recurrence relation

$$s(n+1, r) = s(n, r-1) - ns(n, r) \qquad (iii)$$

By (*i*), $[x]_{n+1} = (x - n)[x]_n$; so that (*ii*) gives

$$\sum_{r=0}^{\infty} s(n+1, r)x^r = x \sum_{r=0}^{\infty} s(n, r)x^r - n \sum_{r=0}^{\infty} s(n, r)x^r$$

$$= \sum_{r=0}^{\infty} [s(n, r-1) - ns(n, r)]x^r$$

and equating coefficients of x^r yields (*iii*).

1.133 The absolute value of $s(n, r)$ is called the **signless** Stirling number of the first kind; we denote it as $s'(n, r)$. Verify that $s(n, r) = (-1)^{n-r}s'(n, r)$.

From the right-hand side of (i) of Problem 1.132 one sees that

$$s(n, r) = \sum i_1 i_2 \cdots i_{(n-1)-(r-1)} \qquad (i)$$

where the summation is over all $(n - r)$ combinations $\{i_1, i_2, \ldots, i_{n-r}\}$ of the set $X = \{-1, -2, \ldots, -(n-1)\}$. Because each summand has algebraic sign $(-1)^{n-r}$, the same is true for $s(n, r)$.

1.134 Construct a triangle of the signless Stirling numbers of the first kind, analogous to Pascal's triangle.

Use the recursion formula

$$s'(n + 1, r) = s'(n, r - 1) + ns'(n, r)$$

[substitute $s(n, r) = (-1)^{n-r}s'(n, r)$ in (iii) of Problem 1.132] and the "edge values" $s'(n, 1) = (n - 1)!$ and $s'(n, n) = 1$ to generate Table 1-1.

Table 1-1

n \ r	1	2	3	4	5	6	...
1	1						
2	1	1					
3	2	3	1				
4	6	11	6	1			
5	24	50	35	10	1		
6	120	274	225	85	15	1	
...	...	...	...	...	...	...	...

1.135 Define the **rising factorial polynomials** by $[x]^0 \equiv 1$ and

$$[x]^n = x(x + 1)(x + 2) \cdots (x + n - 1) \qquad (n = 1, 2, 3, \ldots)$$

Show that

$$[x]^n = \sum_{r=0}^{\infty} s'(n, r)x^r \qquad \text{where} \qquad s'(n, r) = 0 \text{ for } r > n$$

Follows at once from Problem 1.133 [when set X is replaced by $X' = \{+1, +2, \ldots, +(n - 1)\}$, then $s(n, r)$ is replaced by $|s(n, r)| \equiv s'(n, r)$].

1.136 Establish the following analogue to the binomial theorem (Problem 1.38):

$$[x + y]^n = \sum_{r=0}^{n} C(n, r)[x]^{n-r}[y]^r$$

Perform an induction on n. The formula is true for $n = 1$, because

$$[x + y]^1 = x + y = C(1, 0)[x]^1[y]^0 + C(1, 1)[x]^0[y]^1$$

Assume the formula true for $n = m - 1$. Then

$$[x+y]^m = (x+y+m-1)[x+y]^{m-1} = (x+y+m-1)\sum_{r=0}^{m-1} C(m-1,r)[x]^{m-1-r}[y]^r$$

$$= \sum_{r=0}^{m} \{(x+m-r-1)+(y+r)\}C(m-1,r)[x]^{m-1-r}[y]^r$$

$$= \sum_{r=0}^{m} C(m-1,r)[x]^{m-r}[y]^r + \sum_{r=0}^{m} C(m-1,r)[x]^{m-1-r}[y]^{r+1}$$

$$= \sum_{r=0}^{m} C(m-1,r)[x]^{m-r}[y]^r + \sum_{r=0}^{m} C(m-1,r-1)[x]^{m-r}[y]^r$$

$$= \sum_{r=0}^{m} C(m,r)[x]^{m-r}[y]^r$$

and the formula holds for $n = m$.

1.137 Find the number of ways of putting n distinct objects into m distinct boxes, if the (left-to-right) order of objects *within* a box is significant and if empty boxes are permitted. (Note that if $m > n$, at least $m - n$ boxes must be empty.)

Indicate the desired number by $f(n, m)$. Suppose that a distribution of $n - 1$ of the objects—there are $f(n-1, m)$ such distributions—brings i_1 objects into box 1, i_2 objects into box 2, ..., i_m objects into box m; here

$$i_k \geq 0 \quad (k = 1, 2, \ldots, m) \qquad \text{and} \qquad i_1 + i_2 + \cdots + i_m = n - 1$$

Then the nth object can go into box k in $i_k + 1$ ways [leftmost, second from left, ..., $(i_k + 1)$st from left], for a total of

$$(i_1+1)+(i_2+1)+\cdots+(i_m+1) = n-1+m$$

arrangements. Since this number is independent of the particular distribution of the $n - 1$ objects, we have the relation

$$f(n, m) = (n-1+m)f(n-1, m)$$

from which

$$f(n, m) = (m+n-1)(m+n-2)\cdots m = [m]^n$$

1.138 Rework Problem 1.137 if $m \leq n$ and empty boxes are not allowed.

Now each box must be given a leftmost object; this can be done in $P(n, m)$ ways. The remaining $n - m$ objects can, by Problem 1.137, be distributed in $[m]^{n-m}$ ways. So the answer is:

$$P(n,m)\,[m]^{n-m} = \frac{n!}{(n-m)!}\,m(m+1)(m+2)\cdots(n-1)$$

$$= \frac{n!}{(n-m)!}\,\frac{(n-1)!}{(m-1)!} = n!C(n-1, m-1)$$

1.139 If m and n are positive integers, prove that the equation

$$x_1 + x_2 + \cdots + x_m = n \qquad (i)$$

has exactly $[m]^n/n!$ solutions in nonnegative integers x_k. (The result also holds for $n = 0$.)

This is a matter of putting n *identical* objects (1s) into m distinct boxes (an x_1 box, an x_2 box, . . . , an x_m box), empty boxes being allowed. If we temporarily make the 1s distinct by labeling them $1_1, 1_2, \ldots, 1_n$, then Problem 1.137 implies $[m]^n$ arrangements. However, arrangements that differ only with regard to the labels carried by the 1s give the same solution to (i). Thus the answer is $[m]^n/n!$, as stated.

Note: Many books cite the result as $C(n+m-1, m-1)$; equivalence is easily demonstrated.

1.140. Suppose $A = \{a_i : i = 1, 2, \ldots, m\}$ is an alphabet consisting of m letters which are ordered as $a_1 < a_2 < \cdots < a_m$. A word $\theta_1 \theta_2 \cdots \theta_n$ from this alphabet is called an **increasing word** (of length n) if $\theta_1 \le \theta_2 \le \cdots \le \theta_n$. Show that the number of increasing words of length n is $[m]^n/n!$.

An increasing word of length n will consist of x_1 a_1's, followed by x_2 a_2's, . . . , followed by x_m a_m's, where $x_k \ge 0$ $(k = 1, 2, \ldots, m)$ and

$$x_1 + x_2 + \cdots + x_m = n$$

Conversely, any nonnegative integral solution of (i) of Problem 1.139 defines an increasing word of length n.

1.141 A function f whose domain is $N = \{\alpha_1, \alpha_2, \ldots, \alpha_n\}$ and range is $M = \{\beta_1, \beta_2, \ldots, \beta_m\}$ is an **increasing function** (from N to M) if $f(\alpha_i) \le f(\alpha_j)$ whenever $\alpha_i < \alpha_j$. Determine the number of such functions.

We can always suppose the listing of the sets to be such that

$$\alpha_1 < \alpha_2 < \cdots < \alpha_n \quad \text{and} \quad \beta_1 \le \beta_2 \le \cdots \le \beta_m$$

Then an increasing function from N to M will map the first x_1 α's into β_1, the next x_2 α's into β_2, . . . , the last x_m α's into β_m. Here, the x_k $(k = 1, 2, \ldots, m)$ are nonnegative integers whose sum is n. Conversely, any set of x_k with these properties defines an increasing function from N to M. Therefore, by Problem 1.139, the required number is $[m]^n/n! = C(n+m-1, m-1)$.

1.142 For prescribed nonnegative integers $\lambda_1, \lambda_2, \ldots, \lambda_m$, find the number of solutions in integers of the equation $x_1 + x_2 + \cdots + x_m = n$ with $x_i \ge \lambda_i$ for each i.

For each i, let $x_i = \lambda_i + y_i$ and write $\lambda \equiv \lambda_1 + \lambda_2 + \cdots + \lambda_m$. We then have to solve

$$y_1 + y_2 + \cdots + y_m = n - \lambda \qquad y_i \ge 0 \qquad (i = 1, 2, \ldots, m)$$

If $\lambda > n$, there is no solution; otherwise, by Problem 1.139, there are $C(n - \lambda + m - 1, m - 1)$ solutions.

1.143 Use a combinatorial argument to establish the identity

$$[x]^n = \sum_{k=1}^{n} u(n, k)[x]_k \tag{i}$$

where $u(n, k) \equiv (n!/k!)C(n-1, k-1)$.

The number of increasing words of length n afforded by an m-letter alphabet is $[m]^n/n!$, as was shown in Problem 1.140. Consider the set of all increasing words of length n composed from the k-letter subalphabet $a_{i_1} < a_{i_2} < \cdots < a_{i_k}$, *such that each of the k letters appears at least once in the word*. By Problem 1.142—with m replaced by k and with all $\lambda_i = 1$— the cardinality of this set is $C(n-1, k-1)$. Now, there are $C(m, k)$ choices for the subalphabet, and k ranges from 1 to m. Clearly, the corresponding sets constitute a partition of the increasing words of length n; therefore,

$$\frac{[m]^n}{n!} = \sum_{k=1}^{m} C(m, k)C(n-1, k-1) = \sum_{k=1}^{n} C(n-1, k-1)\frac{[m]_k}{k!} \tag{ii}$$

(The reader should verify the second equality.) Define

$$F(x) \equiv [x]^n - n! \sum_{k=1}^{n} C(n-1, k-1) \frac{[x]_k}{k!} = [x]^n - \sum_{k=1}^{n} u(n, k)[x]_k$$

$F(x)$ is a polynomial in x of degree at most $n-1$ which, according to (*ii*), vanishes for $x = 1, 2, \ldots n, \ldots$. Hence, $F(x) \equiv 0$, and (*i*) is proved.

1.144 Find the number of ways of selecting r distinct integers out of the first n positive integers such that the selection does not contain 2 consecutive integers.

Arrange the first n positive integers in a row, in increasing order starting from 1. If a number is chosen, put the symbol Y under that number; otherwise, put the symbol N under the number. Let x_1 be the number of N's preceding the first Y; x_2 the number of N's between the first and second Y's; $\ldots$; x_r the number of N's between the $(r-1)$st and rth Y's; and x_{r+1} the number of N's following the rth Y. Then there is a one-to-one correspondence between acceptable selections and integral solutions of

$$x_1 + x_2 + \cdots + x_{r+1} = n - r \qquad \text{with} \qquad x_1 \geq 0, x_2 \geq 1, \ldots, x_r \geq 1, x_{r+1} \geq 0$$

Problem 1.142 gives the desired number as $C(n - r + 1, r)$.

1.145 Find the number of ways of choosing r positive integers from among the first n positive integers such that no 2 consecutive integers appear in the choice and the choice does not include both 1 and n.

Case 1. The choice includes 1. In the notation of Problem 1.144, $x_1 = 0$ (there is a Y under 1) and $x_{r+1} \geq 1$ (there is an N under n). Thus we count the solutions of

$$x_2 + x_3 + \cdots + x_{r+1} = n - r \qquad \text{with} \qquad x_2 \geq 1, x_3 \geq 1, \ldots, x_{r+1} \geq 1$$

and obtain $C(n - r - 1, r - 1)$.

Case 2. The choice does not include 1. Now $x_1 \geq 1$ (there is an N under 1). Thus we count the solutions of

$$x_1 + x_2 + \cdots + x_{r+1} = n - r \qquad \text{with} \qquad x_1 \geq 1, x_2 \geq 1, \ldots, x_r \geq 1, x_{r+1} \geq 0$$

and obtain $C(n - r, r)$.

The total number of ways is,

$$C(n-r-1, r-1) + C(n-r, r) = \left[1 + \frac{n-r}{r}\right] C(n-r-1, r-1) = \frac{n}{r} C(n-r-1, r-1)$$

1.146 The number of ways of partitioning a set of n elements into m (nonempty) subsets is denoted by $S(n, m)$ and is known as the **Stirling number of the second kind.** By definition, $S(0, 0) = 1$; also, $S(n, m) = 0$ if $m > n$. Show that the number of surjections (onto mappings) from a set of n elements to a set of m elements is $m!\, S(n, m)$.

Given sets $X = \{x_1, x_2, \ldots, x_n\}$ and $Y = \{y_1, y_2, \ldots, y_m\}$, let $X = X_1 \cup X_2 \cup \cdots \cup X_m$ be an arbitrary partition of X into m nonempty subsets. Then any one-to-one correspondence between the y_i and the X_j defines a unique surjection from X to Y; there are precisely $m!$ such one-to-one correspondences. Since there are (by definition) $S(n, m)$ partitions—and since distinct partitions yield distinct surjections—we have $m!\, S(n, m)$ surjections in all.

1.147 Determine the number of ways of distributing n distinct (distinguishable) objects among m identical (indistinguishable) boxes, if (*a*) each box must get at least 1 object; (*b*) not every box need receive an object. (*c*) Repeat (*a*), if the boxes are distinguishable.

Directly from the definition of $S(n, m)$ we have:

$$(a)\quad S(n, m) \qquad (b)\quad S(n, 1) + S(n, 2) + \cdots + S(n, m)$$

(c) This is just the number of surjections from the set of objects to the set of boxes, or $m!\,S(n, m)$.

1.148 Prove: (a) $S(n, 2) = 2^{n-1} - 1$; (b) $S(n, n-1) = C(n, 2)$.

(a) See Problem 1.12. (b) Any partition of a set X with n elements into a class of $n-1$ subsets will contain 1 subset with 2 elements and $n-2$ subsets with 1 element each. Now the 2-element subset can be composed in $C(n, 2)$ ways.

1.149 Show that $S(n+1, m) = S(n, m-1) + mS(n, m)$.

Let $X \equiv \{x_1, x_2, \ldots, x_n\}$, $A \equiv \{x_{n+1}\}$, and $X' \equiv X \cup A$. Then $S(n+1, m)$ is the number of ways of partitioning X' into m subsets. There are two means of effecting such a partition: either take a partition of X into $m-1$ subsets and adjoin the set A or take a partition of X into m subsets and include the element x_{n+1} in any one of these. There are $S(n, m-1)$ ways of accomplishing the former objective and $mS(n, m)$ ways of accomplishing the latter. Thus the relation is established.

1.150 From the recurrence relation of Problem 1.149 and the boundary conditions $S(n, 1) = S(n, n) = 1$ and $S(n, m) = 0$ for $m > n$, construct a triangle of Stirling numbers of the second kind.

The triangle is indicated in Table 1-2.

Table 1-2

n \ m	1	2	3	4	5	6	7
1	1						
2	1	1					
3	1	3	1				
4	1	7	6	1			
5	1	15	25	10	1		
6	1	31	90	65	15	1	
7	1	63	301	350	140	21	1
...	...	...	...	...	...	...	...

1.151 Show that

$$x^n = \sum_{k=1}^{n} S(n, k)\,[x]_k$$

First we establish the formula when x is a positive integer. Let X be a set of n elements and Y a set of m elements. Let A be the set of all mappings f from X to Y. Since each element in X has m possible images in Y, there are m^n functions in the set A. For $k = 1, 2, \ldots, m$ define the subsets

$$A_k \equiv \{f \in A : f(X) \text{ has cardinality } k\}$$

In other words, f is in A_k if and only if f is a surjection from X to some subset of Y with k elements. There are $C(m, k)$ subsets of Y with k elements; hence, by Problem 1.146, the cardinality of A_k is $C(m, k)\,k!\,S(n, k) =$

$[m]_k\, S(n, k)$. Because the sets A_k constitute a partition of A, we have:

$$m^n = \sum_{k=1}^{m} S(n, k)[m]_k = \sum_{k=1}^{n} S(n, k)\, [m]_k$$

The remainder of the proof follows Problem 1.143.

1.152 Prove the "orthogonality relations"

$$(a)\quad \sum_k s(n, k)S(k, m) = S_{nm} \qquad\qquad (b)\quad \sum_k S(n, k)s(k, m) = \delta_{nm}$$

where δ_{nm} is the Kronecker delta.

(*a*) By Problems 1.132 and 1.151,

$$[x]_n \equiv \sum_k s(n, k)x^k \qquad\qquad x^k = \sum_m S(k, m)[x]_m$$

Therefore ,

$$[x]_n = \sum_k s(n, k)\Big(\sum_m S(k, m)[x]_m\Big) = \sum_m \Big(\sum_k s(n, k)S(k, m)\Big)[x]_m$$

By the linear independence of the polynomials $[x]_m$ $(m = 0, 1, \ldots, n)$, any one of them must have the same coefficient on either side of the above equation.

(*b*)

$$x^n = \sum_k S(n, k)[x]_k = \sum_k S(n, k)\Big(\sum_m s(k, m)x^m\Big) = \sum_m \Big(\sum_k S(n, k)s(k, m)\Big)\, x^m$$

1.153 Let $\mathbf{s}_p$ be the $p \times p$ matrix with entries $s(i, j)$ and $\mathbf{S}_p$ be the $p \times p$ matrix with entries $S(i, j)$, where $0 \le i,\ j \le p - 1$. Prove that the two matrices are inverses.

By Problem 1.152, $\mathbf{s}_p\mathbf{S}_p = \mathbf{S}_p\mathbf{s}_p = \mathbf{I}_p$, where $\mathbf{I}_p$ is the $p \times p$ identity matrix.

1.154 Show that $S(n, m) = \sum_k C(n - 1, k)S(k, m - 1)$.

The left-hand side represents the number of partitions of n objects into $m \le n$ cells. Let a denote a particular one of the objects, and in each partition distinguish the cell containing a as the a cell. Partition the partitions (!) according to the composition of the a cell, which must contain r objects $(r = 0, 1, \ldots, n - m)$ besides a. These r objects may be chosen in $C(n - 1, r)$ ways. For each choice, the other $m - 1$ cells may be filled in $S(n - 1 - r, m - 1)$ ways. Thus,

$$S(n, m) = \sum_r C(n - 1, r)S(n - 1 - r, m - 1)$$

and the required formula results on changing the summation index from r to $k = n - 1 - r$.

1.155 Prove that the sequence $\langle S(n, 1), S(n, 2), \ldots, S(n, n)\rangle$ is unimodal (Problem 1.74) for all $n > 2$.

As we see from Problem 1.150, this is certainly true—with a single peak—for small values of n. Making an induction, we suppose the result true for every $k \le n$; i.e., for each $k \le n$, there exists an integer m such that the sequence $\langle S(k, m)\rangle$ peaks at $m = m_k$. Moreover, we include in the induction hypothesis the assertion that m_k is nondecreasing in k (Problem 1.150 bears this out for small k). From Problem 1.149 we have:

$$S(n + 1, m) = S(n, m - 1) + mS(n, m)$$

and

$$S(n + 1, m - 1) = S(n, m - 2) + (m - 1)S(n, m - 1)$$

Hence, by subtraction,

$$S(n+1,m)-S(n+1,m-1)=[S(n,m-1)-S(n,m-2)]+m[S(n,m)-S(n,m-1)]+S(n,m-1)$$

By the induction hypothesis, the right-hand side—and therefore the left-hand side—is positive for all $m \le m_n$, which implies that

$$S(n+1,1)<S(n+1,2)<\cdots<S(n+1,m_n) \qquad (i)$$

Next consider the case $m \ge m_n+2$. By Problem 1.154,

$$S(n+1,m)-S(n+1,m-1)=\sum_{k\le n} C(n,k)[S(k,m-1)-S(k,m-2)]$$

By the induction hypothesis (including the assumed variation of m_k), each bracketed term on the right-hand side is negative. Hence,

$$S(n+1,m_n+1)>S(n+1,m_n+2)>\cdots>S(n+1,n+1) \qquad (ii)$$

Together, (i) and (ii) establish that $\langle S(n+1,m)\rangle$ is unimodal, with either $m_{n+1}=m_n$ or $m_{n+1}=m_n+1$. The induction is complete.

1.156 Prove the following relations:

(a) $C(i+j,i)s(n,i+j)=\sum_k C(n,k)s(k,i)s(n-k,j)$.

(b) $C(i+j,i)S(n,i+j)=\sum_k C(n,k)S(k,i)S(n-k,j)$.

(a) By our definition (Problem 1.132),

$$[x+y]_n=\sum_k s(n,k)(x+y)^k=\sum_k s(n,k)\sum_i C(k,i)x^iy^{k-i} \qquad (i)$$

But, by an induction that exactly parallels Problem 1.136, we have

$$[x+y]_n=\sum_k C(n,k)[x]_k[y]_{n-k} \qquad (ii)$$

In (i), the coefficient of x^iy^j is $s(n,i+j)C(i+j,i)$; in (ii), the coefficient of x^iy^j is $\sum_k C(n,k)s(k,i)s(n-k,j)$.

(b) By Problem 1.151 and formula (ii) above,

$$(x+y)^n=\sum_k S(n,k)[x+y]_k=\sum_k S(n,k)\sum_i C(k,i)[x]_i[y]_{k-i} \qquad (iii)$$

Next we expand $(x+y)^n$ by the binomial theorem and use Problem 1.151 again:

$$(x+y)^n=\sum_k C(n,k)x^ky^{n-k}=\sum_k C(n,k)\sum_i S(k,i)[x]_i\sum_j S(n-k,j)[y]_j \qquad (iv)$$

This time we compare the coefficients of $[x]_i[y]_j$ in (iii) and in (iv) to obtain the result.

1.157 Denote by $S_i(n,k)$ the number of ways of partitioning a set with n elements into k subsets such that each subset has at least i elements. Show that

$$S_i(n,k)=kS_i(n-1,k)+C(n-1,i-1)S_i(n-i,k-1)$$

Let x be a fixed element of a set X with n elements. Any partition of X into k subsets such that each subset has at least i elements falls into one of two categories, according to the cardinality of the subset that contains the element x is (1) greater than i, or (2) equal to i. (Compare Problem 1.154.)

A partition in category 1 is obtained by partitioning the set $X' = X - \{x\}$ into k subsets such that each subset has at least i elements and then including x in one of these k subsets. Thus the number of partitions in category 1 is $S_i(n-1, k)k$.

To generate a category 2 partition, we select $i-1$ elements from the set X' to form a set Y—in $C(n-1, i-1)$ ways. The cardinality of $Z = X' - Y$ is $n - i$, and so Z can be partitioned into $k-1$ subsets of size i or greater in $S_i(n-1, k-1)$ ways. To each such partition we adjoin the set $Y \cup \{x\}$ to obtain a partition of X of the second category. Thus the number of partitions in category 2 is $C(n-1, i-1)S_i(n-i, k-1)$.

1.158 A permutation of a finite set X is a bijective (one-to-one and onto) mapping from X to X. Suppose f is a permutation of X and x is any element of X. Define recursively, $f^1(x) \equiv f(x)$, $f^2(x) \equiv f(f^1(x)), \ldots, f^i(x) \equiv f(f^{i-1}(x)), \ldots$. Since X is finite, there exists a positive integer r such that $f^r(x) = x$. The sequence $\langle x, f^1(x), f^2(x), \ldots, f^{r-1}(x)\rangle$ is called a **cycle of order** (or **length**) r of the permutation f. Obviously, every permutation of X can be represented as a composition of k disjoint cycles, where k is at least 1 and at most the cardinality of X. Let

$$P_{n,k} \equiv \{f : f \text{ is a permutation of an } n\text{-set } X \text{ and } f \text{ has exactly } k \text{ cycles}\}$$

Find the cardinality of $P_{n,k}$.

Let the cardinality of $P_{n,k}$ be $d(n, k)$. The only permutation having n cycles is the identity mapping; so $d(n, n) = 1$. Let X' be the set obtained by adjoining a new element, y, to the set X. Then $d(n+1, k)$ is the number of permutations of X' each of which has k cycles. Now, if f is a permutation of X with $k-1$ cycles, the composition of f with the cycle of order 1 $\langle y\rangle$ defines a permutation of X' with k cycles. Next, consider a permutation of X with k cycles. When y is put into one of these cycles, a permutation of X' with k cycles results. The element y can join a cycle either at the beginning or in between 2 elements of the cycle. (Putting y at the end is equivalent to putting it at the beginning.) Since there are k cycles, y can be included in $k + (n-k)$ ways. Thus we have the relation

$$d(n+1, k) = d(n, k-1) + nd(n, k) \qquad (i)$$

We note at once that (i) is identical to the recurrence relation for the signless Stirling numbers of the first kind (see Problem 1.134); further, the starting values for the $d(n, k)$ are the same as for the $s'(n, k)$. Consequently, $d(n, k) = s'(n, k)$ for all n and k, giving a direct combinatorial significance to the signless Stirling numbers.

1.159 Evaluate $\sum_{k=1}^{n} s'(n, k)$.

From Problem 1.158,

$$\sum_{k=1}^{n} s'(n, k) = \sum_{k=1}^{n} d(n, k) = n!$$

1.160 Find the number of functions from a set X of n elements to a set Y of m elements such that the ranges of these functions each have exactly r elements (generalization of Problem 1.146).

A subset of Y with r elements can be chosen in $C(m, r)$ ways. Once a set is chosen, there are $r!\,S(n, r)$ surjections from X to that set, by Problem 1.146. Thus the total number of functions in this category is

$$C(m, r)\,r!\,S(n, r) = P(m, r)S(n, r)$$

1.161 The number of partitions of a set with n elements is the **Bell number**, B_n; by fiat, $B_0 = 1$. Prove:

$$(a)\quad B_n = \sum_{m=1}^{n} S(n, m) \qquad (b)\quad B_n = \sum_{k=0}^{n-1} C(n-1, k)B_k$$

(*a*) This follows directly from the definitions of the two kinds of numbers.

(*b*) Proof is by the "*a*-cell" argument of Problem 1.154.

1.162 A partition of a nonempty subset Y of a set X is called a **partial partition** of X. Show that the number of partial partitions of a set X with n elements is $B_{n+1} - 1$.

Let a be an element not in X and let X' be a set obtained by adjoining a to X. Excluding the partition of X' into one part, there are $B_{n+1} - 1$ partitions of X'. From each of these partitions delete the set which contains a: this accounts for all the partial partitions of X (still another application of the method of Problem 1.154).

Supplementary Problems

1.163 There are 5 candy bars of different kinds and 4 pieces of different kinds of cakes on a tray. Suppose Eve takes either a candy bar or a piece of cake. Then, Adam takes a candy bar and a piece of cake. In which case has Adam more choices?
Ans. Adam has more choices if Eve takes a candy bar.

1.164 Find the number of 4-character strings (words) that can be formed using the 26 letters of the alphabet.
Ans. $(26)(26)(26)(26) = 456{,}976$

1.165 Count the $m \times n$ binary matrices. *Ans.* 2^{mn}

1.166 If the set A has m elements and the set B has n elements, how many elements have the product sets $A \times B$ and $B \times A$? *Ans.* mn

1.167 If A has m elements and B has n elements, find the number of functions (mappings) from (*a*) A to B, and (*b*) $A \times B$ to B. *Ans.* (*a*) n^m; (*b*) n^{mn}

1.168 There are 4 roads between A and B and 6 roads between B and C. Find the number of ways to go (*a*) from A to C; (*b*) from A to C and back to A; (*c*) from A to C and back to A without using a road more than once.
Ans. (*a*) 24; (*b*) 576; (*c*) 360

1.169 Find the number of ways of placing r objects in $n \geq r$ distinct places so that no place receives more than 1 object, if the r objects are (*a*) distinct (and (*b*) identical. *Ans.* (*a*) $P(n, r)$; (*b*) $C(n, r)$

1.170 There are 5 mathematics students and 7 statistics students in a group. Find the number of ways of selecting 4 students from the group, if (*a*) there are no restrictions; (*b*) all must be mathematics majors; (*c*) all must be statistics majors; (*d*) all must belong to the same discipline; and (*e*) the 2 disciplines must have the same number of representatives.
Ans. (*a*) $C(12, 4)$; (*b*) $C(5, 4)C(7, 0)$; (*c*) $C(5, 0)C(7, 4)$; (*d*) $C(5, 4)C(7, 0) + C(5, 0)C(7, 4)$; (*e*) $C(5, 2)\ C(7, 2)$

1.171 A country club has 8 men and 6 women on its governing board. There is 1 married couple in the board. Find the number of ways of forming a fund-raising committee consisting of 3 men and 3 women from the board such that the committee may include either the husband or the wife but not both.
Ans. $C(7, 2)C(5, 3) + C(7, 3)C(5, 2) + C(7, 3)C(5, 3)$

1.172 In a town council there are 10 Democrats and 11 Republicans. There are 4 women among the Democrats and 3 women among the Republicans. Find the number of planning committees of 8 councillors which have equal numbers of men and women and equal numbers from both parties.
Ans. $C(6, 0)C(4, 4)C(8, 4)C(3, 0) + C(6, 1)C(4, 3)C(8, 3)C(3, 1) + C(6, 2)C(4, 2)C(8, 2)C(3, 2) + C(6, 4)C(4, 0)C(8, 0)C(3, 4)$

1.173 Find the number of ways of scheduling m women speakers and n men speakers so that (*a*) speakers of the same gender are grouped together, and (*b*) the women are grouped together.
Ans. (*a*) $2m!\,n!$; (*b*) $n!\,(n+1)m! = (n+1)!\,m!$

1.174 Find the number, N, of linear arrangements of b identical blue marbles and r identical red marbles such that (*a*) between every pair of marbles of one color there is a marble of the other color; (*b*) between every pair of blue marbles there is exactly 1 red marble; and (*c*) no two red marbles are adjacent.

Ans. (*a*) $N = \begin{cases} 1 \text{ when } |r-b| = 1 \\ 2 \text{ when } r = b \\ 0 \quad \text{otherwise} \end{cases}$ (*b*) $N = \begin{cases} 0 \text{ when } r < b-1 \\ r-b+2 \text{ otherwise} \end{cases}$ (*c*) $N = C(b+1, r)$

1.175 A display case in a jewelry store has 4 shelves. Each shelf can accommodate at most 10 stones. Find the number of ways of displaying 10 diamonds in the case if (*a*) they are indistinguishable, (*b*) they are distinguishable.
Ans. (*a*) $C(13,3) = C(13,10)$; (*b*) $P(13,10)$

1.176 Find the number of ways of assigning 15 distinct paintings to 18 different dormitories so that no dormitory receives more than 1 painting. *Ans.* $P(18,15)$

1.177 Find the number of ways of assigning 18 distinct paintings to 15 different dormitories so that no dormitory receives more than 1 painting and the number of unassigned paintings is a minimum. *Ans.* $P(18,15)$

1.178 Find the number of ways of assigning 15 identical posters to 18 dormitories so that no dormitory receives more than 1 poster. *Ans.* $C(18,15)$

1.179 Find the number of ways of assigning 18 identical posters to 15 dormitories so that no dormitory receives more than 1 poster and the number of unassigned posters is a minimum. *Ans.* 1

1.180 Find the number of ways of assigning 18 identical posters to 15 dormitories with no restrictions on the number of posters a dormitory can receive. [*Hint*: Compare Problem 1.175(*a*).] *Ans.* $C(32,14)$

1.181 There are 15 display cases in an art gallery. Each case can accommodate a row of up to 20 paintings. Find the number of ways of displaying 18 different paintings. *Ans.* $18!\,C(32,14) = P(32,18)$

1.182 There are 12 members in a committee who sit around a table. There is 1 place specially designated for the chairman. Besides the chairman there are 3 people who constitute a subcommittee. Find the number of seating arrangements, if (*a*) the subcommittee sit together as a block, and (*b*) no 2 of the subcommittee sit next to each other. *Ans.* (*a*) $9!\,3!$; (*b*) $8!\,P(9,3)$

1.183 Prove combinatorially that

$$C(3n, n) = 3C(n,3) + 6nC(n,2) + n^3$$

1.184 Derive the result of Problem 1.46 from Pascal's identity,

$$C(n+r+1, r) = C(n+r, r) + C(n+r, n-1)$$

1.185 Show that if p is a prime number and k is any integer, then $k^p - k$ is divisible by p. (This is essentially **Fermat's Little Theorem**.) [*Hint*: The theorem clearly holds for $k = 1$; and we have

$$(k+1)^p - (k+1) = (k^p - k) + [(C(k,1)k^{p-1} + \cdots + C(p, p-1)k]$$

Now use Problem 1.45.]

1.186 A box has 6 blue marbles, 8 red marbles, 11 green marbles, 14 white marbles, and 16 yellow marbles. Find the minimum number of marbles one has to pick to ensure that 10 marbles of the same color are obtained.
Ans. $6 + 8 + (10 - 1)(3) + 1 = 42$

1.187 Find the minimum number of students to be admitted to a college such that at least 1 of the 50 United States is represented by 20 or more students. *Ans.* $(50)(19) + 1 = 951$

1.188 A typical telephone number in the United States is of the form *NXX NXX XXXX*, where the *N*s are digits other than 0 or 1 and the *X*s are any digits. The first 3 digits constitute the area code. Find the minimum number of area codes needed to serve a 23-million-subscriber area. *Ans.* 3

1.189 Four sightseeing buses are to leave the starting point at 1-hour intervals. The numbers of unreserved seats are 8, 10, 13, and 9, respectively. How many additional tickets must be sold by the bus company so that the number of vacant seats is at most 2 in the first bus or at most 3 in the second bus or at most 4 in the third bus or at most 1 in the fourth bus? *Ans.* 27

1.190 There are 20 small towns in a district. A group of 3 people have to be chosen from 1 of these towns. For this purpose volunteers are solicited from all 20 towns. Find the minimum number of volunteers who must come forward. *Ans.* 41

1.191 When Eve was out of town for 14 days, she made 17 long-distance telephone calls to Adam. If she made at least 1 call every day, show that there is a period of consecutive days during which she made exactly 10 calls. (*Hint*: Review Problem 1.85.)

1.192 Show that in any line of 65 people with distinct heights there is a subline of 9 people whose heights strictly increase or strictly decrease. (*Hint*: See Problem 1.90.)

1.193 Show that in any assembly of 924 people there will always be either 7 mutual acquaintances or 7 mutual strangers. (*Hint*: Apply Problem 1.99.)

1.194 Find the number of 5-letter strings with distinct letters in alphabetical order (26-letter alphabet), such that the string (*a*) starts with *H*, and (*b*) ends with *R*. *Ans.* (*a*) $C(18, 4)$; (*b*) $C(17, 4)$.

Chapter 2

Further Basic Tools

2.1 GENERALIZED PERMUTATIONS AND COMBINATIONS

If X is a collection of n objects **that are not necessarily distinct**, any arrangement (or ordering) of $r \leq n$ objects from X is known as a **generalized r-permutation** of X. (If $r = n$, we speak simply of a **generalized permutation** of X.)

Example 1. The collection $X = \{A, A, B, B, B, C, C\}$ has $AABCBBC$ as one of its generalized permutations.

Definition: If n_i $(i = 1, 2, \ldots, k)$, r, and n are $k + 2$ positive integers such that $n_1 + n_2 + \cdots + n_k = r \leq n$, then

$$P(n; n_1, n_2, \ldots, n_k) \equiv \frac{P(n, r)}{n_1! n_2! \cdots n_k!}$$

Since $P(n, r) = P(n, n)/(n - r)!$, it follows from the definition that

$$P(n; n_1, n_2, \ldots, n_k) = P(n; n_1, n_2, \ldots, n_k, n - r)$$

Example 2.

$$P(18; 3, 4, 6) = \frac{P(18, 3 + 4 + 6)}{3!\,4!\,6!} = \frac{18!/5!}{3!\,4!\,6!} = \frac{18!}{3!\,4!\,6!\,5!}$$

$$= \frac{P(18, 3 + 4 + 6 + 5)}{3!\,4!\,6!\,5!} = P(18; 3, 4, 6, 5)$$

Theorem 2.1. The number of generalized permutations of a collection X consisting of n_i identical objects of type i $(i = 1, 2, \ldots, k)$ is $P(n; n_1, n_2, \ldots, n_k)$; here, $n = n_1 + n_2 + \cdots + n_k$.

Proof. Let p be the total number of generalized permutations of X. If the n objects in X were all distinct, there would be $P(n, n)$ permutations of X. Now the n_1 distinct objects of type 1 would give rise to $n_1!$ permutations, the other $n - n_1$ objects being held fixed. This is true for the objects from each of the k types. So, by the product rule, each of the p generalized permutations will define $q \equiv n_1!\, n_2! \cdots n_k!$ permutations of the supposedly distinct elements of X. Thus $pq = P(n, n)$, or

$$p = \frac{P(n, n)}{q} = P(n; n_1, n_2, \ldots, n_k)$$

Example 3. The collection of letters that form the word COMMITTEE is $X = \{C, E, E, I, M, M, O, T, T\}$, with 9 letters belonging to 6 types. The number of generalized permutations of X is

$$P(9; 1, 2, 1, 2, 1, 2) = \frac{9!}{1!\,2!\,1!\,2!\,1!\,2!} = 45{,}360$$

Example 4. Twelve light bulbs (4 identical red, 3 identical white, and 5 identical blue) are to be installed in 18 sockets in a row, leaving 6 empty sockets. Thus we have 18 objects (12 light bulbs and 6 empty sockets) that fall into 4 categories; these may be arranged in $P(18; 4, 3, 5, 6)$ ways. But $P(18; 4, 3, 5, 6) = P(18; 4, 3, 5)$. Thus, without bothering about the empty sockets, one can assert that the number of ways of installing 4 identical red bulbs, 3 identical white bulbs, and 5 identical blue bulbs in 18 sockets in a row is $P(18; 4, 3, 5)$.

Suppose now that X is a collection of n **distinct** objects and suppose that S is any r-subset of X. Then an ordered partition of S is called a **generalized r-combination** of X. Again, if $r = n$, we use the term **generalized combination.**

The number of generalized r-combinations of X having n_1 objects in the first cell, n_2 objects in the second cell, ..., n_k objects in the kth cell, is denoted $C(n; n_1, n_2, \ldots, n_k)$. Since $n_1 + n_2 + \cdots + n_k = r$, the product rule gives

$$C(n; n_1, n_2, \ldots, n_k) = C(n, n_1)C(n - n_1, n_2) \cdots C(n - n_1 - n_2 - \cdots - n_{k-1}, n_k)$$
$$= \frac{n!}{n_1!\, n_2! \cdots n_k!\, (n - r)!} = \frac{P(n, r)}{n_1!\, n_2! \cdots n_k!}$$

and we have proved

Theorem 2.2. $C(n; n_1, n_2, \ldots, n_k) = P(n; n_1, n_2, \ldots, n_k)$, where $n_1 + n_2 + \cdots + n_k = r \leq n$.

Example 5. Seventeen students want to go to a party, and there are 5 vehicles available to them. The numbers of empty seats in these vehicles are 4, 3, 2, 5, and 1. Hence, there are

$$C(17; 4, 3, 2, 5, 1) = \frac{17!\, 2!}{4!\, 3!\, 2!\, 5!\, 1!} \approx 5.15 \times 10^9$$

ways of transporting all but 2 students to the party.

The integers $C(n; n_1, n_2, \ldots, n_k)$ are often referred to as the **multinomial coefficients** (see Problem 2.1).

Theorem 2.3. The number of (unordered) partitions of a set of cardinality n into p_1 subsets of cardinality n_1, p_2 subsets of cardinality n_2, ..., p_k subsets of cardinality n_k (where the n_i are distinct and $\Sigma\, p_i n_i = n$) is given by

$$\frac{C(n; \overbrace{n_1, \ldots, n_1}^{p_1 \text{ terms}}, \overbrace{n_2, \ldots, n_2}^{p_2 \text{ terms}}, \ldots, \overbrace{n_k, \ldots, n_k}^{p_k \text{ terms}})}{p_1!\, p_2! \cdots p_k!} = \frac{n!}{[p_1!\,(n_1!)^{p_1}][p_2!\,(n_2!)^{p_2}] \cdots [p_k!\,(n_k!)^{p_k}]}$$

Proof. The numerator of the given expression counts the *ordered* partitions of the n set into subsets of the required sizes. Now these ordered partitions fall into classes of $p_1!\, p_2! \cdots p_k!$ members each: members in the same class differ only in the order in which like-sized cells are listed. Because each class represents a single unordered partition, and vice versa, the theorem follows.

Example 6. (*i*) The number of ways of placing 12 students in the morning, afternoon, and evening recitation sections so that 4 of them go to each section is $C(12; 4, 4, 4)$. (*ii*) The number of ways of dividing 12 students into groups of 4 is $C(12; 4, 4, 4)/3!$. (*iii*) The number of ways of dividing 12 students into foursomes for bridge (with assignment of N, S, E, W) is

$$\frac{C(12; 4, 4, 4)}{3!}(4!) = 4C(12; 4, 4, 4)$$

2.2. SEQUENCES AND SELECTIONS

Let $X = \{a_1, a_2, \ldots, a_n\}$ be a set of n objects and r a positive integer. Then if X is sampled **with replacement** r times, an **ordered** r-set is obtained which is called an ***r*-sequence** of X. By Chapter 1 there are n^r r sequences of X.

Example 7. In an undergraduate residence hall there are (at least 6) students in each of the 4 years. There is a bench in front of the hall which can accommodate exactly 6 people. Any filling of the bench from left to right by residents (all facing the same way) is a 6-sequence of $X = \{1, 2, 3, 4\}$, where 1 means a first-year student, etc. The number of seating arrangements is $4^6 = 4096$.

Considered **without regard to order**, the r-set obtained above is known as an ***r*-selection** (or ***r*-sample**). In an r selection all that matters are the nonnegative integers x_i giving the numbers of occurrences of the a_i $(i = 1, 2, \ldots, n)$. Problem 1.139 gives at once:

Theorem 2.4. The number of distinct r selections from a collection X of n distinct objects is $[n]^r/r! = C(r + n - 1, n - 1)$.

Example 8. If, in Example 7, the distribution of the years along the bench were irrelevant, there would be only $C(6 + 4 - 1, 4 - 1) = 84$ seating arrangements.

2.3. THE INCLUSION-EXCLUSION PRINCIPLE

The number of elements in a finite set A is denoted by $n(A)$ or by $|A|$. It is easily verified that

$$n(A \cup B) = n(A) + n(B) - n(A \cap B)$$

whenever A and B are finite sets. Thus, to find the number of elements in either A or B, we add $n(A)$ and $n(B)$ (we *include* both sets) and then subtract $n(A \cap B)$ from the sum (we *exclude* what is common to both). This is the idea underlying the inclusion-exclusion principle, which can be formulated in a more general setting involving a finite number of sets.

If A is a subset of X, then the complement of A in X is denoted by A'. If A and B are both subsets of X, then obviously

$$n((A \cup B)') = n(X) - n(A \cup B) = n(X) - [n(A) + n(B)] + n(A \cap B)$$

But $(A \cup B)' = A' \cap B'$, so that

$$n(A' \cap B') = n(X) - [n(A) + n(B)] + n(A \cap B)$$

(It is a little simpler to derive the inclusion-exclusion formula in terms of complements.)

If x is an arbitrary element of X and if A is some subset of X, then the **count of x in $n(A)$** is 1 if x is in A, and 0 if x is not in A.

Example 9. Let $X = \{a, b, c, d, e, f, g, h\}$, $A = \{a, b, c, d\}$, and $B = \{c, d, e\}$. Because $n(X) = 8$, $n(A) = 4$, $n(B) = 3$, $n(A \cap B) = 2$, and $n(A' \cap B') = 3$ the equation $n(A' \cap B') = n(X) - [n(A) + n(B)] + n(A \cap B)$ is seen to hold. The count of a in the left-hand side of this equation is 0, since a is not in A'; the count of a in the right-hand side is $1 - [1 + 0] - 0 = 0$.

Two forms of the inclusion-exclusion principle follow.

***Theorem 2.5* (*Sieve Formula*).** If $A_1, A_2, \ldots, A_m$ are subsets of a finite set X. then

$$n(A'_1 \cap A'_2 \cap \cdots A'_m) = n(X) - s_1 + s_2 - \cdots + (-1)^m s_m$$

where s_k denotes the sum of the cardinalities of all the k-tuple intersections of the given m subsets $(k = 1, 2, \ldots, m)$.

Proof. Let x be an arbitrary element of X. It suffices to show that the count of x is the same on either side of the stated equation. We consider two cases: (i) x is not an element of any of the m subsets; (ii) x is an element of exactly $r \geq 1$ of the m subsets, which we may always suppose to be $A_1, A_2, \ldots, A_r$. In the former case, the count of x is 1 on both sides of the equation. In the latter case, the count of x on the left-hand side is 0. As for the right-hand side, we have

$$s_k = \sum n(A_{i_1} \cap A_{i_2} \cap \cdots \cap A_{i_k}) \qquad (k = 1, 2, \ldots, m)$$

the summation being over all k combinations of $\{1, 2, \ldots, m\}$. In a summand involving a k combination of $\{1, 2, \ldots, r\}$, the count of x will be 1; in all other summands, the count of x will be 0. Hence, the count of x on the right-hand side is

$$1 - \binom{r}{1} + \binom{r}{2} - \binom{r}{3} + \cdots + (-1)^r \binom{r}{r} = (1-1)^r = 0$$

and the proof is complete.

Theorem 2.6. With the notation as in Theorem 2.5,

$$n(A_1 \cup A_2 \cup \cdots \cup A_m) = s_1 - s_2 + \cdots + (-1)^{m-1} s_m$$

Proof. $n(A_1 \cup A_2 \cup \cdots \cup A_m) = n(X) - n(A_1' \cap A_2' \cap \cdots \cap A_m')$; now use the formula from Theorem 2.4.

Example 10 (***Problème des Rencontres***). Let us use the inclusion-exclusion principle, Theorem 2.4, to determine D_m, the number of derangements of m distinct objects (Problem 1.67). Denote by Q the set of all permutations of $\{x_1, x_2, \ldots, x_m\}$, and let A_i $(i = 1, 2, \ldots, m)$ be the subset of Q composed of those permutations that leave x_i fixed. From first principles, $n(Q) = m!$ and, for $k = 1, 2, \ldots, m$,

$$s_k = \sum n(A_{i_1} \cap A_{i_2} \cap \cdots \cap A_{i_k}) = C(m, k)(m-k)! = \frac{m!}{k!}$$

Hence

$$D_m = n(A_1' \cap A_2' \cap \cdots \cap A_m')$$

$$= m!\left[1 - \frac{1}{1!} + \frac{1}{2!} - \cdots + (-1)^m \frac{1}{m!}\right] \approx m!\, e^{-1}$$

2.4. SYSTEMS OF DISTINCT REPRESENTATIVES

Given N sets, not necessarily distinct. If it is possible to choose exactly 1 element from each set, *with the chosen elements distinct*, then the family of N sets has a **system of distinct representatives (SDR)** made up of the selected elements. For an SDR to exist it is obviously necessary that the following **marriage condition** hold for the family: the total number of elements in any subfamily of k sets is at least k $(k = 1, 2, \ldots, N)$.

***Theorem 2.7* (*Philip Hall's Marriage Theorem*).** The marriage condition is also sufficient for the existence of an SDR.

Proof. Assuming that the marriage condition holds for the sets $A_1, A_2, \ldots, A_n$, let the sets be depleted until a family $F' = \{A_1', A_2', \ldots, A_N'\}$ is reached such that removal of 1 more element from any of the A_i' would cause the marriage condition to be violated. We assert that each member of F' consists of a single element; because these elements are distinct (by the marriage condition), F' itself is the required SDR.

Suppose, on the contrary, that A_1' (say) has 2 elements, x and y. Then the minimality of F' requires the existence of subsets P and Q of the set $\{2, 3, \ldots, N\}$ such that

$$X \equiv (A_1' - x) \cup \left(\bigcup_{i \in P} A_i'\right) \qquad \text{and} \qquad Y \equiv (A_1' - y) \cup \left(\bigcup_{i \in Q} A_i'\right)$$

have cardinalities $n(X) \le n(P)$ and $n(Y) \le n(Q)$. Consequently, by addition,

$$n(X) + n(Y) = n(X \cup Y) + n(X \cap Y) \le n(P) + n(Q) \qquad (i)$$

where the first equality follows from the simplest form of the inclusion-exclusion principle. Now, by the definitions of X and Y,

$$X \cup Y = A_1' \cup \left(\bigcup_{i \in P \cup Q} A_i' \right) \quad \text{and} \quad X \cap Y = \bigcup_{i \in P \cap Q} A_i'$$

and the marriage condition gives:

$$n(X \cup Y) \geq 1 + n(P \cup Q) \quad \text{and} \quad n(X \cap Y) \geq n(P \cap Q)$$

By addition and the inclusion-exclusion formula,

$$\begin{aligned} n(X \cup Y) + n(X \cap Y) &\geq 1 + n(P \cup Q) + n(P \cap Q) \\ &= 1 + n(P) + n(Q) \\ &> n(P) + n(Q) \end{aligned} \qquad (ii)$$

The contradiction between (*i*) and (*ii*) establishes the theorem.

Example 11. A family of sets will, in general, have many SDRs. A more refined form of the marriage theorem includes a lower bound on the number of SDRs, expressed in terms of the size of the smallest set in the family.

Companion theorems to Theorem 2.7 play important roles in matrix theory, graph theory, and the theory of partially ordered sets. Some of these will be explored in the Solved Problems.

Solved Problems

GENERALIZED PERMUTATIONS AND COMBINATIONS

2.1 (*The Multinomial Theorem*) Show that the typical term in the expansion of $(x_1 + x_2 + \cdots + x_k)^n$ is

$$C(n; n_1, n_2, \ldots, n_k) x_1^{n_1} x_2^{n_2} \cdots x_k^{n_k} \qquad (n_1 + n_2 + \cdots + n_k = n)$$

The number of ordered partitions of the set

$$S = \{(x_1 + \cdots + x_k)_1, (x_1 + \cdots + x_k)_2, \ldots, (x_1 + \cdots + x_k)_n\}$$

into a cell of n_1 elements, each providing an x_1; ...; a cell of n_k elements, each providing an x_k—is $C(n; n_1, n_2, \ldots, n_k)$.

2.2 There are 20 marbles of the same size but of different colors (1 red, 2 blue, 2 green, 3 white, 3 yellow, 4 orange, and 5 black) in an urn. Find the number of ways of arranging 5 marbles from this urn in a row.

There are 7 distinct cases. (*i*) *All marbles are of the same color.* There is 1 possible 5-sample (the black marbles) and 1 way of arranging it. (*ii*) *Exactly 4 are of the same color.* The number of 5-samples is $C(2, 1)C(6, 1) = 12$. Each sample has $P(5; 4, 1) = 5$ arrangements. So the total number of arrangements is $(12)(5) = 60$. (*iii*) *3 of one color and 2 of another color.* There are $C(4, 1)C(5, 1) = 20$ samples, each yielding $P(5; 3, 2) = 10$ arrangements. So the total here is $(20)(10) = 200$. (*iv*) *3 of one color, 2 of two different colors.* The number of samples is $C(4, 1)C(6, 2) = 60$; each sample gives $P(5; 3, 1, 1) = 20$ arrangements. The total here is $(60)(20) = 1200$. (*v*) *2 of one color, 2 of another color, and 1 of a third color.* The number of samples is $C(6, 2)C(5, 1) = 75$, and each sample yields $P(5; 2, 2, 1) = 30$ arrangements. The total here is $(75)(30) = 2250$. (*vi*) *2 of one color and the other 3 of different colors.* The number of samples is $C(6, 1)C(6, 3) = 120$. Each sample admits $P(5; 2, 1, 1, 1) = 60$ arrangements, for a total of $(120)(60) = 7200$ arrangements. (*vii*) *5 of different colors*. There are $C(7, 5) = 21$ samples, each giving $P(5; 1, 1, 1, 1, 1) = 120$ arrangements. Here the total number of arrangements is $(21)(120) = 2520$.

The grand total of arrangements is $1 + 60 + \cdots + 2520 = 13{,}431$.

2.3 Show that if m and n are positive integers, then $(mn)!$ is divisible by $(m!)^m$.

As a count of generalized permutations,

$$P(mn; \overbrace{m, m, \ldots, m}^{n \text{ terms}}) = \frac{(mn)!}{(m!)^n}$$

is an integer.

2.4 Evaluate

$$\sum_{n_1+n_2+\cdots+n_k=n} C(n; n_1, n_2, \ldots, n_k)$$

By the multinomial theorem, the sum is $(1 + 1 + \cdots + 1)^n = k^n$.

2.5 A particle in the plane is free to move from any lattice point (Problem 1.58) to any of the 4 neighboring lattice points. Find the number of ways that the particle can start from the origin and return to the origin after covering a total distance of $2n$ units.

A path of length $2n$ which returns to its starting point must consist of p rightward steps, p leftward steps, q upward steps, and q downward steps ($2p + 2q = 2n$). Hence the desired number is

$$\sum_{p+q=n} P(2n; p, p, q, q)$$

2.6 A particle starts from a lattice point $U(u_1, u_2, \ldots, u_k)$ in k-dimensional Euclidean space, makes a step of 1 unit parallel to the positive direction of one of the coordinate axes, continues to do likewise at every lattice point en route and stops at the lattice point $V(v_1, v_2, \ldots, v_k)$, creating a path from U to V. Find the number of such paths.

Let $u = u_1 + u_2 + \cdots + u_k$ and $v = v_1 + v_2 + \cdots + v_k$. In any of the considered paths the number of steps parallel to the positive X_i axis is $v_i - u_i$ ($i = 1, 2, \ldots, k$); and so the path length is $v - u$. The required number of paths is therefore $P(v - u; v_1 - u_1, \ldots, v_k - u_k)$.

2.7 Show that $(n!)!$ is divisible by $(n!)^{(n-1)!}$.

Consider a collection of $n!$ objects of $(n - 1)!$ types, with n objects of each type. This collection can be arranged in

$$P(n!; \overbrace{n, n, \ldots, n}^{(n-1)! \text{ terms}}) = \frac{(n!)!}{(n!)^{(n-1)!}}$$

ways.

2.8 Give a noncomputational (combinatorial) proof of Theorem 2.2 in the case $r = n$.

Let X be the n set of Theorem 2.1 and let Y be a set of n distinct locations along a straight line. Any generalized permutation of X determines a unique ordered partition of Y (the ith cell consists of those locations occupied by the n_i identical objects of type i). Conversely, any ordered partition of Y determines a unique generalized permutation of X. This one-to-one correspondence immediately implies Theorem 2.2.

2.9 Obtain from Theorem 2.3 an explicit formula for the Bell number, B_n (Problem 1.161).

First observe that the conclusion of Theorem 2.3 remains true if some (but not all) of the p_i are zero. Make an n-way classification of the partitions of an n set X according to the number p_1 of 1 cells, the number p_2 of 2 cells, . . . , the number p_n of n cells. By Theorem 2.3, a given class will contain

$$\frac{n!}{[p_1!(1!)^{p_1}][p_2!(2!)^{p_2}]\cdots[p_n!(n!)^{p_n}]}$$

partitions of X. The total number of partitions is therefore

$$B_n = n! \sum_{1p_1+2p_2+\cdots+np_n=n} \frac{1}{[p_1!(1!)^{p_1}][p_2!(2!)^{p_2}]\cdots[p_n!(n!)^{p_n}]}$$

2.10 From Problem 2.9 infer that

$$B_n = \frac{d^n}{dx^n} e^{e^x-1}\bigg|_{x=0}$$

Expanding $e^x - 1$ in a Maclaurin series, we have

$$e^{e^x-1} = e^{x^1/1!+x^2/2!+x^3/3!+\cdots} = e^{x^1/1!}e^{x^2/2!}e^{x^3/3!}\cdots$$

$$= \left[\frac{(x^1/1!)^0}{0!} + \frac{(x^1/1!)^1}{1!} + \cdots + \frac{(x^1/1!)^{p_1}}{p_1!} + \cdots\right]$$

$$\times \left[\frac{(x^2/2!)^0}{0!} + \frac{(x^2/2)^1}{1!} + \cdots + \frac{(x^2/2!)^{p_2}}{p_2!} + \cdots\right]$$

$$\cdots\cdots\cdots\cdots\cdots\cdots\cdots\cdots\cdots\cdots\cdots\cdots$$

$$\times \left[\frac{(x^n/n!)^0}{0!} + \frac{(x^n/n!)^1}{1!} + \cdots + \frac{(x^n/n)^{p_n}}{p_n!} + \cdots\right]$$

$$\times \cdots$$

The coefficient of x^n in e^{e^x-1} will be the coefficient of x^n in the product of the first n infinite series on the right; thus this coefficient has the value

$$\sum_{1p_1+2p_2+\cdots+np_n=n} \frac{1}{p_1!(1!)^{p_1}} \frac{1}{p_2!(2!)^{p_2}} \cdots \frac{1}{p_n!(n!)^{p_n}}$$

or $B_n/n!$, by Problem 2.9. The desired result now follows from Taylor's theorem.

SEQUENCES AND SELECTIONS

2.11 Find the number of r sequences of a set X of n distinct elements such that each of the sequences involves every element of X at least once.

Clearly, we must have $r \geq n$. Let P represent a set of r distinct positions along a straight line. Then an r-sequence of the specified kind determines, and is determined by, a surjection from P to X. By Problem 1.146, the required number is $n!\, S(r, n)$.

2.12 Suppose X is a collection of n distinct objects. Find the number of r selections from X such that each selection includes the ith object at least $p_i \geq 0$ times $(i = 1, 2, \ldots, n)$.

Let q_i be the number of times the ith object is included in an n-selection. Then the number of selections is the number of solutions in integers of

$$q_1 + q_2 + \cdots + q_n = r \qquad \text{and} \qquad q_i \geq p_i \qquad (i = 1, 2, \ldots, n)$$

By Problem 1.142, this number is $C(r - p + n - 1, n - 1)$, where $p \equiv \Sigma\, p_i$.

2.13 Find the number of ways of allocating r *identical* objects to n distinct places so that the jth place gets at least $p_j \geq 0$ objects $(j = 1, 2, \ldots, n)$.

This is just the dual of Problem 2.12: each allowable r selection of Problem 2.12 is equivalent to reserving at least p_i of the r identical places (now called ''objects'') for the ith distinct object (now called ''place''). Hence the answer is that found in Problem 2.12, $C(r - p + n - 1, n - 1)$.

2.14 Find the numbers of ways of allocating (*a*) 10 identical VCRs, to dormitories A through F in such a manner that the total number allocated to dormitories A and B does not exceed 4. (*b*) 10 students to their dormitories with the same stipulation as in (*a*).

(*a*) By Problem 1.139 or Problem 2.13 there are $C(r + 1, 1) = r + 1$ ways to distribute $0 \leq r \leq 4$ VCRs between dorms A and B; the remaining $10 - r$ VCRs may then be distributed in $C(10 - r + 3, 3)$ ways. The product rule yields

$$\sum_{r=0}^{4} (r + 1)C(13 - r, 3)$$

(*b*) Choose $0 \leq r \leq 4$ out of the 10 students; this can be done in $C(10, r)$ ways. These can be assigned to dorms A and B in 2^r ways. Now assign the remaining $10 - r$ students, in 4^{10-r} ways. The answer is therefore

$$\sum_{r=0}^{4} C(10, r) 2^r 4^{10-r}$$

[Is it just an accident that we have obtained the first five terms of the binomial expansion of $(2 + 4)^{10}$?]

2.15 Consider a collection of n objects of different types such that any 2 objects of the same type are indistinguishable. Suppose there are x_i objects of type i, where $i = 1, 2, \ldots$. These objects are to be distributed among a group of distinguishable boxes in such manner that box j receives y_j objects $(j = 1, 2, \ldots)$. Denote the number of allowable distributions by $[(x_1, x_2, \ldots) * (y_1, y_2, \ldots)]_n$, where the subscript n is a reminder that $\Sigma\, x_i = \Sigma\, y_j = n$. Interpret and, when possible, evaluate the following expressions:

(*a*) $[(1, 1, \ldots) * (1, 1, \ldots)]_n$

(*b*) $[(1, 1, \ldots) * (r, n - r)]_n$

(*c*) $[(r, n - r) * (1, 1, \ldots)]_n$

(*d*) $[(1, 1, \ldots) * (a_1, a_2, \ldots, a_m)]_n$

(*e*) $[(a_1, a_2, \ldots, a_m) * (1, 1, \ldots)]_n$

(*f*) $[(1, 1, \ldots, 1) * (1, 1, \ldots, 1, n - r)]_n$

(*g*) $[(a_1, a_2, \ldots, a_m) * (r, n - r)]_n$

(*h*) $[(r, n - r) * (a_1, a_2, \ldots, a_m)]_n$

(*a*) There are n distinct objects and they have to be placed in n distinct boxes so that each box gets exactly 1 object. This can be done in $n!$ ways.

(*b*) There are n distinct objects out of which r objects go to box 1 and $n - r$ objects go to box 2. This can be done in $C(n, r)$ ways.

(c) There are r identical objects of one type and $n-r$ identical objects of another type. These n objects are to be put in n distinct boxes so that each box gets exactly one object. This can be done in $C(n, r)$ ways.

(d) There are n distinct objects and they have to be put in m distinct boxes so that box j gets exactly a_j objects ($j = 1, 2, \ldots, m$). This can be done in $C(n; a_1, a_2, \ldots a_m)$ ways (see the definition of a generalized combination).

(e) There are n objects belonging to m types, with a_i objects in type i ($i = 1, 2, \ldots, m$). These n objects have to be put in n distinct boxes such that each box gets exactly 1 object. This can be done in $P(n; a_1, a_2, \ldots, a_m)$ ways (by Theorem 2.1). [Observe that the answers to (d) and (e) are identical.]

(f) This is the special case of (d) corresponding to $m = r + 1$, $a_1 = \cdots = a_r = 1$, $a_{r+1} = n - r$. Hence the answer is $C(n; 1, 1, \ldots, 1, n-r) = P(n, r)$.

(g) Here box 1 must get r objects from among the m categories. Let u_i be the number of objects of the ith category going to box 1; we must have

$$u_1 + u_2 + \cdots + u_m = r \quad \text{and} \quad 0 \le u_i \le a_i \quad (i = 1, 2, \ldots, m) \qquad (*)$$

The number of integral solutions of (*) is the value of the given expression, because to each filling $\langle u_i \rangle$ of box 1 corresponds the *unique* filling $\langle a_i - u_i \rangle$ of box 2. [For a method of handling (*), see Problem 2.21.]

(h) Here the r objects of category 1 must be distributed among m boxes. Let u_j be number of objects of category 1 that go to box j; we must have

$$u_1 + u_2 + \cdots + u_m = r \quad \text{and} \quad 0 \le u_j \le a_j \quad (j = 1, 2, \ldots, m)$$

This is the same system as found in part (g); so the answer is the same. Again note that to each distribution $\langle u_j \rangle$ of the category 1 objects corresponds the *unique* distribution $\langle a_j - u_j \rangle$ of the category 2 objects.

2.16 Prove the **duality principle of distribution**:

$$[(x_1, x_2, \ldots, x_p) * (y_1, y_2, \ldots, y_q)]_n = [(y_1, y_2, \ldots, y_q) * (x_1, x_2, \ldots, x_p)]_n \qquad (i)$$

The proof rests on the simple observation—already exploited in previous problems—that assigning an object to a location is tantamount to assigning the location to the object. More generally, if exactly u_{ij} identical objects of type i are assigned to (the distinct) box j, then exactly u_{ij} identical box numbers j are assigned to (the distinct) object-type i. Consequently, both the left-hand and right-hand sides of (i) are equal to the number of nonnegative integral solutions of the linear system

$$\sum_{i=1}^{p} u_{ij} = y_j \quad (j = 1, 2, \ldots, q-1) \qquad \sum_{j=1}^{q-1} u_{ij} \le x_i \quad (i = 1, 2, \ldots, p)$$

(The system may be solved by the techniques of **integer programming**.)

2.17 Find the value of $[(2, 1, 1, \ldots, 1) * (2, 1, 1, \ldots, 1)]_n$.

There are n objects belonging to $n-1$ types: 2 identical objects, x and x, of type 1 and 1 object of each of the remaining types. Also there are $n-1$ marked boxes. Box 1 will get 2 objects, and the other boxes will get 1 object each. There are three mutually exclusive cases: (i) *Box 1 has the 2 identical objects.* Then the remaining $n-2$ distinct objects can be distributed to the remaining $n-2$ boxes in $(n-2)!$ ways. (ii) *Box 1 has one of the objects x.* The other object x can be distributed in $C(n-2, 1)$ ways. Then we distribute the remaining $n-2$ objects to the other boxes (including box 1) in $(n-2)!$ ways. Thus in this case there are $C(n-2, 1)(n-2)!$ ways. (iii) *Box 1 has neither object x.* In this case the 2 identical objects can be assigned to 2 of the other $n-2$ boxes in $C(n-2, 2)$ ways. Two of the other $n-2$ objects can be assigned to box 1 in $C(n-2, 2)$ ways. There are $n-4$ boxes and $n-4$ objects left. Thus in this case there are $C(n-2, 2)C(n-2, 2)(n-4)!$ ways. The total number of ways is the sum of the ways obtained in the three cases, or

$$\frac{(n^2 - n + 2)(n-2)!}{4}$$

2.18 Consider a set X of n consecutive natural numbers. A ***p*-block** in X is a subset of p consecutive numbers in X. Find the number of ways of forming m pairwise disjoint p blocks in X.

Suppose that the blocks are $B_1, B_2, \ldots, B_m$, where the order is that of the smallest elements of the blocks. Let x_i $(i = 2, 3, \ldots, m)$ be the number of elements of X that are between B_{i-1} and B_i; let x_1 and x_{m+1} be the respective numbers of elements preceding B_1 and succeeding B_m. The number of ways of forming m pairwise disjoint p blocks is the number of solutions in nonnegative integers of $x_1 + x_2 + \cdots + x_{m+1} = n - mp$; i.e., $C(n - mp + m, m)$. (Note that the answer vanishes, as it should, for $mp > n$.)

2.19 Suppose there are x_i identical objects of type i $(i = 1, 2, \ldots, m)$ in a collection. In how many ways may the collection be distributed among n distinct boxes, if box j must receive at least q_{ij} type-i objects $(i = 1, 2, \ldots, m;\ j' = 1, 2, \ldots, n)$?

The fact that there are separate demand conditions on each type of object—and not merely lumped conditions as in Problems 2.15 and 2.16—allows a simple solution, provided the numbers q_{ij} satisfy the necessary conditions

$$\sum_{j=1}^{n} q_{ij} \equiv q_i \leq x_i \qquad (i = 1, 2, \ldots, m)$$

In that case, we may distribute the objects type by type. By Problem 1.142, the ith type can be distributed in $C(x_i - q_i + n - 1, n - 1)$ ways. Therefore, by the product rule, the collection can be distributed in

$$\prod_{i=1}^{m} C(x_i - q_i + n - 1, n - 1)$$

ways.

THE INCLUSION-EXCLUSION PRINCIPLE

2.20 In a dormitory, there are 12 students who take an art course (A), 20 who take a biology course (B), 20 who take a chemistry course (C), and 8 who take a drama course (D). There are 5 students who take both A and B, 7 students who take both A and C, 4 students who take both A and D, 16 students who take both B and C, 4 students who take both B and D, and 3 students who take both C and D. There are 3 who take A, B, and C; 2 who take A, B, and D; 2 who take B, C and D; 3 who take A, C, and D. Finally, there are 2 in all four courses. It is also known that there are 71 students in the dormitory who have not signed up for any of these courses. Find the total number of students in the dormitory.

Let N be the total number of students. Then $71 = N - s_1 + s_2 - s_3 + s_4$, where

$$s_1 = 12 + 20 + 20 + 8 = 60 \qquad s_2 = 5 + 7 + 4 + 16 + 4 + 3 = 39 \qquad s_3 = 3 + 2 + 2 + 3 = 10 \qquad s_4 = 2$$

Thus $71 = N - 29$, or $N = 100$.

2.21 Find the number of solutions in integers of the equation $a + b + c + d = 17$, where $1 \leq a \leq 3$, $2 \leq b \leq 4$, $3 \leq c \leq 5$, $4 \leq d \leq 6$.

Let $a = 1 + \alpha$, $b = 2 + \beta$, $c = 3 + \gamma$, $d = 4 + \delta$. The transformed system is

$$\alpha + \beta + \gamma + \delta = 7 \qquad 0 \leq \alpha, \beta, \gamma, \delta \leq 2 \qquad (i)$$

Let X be the set of all solutions in nonnegative integers of $\alpha + \beta + \gamma + \delta = 7$; and let A be the subset of X for which $\alpha \geq 3$, B be the subset for which $\beta \geq 3$, C be the subset for which $\gamma \geq 3$, D be the subset for which $\delta \geq 3$. Applying Theorem 2.5 and Problem 1.142, we have:

$$n(X) = C(10, 3) \qquad n(A) = n(B) = n(C) = n(D) = C(7, 3)$$

$$n(A \cap B) = n(A \cap C) = \cdots = n(C \cap D) = C(4, 3)$$

and all higher-order intersections are empty. Thus,

$$n(X) = 120 \qquad s_1 = C(4, 1)C(7, 3) = 140 \qquad s_2 = C(4, 2)C(4, 3) = 24$$

and the answer is $120 - 124 + 24 = 4$.

2.22 Prove that the Stirling number of the second kind may be evaluated from the following inclusion-exclusion formula:

$$m!\, S(n, m) = m^n - C(m, 1)(m-1)^n + C(m, 2)(m-2)^n - \cdots + (-1)^{m-1}C(m, m-1)1^n$$

Let M denote the set of all mappings from $X = \{x_1, x_2, \ldots, x_n\}$ to $Y = \{y_1, y_2, \ldots, y_m\}$, and, for $i = 1, 2, \ldots, m$, let A_i be the subset of M consisting of those mappings that fail to cover y_i. We have $n(M) = m^n$ and

$$s_k = C(m, k)(m-k)^n \qquad (k = 1, 2, \ldots, m-1)$$

[There are $C(m, k)$ ways to set apart k elements of Y, and X can be mapped *into* the remaining elements in $(m-k)^n$ ways.] Then Theorem 2.5 gives for the number of mappings that cover every element of Y—i.e., the number of surjections from X to Y—as

$$m^n - C(m, 1)1(m-1)^n + \cdots + (-1)^{m-1}C(m, m-1)1^n$$

But, by Problem 1.146, the number of surjections is $m!\, S(n, m)$.

2.23 Find the number of permutations of the digits 1 through 9 in which (*a*) none of the blocks 2 3, 4 5, and 6 7 8 appears; (*b*) none of the blocks, 3 4, 4 5, and 7 3 8 appears.

Let X be the set of all the permutations; then $n(X) = 9!$.

(*a*) Let A, B, and C be the subsets of permutations which respectively contain 2 3, 4 5, and 6 7 8 as blocks. Then $n(A) = 8!$ [this is the number of permutations of the set $\{1, 2, 3, 4, 5, 6, 7, 8, 9\}$]. Similarly, $n(B) = 8!$, $n(C) = 7!$, $n(A \cap B) = 7!$, $n(A \cap C) = n(B \cap C) = 6!$. Thus the answer is

$$9! - (8! + 8! + 7!) + (7! + 6! + 6!) - 5!$$

(*b*) Let A, B, and C be the subsets of X in which 3 4, 4 5, and 7 3 8 respectively appear as blocks. Then $n(A) = n(B) = 8!$ and $n(C) = 7!$. Observe that $A \cap B$ is the subset of permutations in which 3 4 5 appears as a block; so $n(A \cap B) = 7!$. Here $n(A \cap C) = 0 = n(A \cap B \cap C)$ and $n(B \cap C) = 6!$. Thus the answer is

$$9! - (8! + 8! + 7!) + (7! + 0 + 6!) - 0$$

2.24 A **partition** of a positive integer N is an unordered collection of positive integers (or **parts**) whose sum is N. Let $f(N, r)$ be the number of partitions of N in which each part is repeated fewer than r times ($r = 2, 3, \ldots, N+1$) and let $g(N, r)$ be the number of partitions of N having no part divisible by r. Prove that $f(N, r) = g(N, r)$.

We apply Theorem 2.5 twice to the set X of all partitions of N. First define A_i as the subset of X composed of those partitions in which part i is repeated r or more times ($i = 1, 2, \ldots, \lfloor N/r \rfloor$). Then

$$n\left(\bigcap_i A_i'\right) = f(N, r) = n(X) - s_1 + s_2 - \cdots + (-1)^{\lfloor N/r \rfloor} s_{\lfloor N/r \rfloor}$$

in which the s_k are defined in the usual manner in terms of the intersections of the A_i.

Next define B_i as the subset of X composed of those partitions containing ir as a part ($i = 1, 2, \ldots, \lfloor N/r \rfloor$). We have:

$$n\left(\bigcap_i B_i'\right) = g(N, r) = n(X) - \sigma_1 + \sigma_2 - \cdots + (-1)^{\lfloor N/r \rfloor}\sigma_{\lfloor N/r \rfloor}$$

where the σ_k are defined analogously to the s_k.

But $\sigma_k = s_k$, for all k. Indeed, if $x \in X$ is a partition of N whose parts include $i_1 r, i_2 r, \ldots, i_k r$—so that x has a count of 1 in σ_k—then there exists an $x^* \in X$ in which parts $i_1, i_2, \ldots, i_k$ are each repeated r times—so that x^* has a count of 1 in s_k. Consequently, $f(N, r) = g(N, r)$.

DERANGEMENTS AND OTHER CONSTRAINED ARRANGEMENTS

2.25 Use a combinatorial argument to establish the identity

$$\sum_{r=0}^{n} \frac{D_r}{r!(n-r)!} = 1 \qquad (i)$$

The $n!$ permutations of an n set may be classified according as they have $n, n-1, n-2, \ldots, 0$ fixed points; thus,

$$n! = C(n, n)D_0 + C(n, n-1)D_1 + C(n, n-2)D_2 + \cdots + C(n, 0)D_n$$

Divide through by $n!$ to obtain (i).

2.26 Each of the n children in a class is given a book by the teacher; the books are all distinct. The students are required to return the books after 1 week. The same n books are again distributed for another week. In how many distributions does nobody get the same book twice?

The books can be distributed the first week in $n!$ ways. Each such distribution gives rise to D_n ways of distributing them the second week. So the answer is $n!\, D_n$.

2.27 Given the sequence $X = \langle x_1, x_2, \ldots, x_{2n} \rangle$, find the number of derangements of X such that the first n elements of each derangement are (a) the first n elements of X, and (b) the last n elements of X.

(a) The first n elements can be deranged in D_n ways. The same is true for the last n elements. So the answer is $(D_n)^2$. (b) Each of the $n!$ permutations of the last n elements in the first n places is indeed a derangement. This is true for the other half as well. The answer is $(n!)^2$.

2.28 Each of n women who attend a banquet checks her coat and hat with the receptionist on arrival. Upon leaving, each woman is given a coat and a hat at random. Find the number of ways these coats and hats may be distributed such that (a) nobody gets back either her coat or her hat, and (b) nobody gets back both her coat and her hat.

(a) The number of derangements for coats is D_n and the number of derangements for hats is D_n. Because the handing out of coats is independent of the handing out of hats, there are $(D_n)^2$ distributions with the specified property.

(b) Let A_j ($j = 1, 2, \ldots, n$) be the subset of distributions in which woman j gets back both her coat and her hat. Then, applying Theorem 2.5,

$$n(X) = (n!)^2 \qquad s_r = C(n, r)[(n-r)!]^2 \qquad (r = 1, 2, \ldots, n)$$

and the answer is given by $n(X) - s_1 + \cdots + (-1)^n s_n$.

2.29 Let A denote an r-subset of an n set X. How many permutations of X induce derangements of A?

Let A_i $(i = 1, 2, \ldots, r)$ denote the subset of permutations of X which leave the ith element of A fixed. Then

$$n(X) = n! \qquad s_j = C(r, j)(n-j)! \qquad (j = 1, 2, \ldots, r)$$

and Theorem 2.5 yields the answer as $n(X) - s_1 + \cdots + (-1)^r s_r$.

2.30 Show that $D_n = (n-1)(D_{n-1} + D_{n-2})$, where $n \geq 3$. (Of course, $D_2 = 1$ and $D_1 = 0$.)

Consider those derangements of $X = \{1, 2, \ldots, n\}$ in which r occupies the first position. Then either 1 does not occupy the rth position or it does. There are D_{n-1} derangements of the former type (with respect to $X - \{r\}$, position r functions as position 1) and D_{n-2} of the latter type. The element r can be chosen in $n-1$ ways.

2.31 (*a*) Show that $D_n = nD_{n-1} + (-1)^n$, for $n \geq 2$, and (*b*) Use this result to evaluate D_n. (D_n can be computed without using the inclusion-exclusion principle.)

(*a*) The substitution $E_n = D_n - nD_{n-1}$ takes the recursion formula of Problem 2.30 into $E_n = -E_{n-1}$. Since $E_2 = D_2 - 2D_1 = 1$, the latter recursion formula has the solution $E_n = (-1)^n$, for $n \geq 2$.

(*b*) By iteration,

$$\begin{aligned} D_n &= n((n-1)D_{n-2} + (-1)^{n-1}) + (-1)^n \\ &= n(n-1)D_{n-2} + n(-1)^{n-1} + (-1)^n \\ &= n(n-1)(n-2)D_{n-3} + n(n-1)(-1)^{n-2} + n(-1)^{n-1} + (-1)^n \\ &\qquad \cdots\cdots\cdots\cdots\cdots\cdots\cdots\cdots\cdots\cdots \\ &= n(n-1)\cdots(3) - n(n-1)\cdots(4) + n(n-1)\cdots(5) - \cdots \\ &\qquad + (-1)^{n-2}n(n-1) + (-1)^{n-1}n + (-1)^n \\ &= n!\left[\frac{1}{2!} - \frac{1}{3!} + \cdots + \frac{(-1)^n}{n!}\right] \end{aligned}$$

2.32 Find the number, T_n, of permutations of $Z = \{1, 2, 3, \ldots, n\}$ such that no permutation contains a block of 2 consecutive integers.

Use inclusion-exclusion. Let X denote the set of all permutations of Z; $n(X) = n!$ Let A_i $(i = 1, 2, \ldots, n-1)$ represent the subset of X consisting of all permutations in which the block $i\,i+1$ occurs. If one chooses r *mutually disjoint* blocks, these r objects and the remaining $n - 2r$ elements of Z together generate $(n-r)!$ permutations. *This same number*, $(n-r)!$, *is obtained for any selection of r blocks, disjoint or not* (prove it!). Hence, $s_r = C(n-1, r)(n-r)!$, for $r = 1, 2, \ldots, n-1$, and

$$\begin{aligned} T_n &= \sum_{r=0}^{n-1} (-1)^r C(n-1, r)(n-r)! = \sum_{r=0}^{n-1} (-1)^r \left[\frac{n!}{r!} - \frac{(n-1)!}{(r-1)!}\right] \\ &= n! \sum_{r=0}^{n-1} \frac{(-1)^r}{r!} + (n-1)! \sum_{s=0}^{n-2} \frac{(-1)^s}{s!} \end{aligned}$$

Referring to Example 10, we see that

$$T_n = [D_n - (-1)^n] + [D_{n-1} - (-1)^{n-1}] = D_n + D_{n-1}$$

or, by Problem 2.30, $T_n = D_{n+1}/n$.

2.33 Tabulate D_n and T_n for $n = 1(1)10$.

Start with $D_1 = 0$ and $T_1 = 1$. To compute D_n use the relation $D_n = (n-1)T_{n-1}$ (Problem 2.32); to compute T_n use the relation $T_n = D_n + D_{n-1}$ (Problem 2.32). See Table 2-1.

Table 2-1

n	D_n	T_n
1	0	1
2	1	1
3	2	3
4	9	11
5	44	53
6	265	309
7	1 854	2 119
8	14 833	16 687
9	133 496	148 329
10	1 334 961	1 468 457

2.34 Find the number of ways the integers $0, 1, 2, \ldots, n-1$ can be arranged in a circle so that no arrangement, when read clockwise, has a block of 2 consecutive integers or the block $[n-1\ 0]$.

Let X be the set of circular permutations of $\{0, 1, 2, \ldots, n-1\}$; by Problem 1.34, $n(X) = (n-1)!$. If we operate modulo n, the n excluded blocks can be notated as $\underline{i\ i+1}$ $(i = 0, 1, 2, \ldots, n-1)$. Let A_i be the subset of X in which each circular permutation contains the block $i\ i+1$. To evaluate s_r, for $r < n$, we may suppose the r blocks selected to be disjoint (see Problem 2.32), thus leading to $(r + n - 2r - 1)! = (n - r - 1)!$ *circular* permutations; i.e.,

$$s_r = C(n, r)(n-r-1)! \qquad (r = 1, 2, \ldots, n-1)$$

Clearly, $s_n = 1$. Theorem 2.5 now yields the answer

$$\sum_{r=0}^{n-1} (-1)^r C(n, r)(n-r-1)! \ + \ (-1)^n = n! \sum_{r=0}^{n} \frac{(-1)^r}{r!\,(n-r)}$$

2.35 There are 8 letters to different people to be placed in 8 different addressed envelopes. Find the number of ways of doing this so that at least 1 letter gets to the right person.

The answer is, from Table 2-1, $8! - D_8 = 40\,320 - 14\,833 = 25\,487$.

2.36. Find the number of 4-digit positive integers the sum of the digits of which is 31.

We wish to count the solutions in integers of the system

$$a + b + c + d = 31 \qquad 0 \le a, b, c, d \le 9$$

By the method of Problem 2.21 the answer is

$$C(34, 3) - C(4, 1)C(24, 3) + C(4, 2)C(14, 3) - C(4, 3)C(4, 3) + 0$$

2.37 Find the number of **injections** (one-to-one mappings) from a finite set X of n elements to X such that each has at least 1 fixed point.

The total number of injections is $n!$. Any derangement of X defines a unique injection which has no fixed point, and vice versa. Thus the answer is $n! - D_n$.

2.38 Out of 30 students in a dormitory, 15 take an art course, 8 take a biology course, and 6 take a chemistry course. It is known that 3 students take all 3 courses. Show that 7 or more students take none of the courses.

Let A be the set of students who take the art course, B be the set for biology, and C be the set for chemistry. Then $s_1 = 15 + 8 + 6 = 29$ and $s_3 = 3$ (given). Let x be the number who are not in any of these courses; by Theorem 2.5,

$$x = 30 - 29 + s_2 - 3 = s_2 - 2$$

Now the intersection of the 3 sets is a subset of the intersection of any 2 of the sets. Therefore,

$$s_2 \geq 3 + 3 + 3 = 9 \qquad \text{or} \qquad x \geq 9 - 2 = 7$$

2.39 There are 6 pairs of children's gloves in a box. Each pair is of a different color. Suppose the right gloves are distributed at random to 6 children, and then the left gloves also are distributed to them at random. Find the probability that (*a*) no child gets a matching pair, (*b*) everybody gets a matching pair, (*c*) exactly 1 child gets a matching pair, and (*d*) at least 2 children get matching pairs.

The right gloves can be distributed in 6! ways, after which the left gloves can be distributed in 6! ways. Thus there are (6!)(6!) equiprobable outcomes.

(*a*) For each of the 6! distributions of right gloves there are D_6 distributions of left gloves that result in no matching pairs. The required probability is therefore $(6!)D_6/(6!)^2 = D_6/6!$.

(*b*) For each of the 6! distributions of right gloves there is 1 distribution of left gloves that yields 6 matching pairs. The required probability is $(6!)(1)/(6!)^2 = 1/6!$.

(*c*) For each of the 6! distributions of right gloves there are $(1)D_5$ distributions of left gloves that give Annie—and Annie alone—a matching pair. Hence the required probability is

$$\frac{6[(6!)(1)D_5]}{(6!)^2} = \frac{D_5}{5!}$$

(*d*) Using the results of (*a*) and (*c*),

$$\text{Probability} = 1 - \frac{D_6}{6!} - \frac{D_5}{5!}$$

COMBINATORIAL NUMBER THEORY

2.40 Find the number of positive integers less than 601 that are not divisible by 3 or 5 or 7.

Let $X = \{1, 2, \ldots, 600\}$; then $n(X) = 600$. If A, B, and C are the subsets of integers in X that are divisible by 3, 5, and 7, respectively,

$$s_1 = n(A) + n(B) + n(C) = \left(\frac{600}{3}\right) + \left(\frac{600}{5}\right) + \left\lfloor\frac{600}{7}\right\rfloor = 405$$

$$s_2 = n(A \cap B) + n(A \cap C) + n(B \cap C)$$
$$= \left(\frac{600}{15}\right) + \left\lfloor\frac{600}{21}\right\rfloor + \left\lfloor\frac{600}{35}\right\rfloor = 85$$

$$s_3 = n(A \cap B \cap C) = \frac{600}{105} = 5$$

Thus, $n(A' \cap B' \cap C') = 600 - 405 + 85 - 5 = 275$.

2.41 (*Sieve of Eratosthenes*) Derive an expression for the counting function

$$\pi(n) \equiv \text{number of primes not exceeding the positive integer } n$$

Eratosthenes' method is based on the observation that an integer $k \geq 2$ is composite if and only if it is divisible by a prime $p \leq k^{1/2}$. Thus let $X = \{2, 3, \ldots, n\}$ and let r be the number of primes that do not exceed $n^{1/2}$; i.e.,

$$2 = p_1 < p_2 < \cdots < p_r \leq n^{1/2} < p_{r+1} \qquad (i)$$

Then, if A_i $(i = 1, 2, \ldots, r)$ represents the subset of X composed of multiples of p_i, the union $\cup_i A_i$ will consist of all composite integers in X and the first r primes.

We compute $n(\cup_i A_i)$ by means of Theorem 2.5. Here,

$$s_1 = \left\lfloor \frac{n}{p_1} \right\rfloor + \left\lfloor \frac{n}{p_2} \right\rfloor + \cdots + \left\lfloor \frac{n}{p_r} \right\rfloor = \sum_{i=1}^{r} \left\lfloor \frac{n}{p_i} \right\rfloor$$

and, in general,

$$s_j = \sum_{1 \leq i_1 < i_2 < \cdots < i_j \leq r} \left\lfloor \frac{n}{p_{i_1} p_{i_2} \cdots p_{i_j}} \right\rfloor \qquad (j = 1, \ldots, r) \qquad (ii)$$

the summation in (*ii*) being over all j combinations of $\{1, 2, \ldots, r\}$. Hence,

$$n\left(\bigcup_i A_i\right) = s_1 - s_2 + \cdots + (-1)^{r-1} s_r$$

and so

$$\pi(n) = n - 1 + r - s_1 + s_2 - \cdots + (-1)^r s_r \qquad (iii)$$

where r is given (*i*) and the s_j by (*ii*). Note that if the π function is extended to arbitrary real arguments, r can be expressed as $\pi(n^{1/2})$.

2.42 Show that 97 is the twenty-fifth prime.

Since 98, 99, and 100 are composite, it suffices to show that $\pi(100) = 25$. In the notation of Problem 2.41, $r = 4$ $(p_1 = 2,\ p_2 = 3,\ p_3 = 5,\ p_4 = 7)$;

$$s_1 = \left\lfloor \frac{100}{2} \right\rfloor + \left\lfloor \frac{100}{3} \right\rfloor + \left\lfloor \frac{100}{5} \right\rfloor + \left\lfloor \frac{100}{7} \right\rfloor = 117$$

$$s_2 = \left\lfloor \frac{100}{(2)(3)} \right\rfloor + \left\lfloor \frac{100}{(2)(5)} \right\rfloor + \left\lfloor \frac{100}{(2)(7)} \right\rfloor + \left\lfloor \frac{100}{(3)(5)} \right\rfloor + \left\lfloor \frac{100}{(3)(7)} \right\rfloor + \left\lfloor \frac{100}{(5)(7)} \right\rfloor = 45$$

$$s_3 = \left\lfloor \frac{100}{(3)(5)(7)} \right\rfloor + \left\lfloor \frac{100}{(2)(5)(7)} \right\rfloor + \left\lfloor \frac{100}{(2)(3)(7)} \right\rfloor + \left\lfloor \frac{100}{(2)(3)(5)} \right\rfloor = 6$$

$$s_4 = \left\lfloor \frac{100}{(2)(3)(5)(7)} \right\rfloor = 0$$

whence

$$\pi(100) = 100 - 1 + 4 - 117 + 45 - 6 + 0 = 25$$

2.43 A positive integer is **squarefree** if its prime factorization involves no power higher than the first. (We agree that 1 is squarefree.) Show how to compute the number of squarefree integers not exceeding a given integer n.

Let $X = \{1, 2, \ldots, n\}$ and, as in Problem 2.41, let r be the number of primes not exceeding the square root of n. Now define A_i $(i = 1, 2, \ldots, r)$ as the subset of X composed of multiples of p_i^2. The sought number is just $n(\bigcap_i A'_i)$, which is given by Theorem 2.5.

2.44 Find the number of squarfree integers not exceeding 100.

Follow Problem 2.43 (and Problem 2.42):

$$s_1 = \left\lfloor \frac{100}{2^2} \right\rfloor + \left\lfloor \frac{100}{3^2} \right\rfloor + \left\lfloor \frac{100}{5^2} \right\rfloor + \left\lfloor \frac{100}{7^2} \right\rfloor = 42$$

$$s_2 = \left\lfloor \frac{100}{2^2 3^2} \right\rfloor + \left\lfloor \frac{100}{2^2 5^2} \right\rfloor + 0 + \cdots + 0 = 3$$

$$s_3 = 0$$

$$s_4 = 0$$

and
$$n\left(\bigcap_i A'_i\right) = 100 - 42 + 3 = 61$$

2.45 Two positive integers are **relatively prime** if the only positive divisor they have in common is 1. The number of positive integers not exceeding n and relatively prime to n is denoted by $\phi(n)$, where ϕ is **Euler's phi** (or **totient**) **function**. Obtain a formula for Euler's phi function.

Let $p_1, p_2, \ldots, p_k$ be the distinct prime divisors of the positive integer n, and let A_i $(i = 1, 2, \ldots, k)$ be the set of all positive integers from 1 to n that are divisible by p_i. Then, in Theorem 2.5,

$$s_j = \sum \frac{n}{p_{i_1} p_{i_2} \cdots p_{i_j}} \qquad (j = 1, 2, \ldots, k)$$

where the summation is over all j combinations of $\{1, 2, \ldots, k\}$. Thus,

$$\begin{aligned}
\phi(n) &= n - s_1 + s_2 - \cdots + (-1)^k s_k \\
&= n\left[1 - \left(\frac{1}{p_1} + \cdots\right) + \left(\frac{1}{p_1 p_2} + \cdots\right) - \left(\frac{1}{p_1 p_2 p_3} + \cdots\right) + \cdots\right] \\
&= n\left(-\frac{1}{p_1}\right)\left(1 - \frac{1}{p_2}\right)\cdots\left(1 - \frac{1}{p_k}\right) \\
&= \frac{n}{p_1 p_2 \cdots p_k}(p_1 - 1)(p_2 - 1)\cdots(p_k - 1)
\end{aligned}$$

2.46 Evaluate $\phi(3528)$.

Since $3528 = (2^3)(3^2)(7^2)$, the distinct prime divisors are 2, 3, and 7. Problem 2.45 gives:

$$\phi(3528) = \frac{3528}{(2)(3)(7)}(2 - 1)(3 - 1)(7 - 1) = 1008$$

2.47 Show that $\phi(p) = p - 1$ if and only if p is prime.

If p is prime, $\phi(p) = p - 1$, by Problem 2.45. Conversely, if p is not prime, there exists a positive integer $1 < d < p$ such that d divides p. Therefore p and d are not relatively prime, and the definition of the phi function implies $\phi(p) \le p - 2$. Certainly, then, $\phi(p) \ne p - 1$.

2.48 A function whose domain is the set of positive integers is called a **number-theoretic function**. A number-theoretic function f is **multiplicative** if the range of f is closed under multiplication and if $f(mn) = f(m)f(n)$ whenever m and n are relatively prime. Show that Euler's phi function is multiplicative.

Let m and n be relatively prime. If the distinct prime factors of m form the set $M = \{p_1, p_2, \ldots, p_r\}$ and the distinct prime factors of n form the set $N = \{q_1, q_2, \ldots, q_s\}$, then $M \cap N = \emptyset$. Let p be the product of the r primes in M and q be the product of the s primes in N. By Problem 2.45,

$$\begin{aligned}
\phi(mn) &= \frac{mn}{pq}(p_1 - 1)\cdots(p_r - 1)(q_1 - 1)\cdots(q_s - 1) \\
&= \left[\frac{m}{p}(p_1 - 1)\cdots(p_r - 1)\right]\left[\frac{n}{q}(q_1 - 1)\cdots(q_s - 1)\right] \\
&= \phi(m)\phi(n)
\end{aligned}$$

2.49 Suppose $1 = d_1, d_2, \ldots, d_r = n$ are the distinct positive divisors of the positive integer n. Show that $\Sigma\, \phi(d_i) = n$.

First observe that the summation can also be written as $\Sigma\, \phi(n/d_i)$; for as d_i runs through the divisors of n (in increasing order), n/d_i runs through the divisors of n (in decreasing order). Let $X = \{1, 2, \ldots, n\}$ and, for $i = 1, 2, \ldots, r$, let $X_i = \{m \in X : \text{the g.c.d of } m \text{ and } n \text{ is } d_i\}$. Because any pair of positive integers has a unique g.c.d., and because $d_i \in X_i$ for each i, it is clear that $\{X_1, X_2, \ldots, X_r\}$ is a partition of X. Now, m is in X_i if and only if m/d_i and n/d_i are relatively prime. Thus the number of elements in X_i is the number of positive integers not exceeding n/d_i and relatively prime to it; i.e. $\phi(n/d_i)$. The desired result now follows from the fact that $\{X_i\}$ is a partition.

2.50 Verify the result of Problem 2.49 for $n = 12$.

The distinct divisors of 12 are 1, 2, 3, 4, 6, and 12.

$$\begin{aligned}
X_1 &= \{1, 5, 7, 11\} &&\text{and}&& n(X_1) = \phi(12/1) = 4 \\
X_2 &= \{2, 10\} &&\text{and}&& n(X_2) = \phi(12/2) = 2 \\
X_3 &= \{3, 9\} &&\text{and}&& n(X_3) = \phi(12/3) = 2 \\
X_4 &= \{4, 8\} &&\text{and}&& n(X_4) = \phi(12/4) = 2 \\
X_6 &= \{6\} &&\text{and}&& n(X_6) = \phi(12/6) = i \\
X_{12} &= \{12\} &&\text{and}&& n(X_{12}) = \phi(12/12) = 1
\end{aligned}$$

and $4 + 2 + 2 + 2 + 1 + 1 = 12$.

2.51 The **Möbius function,** $\mu(n)$, of a positive integer n is defined by $\mu(n) = 1$, if $n = 1$ or n is the product of an even number of distinct primes; $\mu(n) = -1$, if n is the product of an odd number of distinct primes; and $\mu(n) = 0$ for all other n. Show that the Euler phi function and the Möbius function are related as follows:

$$\phi(n) = n \sum_{d \mid n} \frac{\mu(d)}{d}$$

where the sum is over all the divisors of n, including 1 and n.

This follows immediately from the second equation of Problem 2.45.

2.52 Rewrite the result of Problem 2.41 in terms of the Möbius function.

From the formula (*ii*) for the s_j it is apparent that

$$n - s_1 + s_2 - s_3 + \cdots = \sum_{d|p_1p_2\cdots p_r} \mu(d)\left\lfloor \frac{n}{d} \right\rfloor$$

(The term n on the left side corresponds to $d = 1$.) Hence (*iii*) becomes

$$\pi(n) - \pi(n^{1/2}) = -1 + \sum_{d|p_1p_2\cdots p_r} \mu(d)\left\lfloor \frac{n}{d} \right\rfloor$$

2.53 Show that the Möbius function is multiplicative (Problem 2.48).

We have to show that $\mu(mn) = \mu(m)\mu(n)$ whenever m and n are relatively prime. This is true when $m = 1$ or $n = 1$. Suppose $m > 1$ and $n > 1$. It is also true $(0 = 0)$ when either m or n is divisible by p^k, where p is a prime and $k > 1$. The only remaining case is when both m and n are products of distinct primes: $m = p_1p_2 \cdots p_s$ and $n = q_1q_2 \cdots q_t$. Then

$$\mu(mn) = (-1)^{s+t} = (-1)^s(-1)^t = \mu(m)\mu(n)$$

2.54 Let f be a multiplicative function and let

$$g(n) = \sum_{d|n} f(d)$$

where the sum is over all the divisors d of n. Show that g is multiplicative.

Let $d_1, d_2, \ldots, d_s$ be the distinct divisors of the positive integer m and let $e_1, e_2, \ldots, e_t$ be the distinct divisors of the positive integer n. Then,

$$g(m) = \sum_{i=1}^{s} f(d_i) \qquad \text{and} \qquad g(n) = \sum_{j=1}^{t} f(e_j)$$

If m and n are relatively prime, the distinct divisors of mn are the st numbers d_ie_j, with d_i and e_j relatively prime for all i and j. Therefore,

$$g(mn) = \sum_{i=1}^{s}\sum_{j=1}^{t} f(d_ie_j) = \sum_{i=1}^{s} f(d_i) \sum_{j=1}^{t} f(e_j) = g(m)g(n)$$

2.55 If $n \geq 2$, evaluate

$$(a)\ \sum_{d|n} \mu(d) \qquad \text{and} \qquad (b)\ \sum_{d|n} |\mu(d)|$$

We need consider only those divisors of n which are squarefree (including 1). Suppose n has k prime factors.

(*a*) The sum of the Möbius-function values of the squarefree divisors of n is

$$C(k, 0) - C(k, 1) + C(k, 2) - \cdots + (-1)^k C(k, k) = 0$$

[by Problem 1.39(*b*)].

(*b*) 2^k [by Problem 1.39(*a*)].

2.56 Let n be a positive integer and let its divisors be

$$1 = d_1 < d_2 < \cdots < d_t = n$$

Prove that for all i and j,

$$d_j \mid d_{t-i+1} \quad \text{if and only if } d_i \mid d_{t-j+1}$$

We have $d_i d_{t-i+1} = n$ and $d_j d_{t-j+1} = n$. Therefore,

$$d_i d_{t-i+1} = d_j d_{t-j+1} \qquad \text{or} \qquad \frac{d_{t-i+1}}{d_j} = \frac{d_{t-j+1}}{d_i}$$

One side of this last equation is an integer if and only if the other side is.

2.57 (*a*) (*Classical Möbius Inversion Formula*) Let f be a number-theoretic function and let

$$g(n) = \sum_{d|n} f(d)$$

Show that

$$f(n) = \sum_{d|n} \mu(d) g\left(\frac{n}{d}\right) = \sum_{d|n} \mu\left(\frac{n}{d}\right) g(d)$$

(*b*) Prove the converse of (*a*).

(*a*) Denote the divisors of n as in Problem 2.56. We have:

$$\sum_{i=1}^{t} \mu(d_i) g\left(\frac{n}{d_i}\right) = \sum_{i=1}^{t} \mu(d_i) g(d_{t-i+1})$$

$$= \sum_{i=1}^{t} \mu(d_i) \sum_{d_j | d_{t-i+1}} f(d_j)$$

Problem 2.56 may be used to invert the order of the double summation:

$$\sum_{i=1}^{t} \mu(d_i) \sum_{d_j | d_{t-i+1}} f(d_j) = \sum_{j=1}^{t} f(d_j) \sum_{d_i | d_{t-j+1}} \mu(d_i)$$

On the right-hand side the inner sum vanishes for $j = 1, 2, \ldots, t-1$—in consequence of Problem 2.55(*a*)—while for $j = t$ it has the value $\mu(d_1) = 1$. We see then that

$$\sum_{i=1}^{t} \mu(d_i) g\left(\frac{n}{d_i}\right) = f(d_t) = f(n)$$

which is the Möbius formula.

(*b*) In the notation of Problems 2.56 and 2.57(*a*), we have

$$\sum_{j=1}^{t} f(d_j) = \sum_{j=1}^{t} \sum_{d_i | d_j} \mu(d_i) g\left(\frac{d_j}{d_i}\right) = \sum_{j=1}^{t} \sum_{d_i | d_j} \mu\left(\frac{d_j}{d_i}\right) g(d_i)$$

$$= \sum_{i=1}^{t} g(d_i) \sum_{d_{t-j+1} | d_{t-i+1}} \mu\left(\frac{d_{t-i+1}}{d_{t-j+1}}\right)$$

where the last step follows from Problem 2.56. For simplicity, replace the summation index $t - j + 1$ by k, obtaining

$$\sum_{j=1}^{t} f(d_j) = \sum_{i=1}^{t} g(d_i) \sum_{d_k | d_{t-i+1}} \mu(d_k)$$

For $i = 1, 2, \ldots, t-1$, the coefficient of $g(d_i)$ is zero [Problem 2.55(*a*)]; for $i = t$, the coefficient is $\mu(d_1) = 1$. Thus,

$$\sum_{j=1}^{t} f(d_j) = g(d_t) = g(n)$$

and the proof is complete.

2.58 With reference to Problem 2.57(*a*), show that f is multiplicative if and only if g is multiplicative.

If f is multiplicative, then g is multiplicative, by Problem 2.54.

Suppose g is multiplicative. Let m and n be relatively prime and let the divisors of m and n be $d_1, d_2, \ldots, d_a$ and $e_1, e_2, \ldots, e_t$, respectively. The divisors of mn are d_ie_j, where $1 \leq i \leq s$ and $1 \leq j \leq t$; hence, by the Möbius formula,

$$f(mn) = \sum_i \sum_j \mu(d_ie_j) g\left(\frac{mn}{d_ie_j}\right)$$

But μ is multiplicative (Problem 2.53), and d_i and e_j are relatively prime, so that

$$\mu(d_ie_j) = \mu(d_i)\mu(e_j)$$

Also (by assumption) g is multiplicative, and m/d_i and n/e_j are relatively prime, so that

$$g\left(\frac{mn}{d_ie_j}\right) = g\left(\frac{m}{d_i}\right) g\left(\frac{n}{e_j}\right)$$

Thus

$$f(mn) = \sum_i \mu(d_i) g\left(\frac{m}{d_i}\right) \sum_j \mu(e_j) g\left(\frac{n}{e_j}\right) = f(m)f(n)$$

and f is multiplicative.

2.59 The number of positive divisors of the positive integer n is denoted by $\tau(n)$, and the sum of the positive divisors of n is denoted by $\sigma(n)$. (*a*) Prove that these two functions are multiplicative. (*b*) Show how to compute their values for a given n.

(*a*) If f is multiplicative and if

$$g(n) = \sum_{d|n} f(d)$$

then by Problem 2.54, g is multiplicative. Now the function $f(n) \equiv 1$ is clearly multiplicative; to it corresponds $g(n) = \tau(n)$, which is therefore also multiplicative. The same argument, using $f(n) = n$, shows that $\sigma(n)$ is multiplicative.

(*b*) If p is prime, the divisors of p^k are $1, p, p^2, \ldots, p^k$. Hence,

$$\tau(p^k) = 1 + k \qquad \text{and} \qquad \sigma(p^k) = 1 + p + p^2 + \cdots + p^k = \frac{p^{k+1} - 1}{p - 1}$$

These results, the prime factorization theorem, and the multiplicative property of $\tau(n)$ and $\sigma(n)$ allow the computation of the two functions. For example,

$$\tau(4608) = \tau(2^3 3^2 4^3) = \tau(2^3)\tau(3^2)\tau(4^3) = (1+3)(1+2)(1+3) = 48$$

$$\sigma(4608) = \left(\frac{2^{3+1} - 1}{2 - 1}\right)\left(\frac{3^{2+1} - 1}{3 - 1}\right)\left(\frac{4^{3+1} - 1}{4 - 1}\right) = (15)(26)(85) = 33{,}150$$

2.60 Show that for any positive integer n,

$$\sum_{d|n} \mu(d)\tau\left(\frac{n}{d}\right) = 1$$

Apply the Möbius inversion formula to the relation

$$\tau(n) = \sum_{d|n} 1$$

2.61 Consider the formal series $L(s) = \sum_{n=1}^{\infty} f(n)n^{-s}$ defined by a number-theoretic function f. Show that if f is multiplicative, this series can be represented as an infinite product:

$$L(s) = \prod_{p \text{ prime}} F(p, s) \qquad \text{where } F(p, s) = 1 + f(p)p^{-s} + f(p^2)p^{-2s} + f(p^3)p^{-3s} + \cdots$$

[$F(p, s)$ is also a *formal* series.]

It suffices to show that, for every positive integer n, the term n^{-s} arises in the infinite product with the same coefficient that it has in the infinite series; namely, $f(n)$. Let n have the unique prime factorization

$$n = \prod_{i=1}^{k} p_i^{a_i} \qquad \text{whence } n^{-s} = \prod_{i=1}^{k} p_i^{-a_i s}$$

It is seen that when the infinite product is multiplied out, one and only one product gives a term in n^{-s}: this is the product of $f(p_1^{a_1})p_1^{-a_1 s}$ from $F(p_1, s)$, $f(p_2^{a_2})p_2^{-a_2 s}$ from $F(p_2, s), \ldots, f(p_k^{a_k})p_k^{-a_k s}$ from $F(p_k, s)$, and 1s from all other $F(p, s)$. Thus, n^{-s} appears with coefficient

$$f(p_1^{a_1})f(p_2^{a_2}) \cdots f(p_k^{a_k}) = f(p_1^{a_1} p_2^{a_2} \cdots p_k^{a_k}) = f(n)$$

f being multiplicative.

2.62 The celebrated **Riemann zeta function** is defined by

$$\zeta(s) = \sum_{n=1}^{\infty} \frac{1}{n^s}$$

Prove that

$$\frac{1}{\zeta(s)} = \sum_{n=1}^{\infty} \frac{\mu(n)}{n^s}$$

By Problem 2.61, with $f(n) \equiv 1$,

$$\zeta(s) = \prod \left(\frac{1}{1 - p^{-s}}\right) = \frac{1}{\prod (1 - p^{-s})}$$

On the other hand, for $f(n) = \mu(n)$, Problem 2.61 gives

$$\sum_{n=1}^{\infty} \frac{\mu(n)}{n^s} = \prod (1 - p^{-s}) = \frac{1}{\zeta(s)}$$

2.63 Let $A = \{a_1, a_2, \ldots, a_m\}$ be an alphabet with m distinct letters. A **circular word of length n** from A is an n sequence of A arranged clockwise around a circle, as in Fig. 2-1(a). In a circular word only the clockwise order of the letters matters: circular words f and f' are identical if some rotation of the diagram for f takes it into the diagram for f'. Count the *distinct* circular words of length n from A.

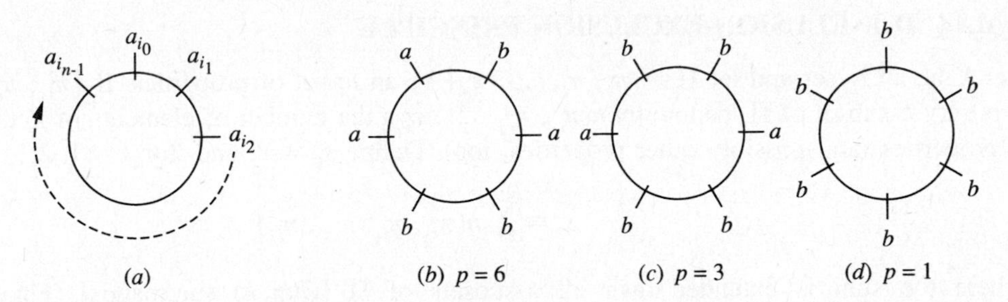

Fig. 2-1

It is apparent from Fig. 2-1(*a*) that when a circular word of length n is "spelled"— with *any* of its letters chosen as the initial letter—the clockwise sequence repeats after n letters. Hence, the set of *repetition-distances* for the circular word is nonempty. The smallest element, p, of this set is called the **period** of the circular word. Figures 2-1(*b*)–(*d*) show three circular words of length $n = 6$ ($m \geq 2$), and their corresponding periods.

It is easy to see that any repetition-distance—in particular, the period p—must divide n. Furthermore, a circular word of period p determines precisely p distinct n sequences of A. [For example, the circular word of Fig. 2-1(*c*) determines the 3 sequences $babbab$, $abbabb$, $bbabba$.] Consequently, the $M(p)$ (say) circular words with a given period p account for $pM(p)$ n sequences. Because there are m^n n-sequences in all, and because p is a divisor of n, we must have

$$\sum_{p|n} pM(p) = m^n$$

Application of the Möbius inversion formula to this last equation yields

$$xM(x) = \sum_{d|x} \mu(d)m^{x/d} \qquad \text{or} \qquad M(x) = \frac{1}{x}\sum_{d|x} \mu(d)m^{x/d}$$

and the answer to our problem is $\sum_{p|n} M(p)$.

2.64 How many circular words of length 12 may be drawn from a 5-letter alphabet?

The periods, or divisors of 12, are 1, 2, 3, 4, 6, and 12; the corresponding Möbius-function values are 1, −1, −1, 0, 1, and 0, respectively. The integers $M(p)$ are given by the expression of Problem 2.63:

$$M(1) = \frac{1}{1}[(1)5^{1/1}] = 5$$

$$M(2) = \frac{1}{2}[(1)5^{2/1} + (-1)5^{2/2}] = 10$$

$$M(3) = \frac{1}{3}[(1)5^{3/1} + (-1)5^{3/3}] = 40$$

$$M(4) = \frac{1}{4}[(1)5^{4/1} + (-1)5^{4/2} + (0)5^{4/4}] = 150$$

$$M(6) = \frac{1}{6}[(1)5^{6/1} + (-1)5^{6/2} + (-1)5^{6/3} + (1)5^{6/6}] = 2580$$

$$M(12) = \frac{1}{12}[(1)5^{12/1} + (-1)5^{12/2} + (-1)5^{12/3} + (0)5^{12/4} + (1)5^{12/6} + (0)5^{12/12}]$$
$$= 20{,}343{,}700$$

The number of circular words is the sum of these 6 integers, or 20,346,485. (For comparison, the number of 12 sequences is $5^{12} = 244{,}140{,}625$.)

GENERALIZED INCLUSION-EXCLUSION PRINCIPLE

2.65 Let X be an N set and let $\Pi = \{\pi_1, \pi_2, \ldots, \pi_m\}$ be an m-set of properties. If $\{\pi_{i_1}, \pi_{i_2}, \ldots, \pi_{i_k}\}$ is an arbitrary k subset of Π, denote by $n(\pi_{i_1}, \pi_{i_2}, \ldots, \pi_{i_k})$ the number of elements of X that possess these k properties (and possibly other properties, too). Define $s_0 \equiv N$, and, for $k = 1, 2, \ldots, m$,

$$s_k \equiv \sum n(\pi_{i_1}, \pi_{i_2}, \ldots, \pi_{i_k})$$

where the sum is extended over all k-subsets of Π [$C(m, k)$ summands]. Finally, let e_j ($j = 0, 1, \ldots, m$) be the number of elements of X that have *exactly* j of the m properties, and let f_j ($j = 1, 2, \ldots, m$) be the number of elements that have *at least* j properties. (*a*) Prove

Theorem 2.8.

$$e_j = s_j - C(j+1, 1)s_{j+1} + C(j+2, 2)s_{j+2} - \cdots + (-1)^{m-j}C(m, m-j)s_m$$

(*b*) Prove

Theorem *2.9.*

$$f_j = s_j - C(j, 1)s_{j+1} + C(j+1, 2)s_{j+2} - \cdots + (-1)^{m-j}C(m-1, m-j)s_m$$

(*a*) We show that every element of X has the same count on (makes the same contribution to) either side of the equation of Theorem 2.8.

If an element has fewer than j properties, its count is 0 on either side.

If an element has exactly j properties, its count is 1 on the left, and $(C_j, j) = 1$ (in s_j) on the right.

If an element has exactly $j + l$ properties ($1 \leq l \leq m - j$), its count on the left is 0. On the right, it has the count $C(j+l, j)$ in s_j; $C(j+l, j+1)$ in s_{j+1}; $\ldots$; $C(j+l, j+l)$ in s_{j+l}. The total count on the right is therefore

$$C(j+l, j) - C(j+1, 1)C(j+l, j+1) + C(j+2, 2)C(j+l, j+2) - \cdots + (-1)^l C(j+l, l)C(j+l, j+l)$$
$$= C(j+l, j) - C(l, 1)C(j+l, j) + C(l, 2)C(j+l, j) - \cdots + (-1)^l C(l, l)C(j+l, j)$$
$$= C(j+l, j)(1-1)^l = 0$$

This completes the proof.

Observe that, for $j = 0$, Theorem 2.8 is just the sieve formula [in Theorem 2.5 let A_i ($i = 1, 2, \ldots, m$) be the subset of elements possessing property π_i].

(*b*) It is enough to show that the given series for f_j satisfies the difference equation $f_j - f_{j+1} = e_j$ and the end condition $f_m = e_m$. Now, using Pascal's identity,

$$f_j - f_{j+1} = s_j - [C(j, 1) + C(j, 0)]s_{j+1} + [C(j+1, 2) + C(j+1, 1)]s_{j+2}$$
$$- \cdots + (-1)^{m-j}[C(m-1, m-j) + C(m-1, m-j-1)]s_m$$
$$= s_j - C(j+1, 1)s_{j+1} + C(j+2, 2)s_{j+2} - \cdots + (-1)^{m-j}C(m, m-j)s_m$$
$$= e_j$$

Further, $f_m = s_m = e_m$.

2.66 With X and Π as in Problem 2.65, find the number of elements of X having an even number of the properties and the number having an odd number of the properties.

If we define $E(x) \equiv e_0 + e_1x + e_2x^2 + \cdots + e_mx^m$, then the numbers of elements having an even and an odd number of properties will be $\frac{1}{2}[E(1) \pm E(-1)]$. Now, by Theorem 2.8,

$$E(x) = \sum_{j=0}^{m} e_j x^j = \sum_{j=0}^{m} \left[\sum_{k=j}^{m} (-1)^{k-j} C(k, j) s_k \right] x^j$$

$$= \sum_{k=0}^{m} (-1)^k s_k \sum_{j=0}^{k} C(k, j)(-x)^j = \sum_{k=0}^{m} (-1)^k s_k (1-x)^k$$

$$= \sum_{k=0}^{m} s_k (x-1)^k$$

Therefore,
$$\tfrac{1}{2}[E(1) \pm E(-1)] = \tfrac{1}{2}\left[s_0 \pm \sum_{k=0}^{m} (-2)^k s_k \right]$$

2.67 Find the number of m-letter words that can be formed using the letters A, B, C, and D, such that each word has an odd number of A's.

In the notation of Problem 2.65 let π_i ($i = 1, 2, \ldots, m$) be the property that the ith letter of the word is A. We have $s_0 = 4^m$ possible words, and

$$s_k = C(m, k)4^{m-k} \qquad (k = 1, 2, \ldots, m)$$

By Problem 2.66 the desired number is

$$\tfrac{1}{2}\left[s_0 - \sum_{k=0}^{m} (-2)^k s_k \right] = \tfrac{1}{2}\left[4^m - \sum_{k=0}^{m} C(m, k)(-2)^k 4^{m-k} \right]$$

$$= \tfrac{1}{2}(4^m - 2^m) = 2^{m-1}(2^m - 1)$$

2.68 Check Problem 2.67 by direct use of the generating function $E(x)$.

Since $e_j = C(m, j)3^{m-j}$ for all j,

$$E(x) = \sum_{j=0}^{m} C(m, j) x^j 3^{m-j} = (x+3)^m$$

and so $\frac{1}{2}[E(1) - E(-1)] = \frac{1}{2}(4^m - 2^m)$.

2.69 Of the 100 students in Problem 2.20, how many take (*a*) exactly 1 course? (*b*) exactly 2 courses? (*c*) exactly 3 courses? (*d*) at least 1 course? (*e*) at least 2 courses? (*f*) at least 3 courses?

With the s_k as computed in Problem 2.20, Theorems 2.7 and 2.8 give:

(*a*) $e_1 = 60 - C(2, 1)(39) + C(3, 2)(10) - C(4, 3)(2) = 4$

(*b*) $e_2 = 39 - C(3, 1)(10) + C(4, 2)(2) = 21$

(*c*) $e_3 = 10 - C(4, 1)(2) = 2$

(*d*) $f_1 = 60 - C(1, 1)(39) + C(2, 2)(10) - C(3, 3)(2) = 29$

(*e*) $f_2 = 39 - C(2, 1)(10) + C(3, 2)(2) = 25$

(*f*) $f_3 = 10 - C(3, 1)(2) = 4$

2.70 A function w from a set X to the set of real numbers is called a **weight function** on X. If X is finite and A is a subset of X, the **weight of A**, denoted $w(A)$, is the sum of all $w(x)$ for $x \in A$. If Π is a set of m properties (cf. Problem 2.65), let A_j ($j = 0, 1, 2, \ldots, m$) be the set of all elements in X having *exactly j* properties and j let B_j $(j = 1, 2, \ldots, m)$ be the set of all elements in X having *at least j* properties. For each j, write $E_j \equiv w(A_j)$ and $F_j \equiv w(B_j)$. If Q is a subset of Π, $w(Q)$ is defined as the sum of the weights of all the elements in X which have every property in Q. Finally, in analogy to the s_k of Problem 2.65, define

$$S_k = \begin{cases} w(X) & k = 0 \\ \sum\limits_{\substack{Q \subset \Pi \\ n(Q)=k}} w(Q) & k = 1, 2, \ldots, m \end{cases} \tag{i}$$

Prove that Theorems 2.8 and 2.9 remain valid if e_j, f_j, and s_j are replaced, respectively, by E_j, F_j, and S_j.

In the proof of Problem 2.65(*a*), simply replace "count of x" (1 or 0) by "weight of x" [$w(x)$ or 0]. The proof of Problem 2.65(*b*) extends without change. Note that for the particular weight function $w(x) \equiv 1$, we have $S_j = s_j$, etc.

2.71 Let $X = \{a, b, c, d, e, f\}$ be a set with weights 2, 3, 4, 5, 6, and 7, respectively, and let $\Pi = \{\alpha, \beta, \gamma, \delta\}$ be a set of properties. Given: (*i*) a, b, c, e, and f have property α; (*ii*) b, c, d, and f have property β; (*iii*) a, d, e, and f have property γ; (*iv*) b, c, d, and e have property δ. Compute all E_j, F_j, and S_j.

In the notation of Problem 2.70, A_0, A_1, and A_4 are empty; $A_2 = \{a\}$; $A_3 = \{b, c, d, e, f\}$. Hence,

$$E_0 = 0 \qquad E_1 = 0 \qquad E_2 = 2 \qquad E_3 = 25 \qquad E_4 = 0$$

Further, $B_1 = B_2 = X$, $B_3 = A_3$, and B_4 is empty; hence

$$F_1 = F_2 = 27 \qquad F_3 = 25 \qquad F_4 = 0$$

By (*i*) of Problem 2.70, $S_0 = w(X) = 2 + 3 + 4 + 5 + 6 + 7 = 27$.

$$w(\{\alpha\}) = w(a) + w(b) + w(c) + w(e) + w(f) = 22$$

$$w(\{\beta\}) = w(b) + w(c) + w(d) + w(f) = 19$$

$$w(\{\gamma\}) = w(a) + w(d) + w(e) + w(f) = 20$$

$$w(\{\delta\}) = w(b) + w(c) + w(d) + w(e) = 18$$

$$S_1 = 22 + 19 + 20 + 18 = 79$$

$$w(\{\alpha, \beta\}) = w(b) + w(c) + w(f) = 14$$

$$w(\{\alpha, \gamma\}) = w(a) + w(e) + w(f) = 15$$

$$w(\{\alpha, \delta\}) = w(b) + w(c) + w(e) = 13$$

$$w(\{\beta, \gamma\}) = w(d) + w(f) = 12$$

$$w(\{\beta, \delta\}) = w(b) + w(c) + w(d) = 12$$

$$w(\{\gamma, \delta\}) = w(d) + w(e) = 11$$

$$S_2 = 14 + 15 + 13 + 12 + 12 + 11 = 77$$

$$w(\{\alpha, \beta, \gamma\}) = w(f) = 7$$

$$w(\{\alpha, \beta, \delta\}) = w(b) + w(c) = 7$$

$$w(\{\beta, \gamma, \delta\}) = w(d) = 5$$

$$w(\{\alpha, \gamma, \delta\}) = w(e) = 6$$

$$S_3 = 7 + 7 + 5 + 6 = 25$$

Finally, $S_4 = E_4 = 0$.

2.72 Verify the generalized Theorems 2.8 and 2.9 for the data of Problem 2.71.

Checking Theorem 2.7 for $j = 0, 1, \ldots, 4$:

$$S_0 - C(1,1)S_1 + C(2,2)S_2 - C(3,3)S_3 + C(4,4)S_4 = 27 - 79 + 77 - 25 + 0 = 0 = E_0$$

$$S_1 - C(2,1)S_2 + C(3,2)S_3 - C(4,3)S_4 = 79 - 154 + 75 - 0 = 0 = E_1$$

$$S_2 - C(3,1)S_3 + C(4,2)S_4 = 77 - 75 + 0 = 2 = E_2$$

$$S_3 - C(4,1)S_4 = 25 - 0 = 25 = E_3$$

$$S_4 = 0 = E_4$$

Checking Theorem 2.9 for $j = 1, \ldots, 4$:

$$S_1 - C(1,1)S_2 + C(2,2)S_3 - C(3,3)S_4 = 79 - 77 + 25 - 0 = 27 = F_1$$

$$S_2 - C(2,1)S_3 + C(3,2)S_4 = 77 - 50 + 0 = 27 = F_2$$

$$S_3 - C(3,1)S_4 = 25 - 0 = 25 = F_3$$

$$S_4 = 0 = F_4$$

2.73 (*Problème des Ménages*) Find the number of ways of seating $n \geq 3$ married couples at a circular table (having $2n$ numbered seats) so that the sexes alternate and no husband and wife sit side by side.

Let the wives be seated first: they can occupy the odd-numbered seats in $n!$ ways, or they can occupy the even-numbered seats in $n!$ ways. For each of these $2n!$ arrangements, there is a fixed number, M_n, of ways in which the husbands can be seated, no husband next to his wife. (M_n is called the **ménage number**.) This being the case, we may assume the wives disposed and the empty seats renumbered as in Fig. 2-2(*a*). The spouse of W_i will be notated H_i $(i = 1, 2, \ldots, n)$.

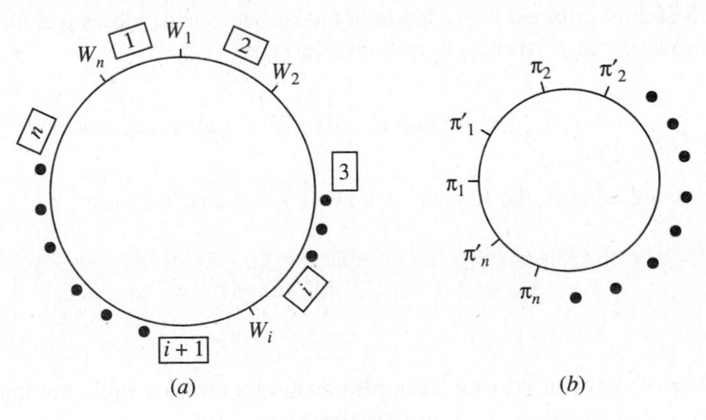

Fig. 2-2

There are $n!$ (unrestricted) arrangements of the husbands. We distinguish $2n$ properties of the arrangements:

π_i $(i = 1, 2, \ldots, n)$: H_i is put in seat i

π_i' $(i = 1, 2, \ldots, n-1)$: H_i is put in seat $i + 1$

π_n': H_n is put in seat 1

In using Theorem 2.8 to evaluate $e_0 = M_n$, one need only consider k subsets of *compatible* properties. (A k-subset containing incompatible properties—e.g., π'_1 and π_2—contributes zero to s_k.) Now, to say that an arrangement of husbands exhibits certain k compatible properties means that certain k husbands are sitting in forbidden seats; there exist $(n-k)!$ such arrangements. In other words, $s_k = p_k(n-k)!$, where p_k is the number of ways of choosing k compatible properties from among the given $2n$ properties.

All that is left is to determine the numbers p_k $(k = 1, 2, \ldots, 2n)$. To this end, dispose the properties around a circle [Fig. 2-2(*b*)]; in this configuration consecutive properties, and these only, are incompatible. Let a property be colored *red* if it is included in the k sample and *blue* if not included.

Case 1. π_1 is red. Then π'_1 and π'_n are blue, leaving a linear sequence $\pi_2\pi'_2 \cdots \pi_n$ from which the remaining $k-1$ compatible properties must be selected. By Problem 1.174(*c*), this can be accomplished in $C(2n-k-1, k-1)$ ways.

Case 2. π_1 is blue. Again by Problem 1.174(*c*), the selection can be made in $C(2n-k, k)$ ways.

Thus we have found

$$p_k = C(2n-k-1, k-1) + C(2n-k, k) = \frac{2n}{2n-k}C(2n-k, k)$$

(which expression properly vanishes for $k > n$). Consequently, the ménage problem has the solution $2n!\,M_n$, with

$$M_n = n! - \frac{2n}{2n-1}C(2n-1, 1)(n-1)! + \frac{2n}{2n-2}C(2n-2, 2)(n-2)!$$
$$- \cdots + (-1)^n \frac{2n}{2n-n}C(2n-n, n)(n-n)!$$

2.74 Solve the ménage problem if the alternation of sexes is no longer required. (Now there will be a positive solution for the case $n = 2$.)

Distinguish k couples and consider an arrangement in which each of these couples sit side by side. Each couple may be treated as a unit; the 2 seats occupied by each couple may also be taken as a unit. There are $2n-2k$ more individuals and $2n-2k$ more seats. Thus, in effect, there are $2n-k$ units of people to be assigned to $2n-k$ units of seats. Since the seats are labeled, the number of possible assignments is $(2n-k)!$. Now, our k couples will still each be adjacent if any subset of the husbands trade places with their wives. This leads to a total of $2^k(2n-k)!$ arrangements, whence, in the sieve formula,

$$s_k = C(n, k)2^k(2n-k)! \qquad (k = 1, 2, \ldots, n)$$

Because the seats are labeled, the formula extends to $k = 0$, and we have

$$M'_n = \sum_{k=0}^{n} (-1)^k C(n, k)2^k(2n-k)! \qquad (n = 2, 3, \ldots)$$

2.75 Find the number of ways of seating 5 couples around a circular table having 10 numbered seats, under the conditions of (*a*) Problem 2.73 and (*b*) Problem 2.74.

(*a*)
$$M_5 = 5! - \tfrac{10}{9}C(9, 1)4! + \tfrac{10}{8}C(8, 2)3! - \tfrac{10}{7}C(7, 3)2! + \tfrac{10}{6}C(6, 4)1! - \tfrac{10}{5}C(5, 5)0!$$
$$= 13$$

(*b*)
$$M'_5 = \sum_{k=0}^{5} (-1)^k C(5, k)2^k(10-k)! = 1{,}263{,}360$$

THE PERMANENT OF A MATRIX

2.76 Let $A = (a_{ij})$ be a matrix with m rows and n columns, where $m \leq n$. The **permanent** of A is defined as

$$\text{per}(A) = \sum a_{1i_1} a_{2i_2} \cdots a_{mi_m}$$

where the summation is over all m-permutations of the set $N = \{1, 2, \ldots, n\}$. Compute the permanents of the following matrices:

$$(a)\quad A = \begin{bmatrix} 1 & 1 \\ 1 & 1 \end{bmatrix} \qquad (b)\quad B = \begin{bmatrix} 1 & 2 & 3 \\ 1 & 2 & 3 \end{bmatrix} \qquad (c)\quad C = \begin{bmatrix} 1 & 1 & 1 & 1 \\ 1 & 2 & 3 & 4 \\ 1 & 1 & 1 & 1 \end{bmatrix}$$

(*a*)

$$\text{per}(A) = (1)(1) + (1)(1) = 2$$

The permanent of a *square* matrix is made up of the same products that compose the determinant, but without the prefixed algebraic signs.

(*b*)

$$\text{per}(B) = (1)(2) + (2)(1) + (1)(3) + (3)(1) + (2)(3) + (3)(2)$$

$$= 22$$

(*c*) By its very definition, the permanent of an $m \times n$ matrix A is the sum of the permanents of all $m \times m$ submatrices of A. For the given matrix C,

Columns 1, 2, and 3:

$$(1)(2)(1) + (1)(3)(1) + (1)(1)(1) + (1)(3)(1) + (1)(1)(1) + (1)(2)(1)$$
$$= 2 + 3 + 1 + 3 + 1 + 2 = 19$$

Columns 1, 2, and 4:

$$(1)(2)(1) + (1)(4)(1) + (1)(1)(1) + (1)(4)(1) + (1)(1)(1) + (1)(2)(1)$$
$$= 2 + 4 + 1 + 4 + 1 + 2 = 14$$

Columns 1, 3, and 4:

$$(1)(3)(1) + (1)(4)(1) + (1)(1)(1) + (1)(4)(1) + (1)(1)(1) + (1)(3)(1)$$
$$= 3 + 4 + 1 + 4 + 1 + 3 = 16$$

Columns 2, 3, and 4:

$$(1)(3)(1) + (1)(4)(1) + (1)(2)(1) + (1)(4)(1) + (1)(2)(1) + (1)(3)(1)$$
$$= 3 + 4 + 2 + 4 + 2 + 3 = 18$$

Thus, $\text{per}(C) = 12 + 14 + 16 + 18 = 60$.

2.77 Let I_n and J_n respectively denote the $n \times n$ identity matrix and the $n \times n$ matrix of 1s. Evaluate

$$(a)\quad \text{per}\left(\frac{1}{n} J_n\right) \qquad (b)\quad \text{per}(J_n - I_n)$$

(*a*) The required permanent is the sum of $P(n, n) = n!$ terms, each equal to $(1/n)^n$. Thus,

$$\text{per}\left(\frac{1}{n} J_n\right) = \frac{n!}{n^n}$$

(*b*) The $n \times n$ matrix $J_n - I_n$ has 0s on the main diagonal and 1s elsewhere. Hence, each derangement of $\{1, 2, \ldots, n\}$ contributes 1 to per $(J_n - I_n)$, while every other permutation of the set contributes 0. That is, per $(J_n - I_n) = D_n$.

2.78 Let the product of the row sums of any matrix B be denoted by $P(B)$. Let $B = (b_{ij})$ be an $m \times n$ matrix with $m \le n$. An $m \times n$ matrix obtained from B by replacing the elements of exactly r columns by 0s shall be denoted as $B_r = (b'_{ij})$; there are $C(n, r)$ such matrices, for each r. Define

$$P_r \equiv \sum_{C(n,\,r)\text{ terms}} P(B_r) \qquad (r = 0, 1, 2, \ldots, n)$$

[Clearly, $P_0 = P(B)$ and $P_n = 0$.] Prove:

Theorem 2.10.

$$\text{per}\,(B) = P_{n-m} - C(n-m+1, 1)P_{n-m+1} + C(n-m+2, 2)P_{n-m+2} - \cdots + (-1)^{m-1}C(n-1, m-1)P_{n-1}$$

We shall apply inclusion-exclusion—as developed in Problem 2.70—to the set X of *m-sequences* (Section 2.2) of $N = \{1, 2, \ldots, n\}$; we know that $|X| = n^m$. Define a weight function w on X as follows: If $x = \langle j_1, j_2, \ldots, j_m \rangle$ belongs to X, then

$$w(x) = b_{1j_1} b_{2j_2} \cdots b_{mj_m}$$

Cleverly consider the n-set of properties $\Pi = \{\pi_1, \pi_2, \ldots, \pi_n\}$, where π_i $(i = 1, 2, \ldots, n)$ is the property that an element of X does not contain i. Then an m-sequence x is an *m-permutation* if and only if x possesses exactly $n - m$ of the properties. Hence (going over to the notation of Problem 2.70) if A_{n-m} is the set of all such x, we have $E_{n-m} = \text{per}\,(B)$.

Now observe that, by the rule for multiplying series,

$$P(B_r) = \Big(\sum_{j_1} b'_{1j_1}\Big)\Big(\sum_{j_2} b'_{2j_2}\Big) \cdots \Big(\sum_{j_m} b'_{mj_m}\Big) = \sum_{j_1, j_2, \ldots, j_m} b'_{1j_1} b'_{2j_2} \cdots b'_{mj_m}$$

Since $b'_{ij} = 0$ for r given values of j, and since $b'_{ij} = b_{ij}$ for the remaining $n - r$ values of j, the multiple summation need be extended only over those sequences $x = \langle j_1, j_2, \ldots, j_m \rangle$ which possess r specified properties in Π; moreover, the primes may be dropped from the summand. Therefore—still following Problem 2.70—

$$P(B_r) = \sum_{x \in Q} w(x) = w(Q)$$

so that (*i*) of Problem 2.70 gives $P_r = S_r$ $(r = 0, 1, \ldots, n)$. Expressing E_{n-m} in terms of these S_r via the generalized Theorem 2.8 yields Theorem 2.10.

2.79 A simple example of a **circulant matrix** (of order 3) is

$$C = \begin{bmatrix} x & y & z \\ y & z & x \\ z & x & y \end{bmatrix}$$

By calculating per (C) in 2 ways, establish the identity

$$x^3 + y^3 + z^3 = (x + y + z)^3 - 3(x + y)(y + z)(z + x)$$

From the definition of the permanent,

$$\begin{aligned}\text{per}(C) &= (xyz + x^3) + (yxz + y^3) + (zyx + z^3) \\ &= x^3 + y^3 + z^3 + 3xyz \end{aligned} \qquad (i)$$

For Theorem 2.10, $P_0 = (x + y + z)^3$,

$$\begin{aligned}P_1 &= (y + z)(z + x)(x + y) + (x + z)(y + x)(z + y) + (x + y)(y + z)(z + x) \\ &= 3(x + y)(y + z)(z + x)\end{aligned}$$

and $P_2 = 3xyz$. Therefore,

$$\text{per}(C) = (x + y + z)^3 - 3(x + y)(y + z)(z + x) + 3xyz \qquad (ii)$$

Comparison of (*i*) and (*ii*) yields the identity.

ROOK POLYNOMIALS AND HIT POLYNOMIALS

2.80 In the game of chess a **rook** is a piece that can capture another piece or pawn situated in the same row or same column, provided that there is no other piece or pawn between the two. The **problem of nontaking rooks** asks for the number of ways, $r_k(\mathscr{C})$, of positioning k identical rooks on a chessboard, $\mathscr{C}$, with n squares, so that no rook can capture another rook. The polynomial

$$R(x, \mathscr{C}) \equiv 1 + r_1x + r_2x^2 + \cdots + r_nx^n$$

is called the **rook polynomial** of $\mathscr{C}$. Find the rook polynomial of the standard 8×8 chessboard.

One can choose k rows out of the 8 rows in $C(8, k)$ ways. After that the k rooks can be placed in these k rows so that there is exactly 1 rook in each row and no 2 rooks are in the same column, in $P(8, k)$ ways. Hence, for $k = 1, 2, \ldots, 8$ (and also for $k = 0$ and $k > 8$), $r_k = C(8, k)P(8, k)$, giving

$$R(x, \mathscr{C}_{8\times 8}) = 8! \sum_{k\geq 0} C(8, k)^2 x^k$$

2.81 (*Restricted Permutations*) For each element i of $N = \{1, 2, \ldots, n\}$ let there be specified a (possibly empty) subset, A_i, of N. Required is the number of bijective functions f from N to N such that, for every i, $f(i)$ does not belong to A_i. (Because f is a permutation of N, the elements of A_i are called the **forbidden positions** for $i \in N$.)

This problem may be transformed into one of nontaking rooks on a subboard $\mathscr{C}'$ of an $n \times n$ chessboard $\mathscr{C}$, *provided* that one agrees that 2 rooks on $\mathscr{C}'$ are nontaking if and only if they are nontaking on $\mathscr{C}$. (On this understanding, there are 2 nontaking rooks on the subboard of Fig. 2-3.) Specifically, let square (i, j) of $\mathscr{C}$ be colored *red* if $j \in A_i$; otherwise, let it be colored *green*. The green squares compose our subboard $\mathscr{C}'$: it is evident that every placement of n nontaking rooks on $\mathscr{C}'$ corresponds to a permutation f of N with the desired property, and vice versa. Hence there exist $r_n(\mathscr{C}')$ such permutations.

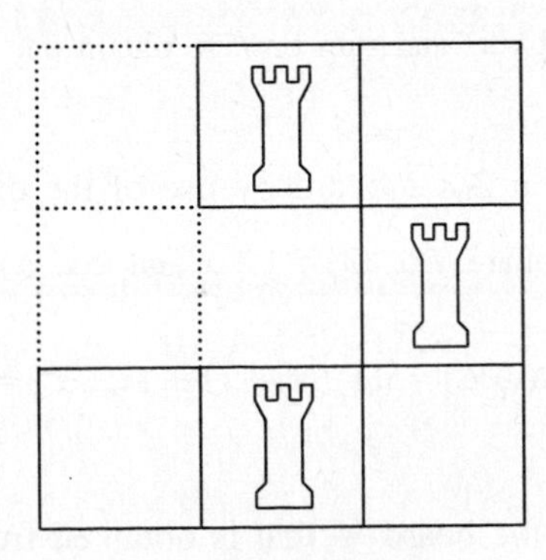

Fig. 2-3

An alternative solution follows from consideration of the red subboard (call it $\mathscr{C}''$). A placement of k nontaking rooks on $\mathscr{C}''$ corresponds to a family of $(n-k)!$ permutations with k elements in forbidden positions. Thus, in the sieve formula,

$$s_k = r_k(\mathscr{C}'')(n-k)! \qquad (k = 0, 1, \ldots, n)$$

and the required number is

$$r_n(\mathscr{C}') = \sum_{k=0}^{n} (-1)^k r_k(\mathscr{C}'')(n-k)! \tag{i}$$

2.82 Show that (i) of Problem 2.81 is self-inverse; i.e.,

$$r_n(\mathscr{C}'') = \sum_{k=0}^{n} (-1)^k r_k(\mathscr{C}')(n-k)!$$

In Problem 2.81 replace each subset A_i by its complement in N. This interchanges $\mathscr{C}'$ and $\mathscr{C}''$.

2.83 Reformulate the ménage problem—the wives already having been seated as in Fig. 2-2(a)—as a problem of nontaking rooks.

In the notation of Problem 2.81 choose $A_i = \{i, i+1\}$, for $i = 1, 2, \ldots, n-1$, and $A_n = \{1, n\}$. Then $r_n(\mathscr{C}') = M_n$.

2.84 (*Expansion Formula for the Rook Polynomial*) Given a chessboard $\mathscr{C}$ in which a **special square** has been distinguished. Let $\mathscr{D}$ be the board obtained from $\mathscr{C}$ by deleting the row and the column of the special square, let $\mathscr{E}$ be the board obtained from $\mathscr{C}$ by deleting only the special square. Prove that

$$R(x, \mathscr{C}) = xR(x, \mathscr{D}) + R(x, \mathscr{E}) \tag{i}$$

Either there is a rook in the special square or there is no rook in that square. The number of ways of placing k nontaking rooks in $\mathscr{C}$, one of which is in the special square, is $r_{k-1}(\mathscr{D})$; the number of ways of placing k nontaking rooks so that no rook is in the special square is $r_k(\mathscr{E})$. Consequently,

$$r_k(\mathscr{C}) = r_{k-1}(\mathscr{D}) + r_k(\mathscr{E})$$

Multiply this equation through by x^k and sum on k to obtain (i).

2.85 Find the rook polynomial of a 2×2 board by use of the expansion formula.

For any choice of the special square, $R(x, \mathscr{D}) = 1 + x$ and $R(x, \mathscr{E}) = 1 + 3x + x^2$. Hence,

$$R(x, \mathscr{C}) = x(1+x) + (1 + 3x + x^2) = 1 + 4x + 2x^2$$

2.86 Find the rook polynomial of the board $\mathscr{C}$ that is obtained from a 3×3 board by deleting the middle square in the first row and the first and the last squares in the third row. [Fig. 2-4(a)].

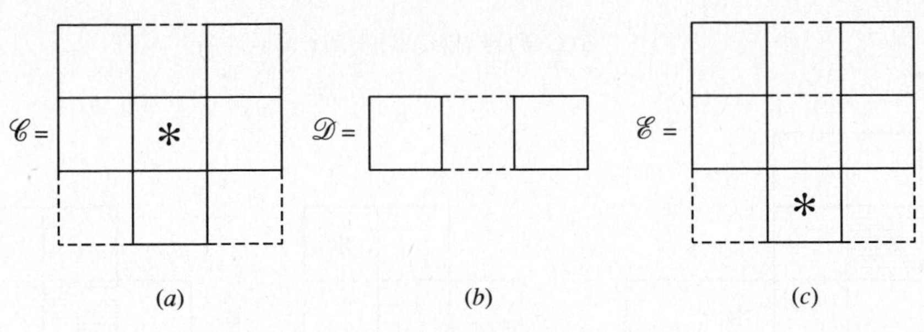

Fig. 2-4

Choosing the central square of $\mathscr{C}$ as the special square, we have

$$R(x, \mathscr{C}) = xR(x, \mathscr{D}) + R(x, \mathscr{E}) = x(1 + 2x) + R(x, \mathscr{E})$$

Further decomposing $\mathscr{E}$ relative to its third-row square,

$$R(x, \mathscr{E}) = x(1 + 4x + 2x^2) + (1 + 4x + 2x^2)$$

Thus,
$$R(x, \mathscr{C}) = x(1 + 2x) + (x + 1)(1 + 4x + 2x^2) = 1 + 6x + 8x^2 + 2x^3$$

2.87 In how many ways may nontaking rooks be placed on the board $\mathscr{C}$ of Problem 2.86?

$R(1, \mathscr{C}) - 1 = 16$ ways.

2.88 Let $\mathscr{A}$ and $\mathscr{B}$ be two **disjunct** (or **disjoint**) subboards of a given chessboard $\mathscr{C}$; this means that no row or column of $\mathscr{C}$ contains both a square of $\mathscr{A}$ and a square of $\mathscr{B}$. Show that the rook polynomial of the subboard $\mathscr{A} \cup \mathscr{B}$ is given by

$$R(x, \mathscr{A} \cup \mathscr{B}) = R(x, \mathscr{A})R(x, \mathscr{B})$$

If k nontaking rooks are placed on $\mathscr{A} \cup \mathscr{B}$, there must exist a nonnegative integer t such that t (nontaking) rooks are on $\mathscr{A}$ and $k - t$ rooks are on $\mathscr{B}$. Since $\mathscr{A}$ and $\mathscr{B}$ are disjunct, the way the t rooks are placed is independent of the way the $k - t$ rooks are placed. Hence, the number of ways of placing the k rooks is

$$r_k(\mathscr{A} \cup \mathscr{B}) = \sum_{t=0}^{k} r_t(\mathscr{A})r_{k-t}(\mathscr{B}) \qquad (i)$$

The left side of (i) is the coefficient of x^k in $R(x, \mathscr{A} \cup \mathscr{B})$; the right side is the coefficient of x^k in $R(x, \mathscr{A})R(x, \mathscr{B})$.

2.89 Find the rook polynomial of the forbidden subboard (squares marked with $\times$) in Fig. 2-5.

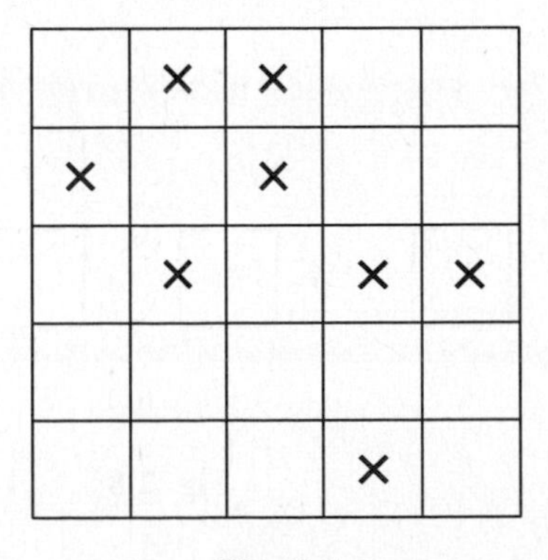

Fig. 2-5

Figure 2-6 shows the decomposition of the forbidden board relative to the indicated special square:

$$R(x, \mathscr{C}) = xR(x, \mathscr{D}) + R(x, \mathscr{E}) \tag{i}$$

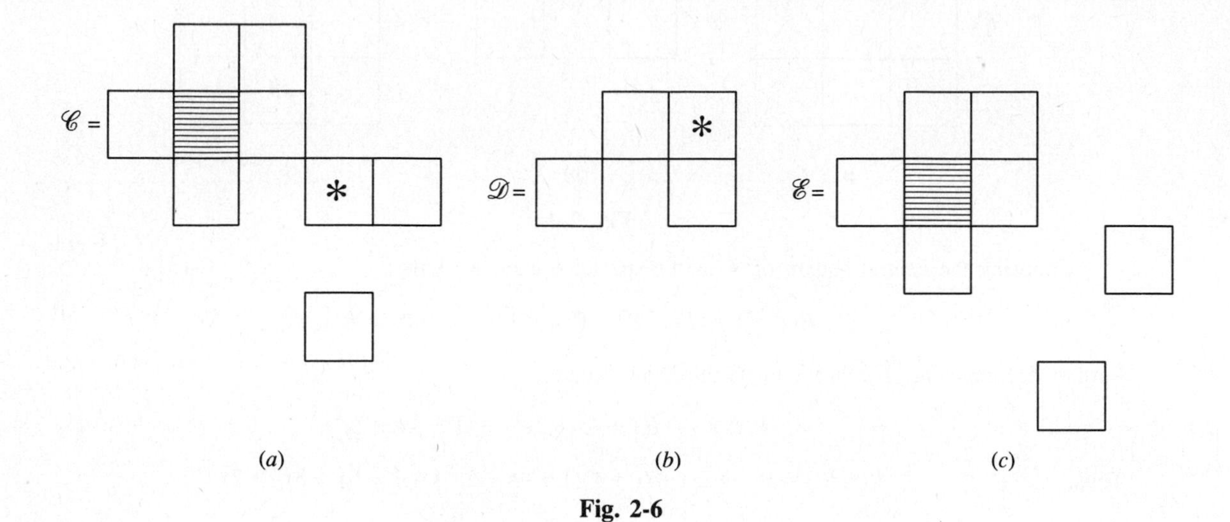

Fig. 2-6

Now decompose $\mathscr{D}$ as indicated in Figs. 2-6(*b*) and 2-7:

$$R(x, \mathscr{D}) = x(1 + x) + (1 + 3x + 2x^2) = 1 + 4x + 3x^2 \tag{ii}$$

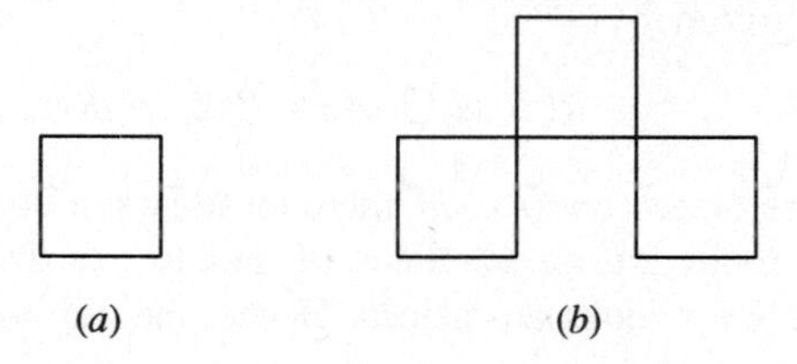

Fig. 2-7

The board $\mathscr{E}$ is the union of 2 disjunct subboards, one of which is a single square and the other is the board $\mathscr{F}$ of Fig. 2-8(*a*). Thus,

$$R(x, \mathscr{E}) = (1 + x)R(x, \mathscr{F})$$

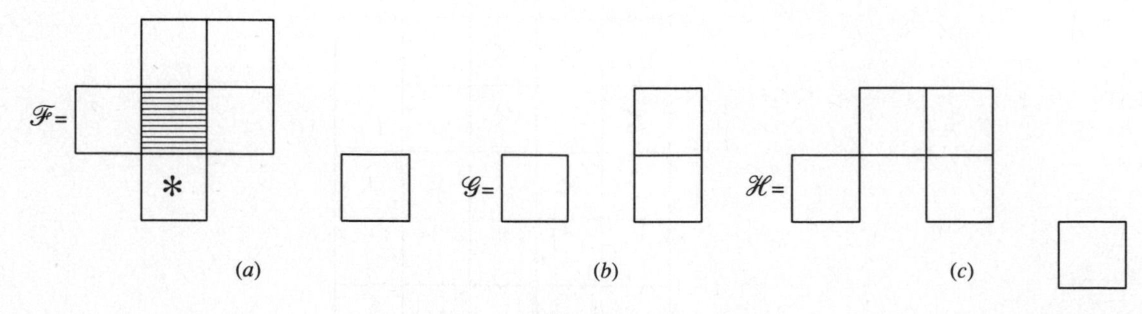

Fig. 2-8

Board $\mathscr{F}$ may be decomposed into boards $\mathscr{G}$ and $\mathscr{H}$ (Fig. 2-8), where $\mathscr{H}$ is itself the union of 2 disjunct parts, $\mathscr{D}$ and a single square. By use of Problem 2.88,

$$R(x, \mathscr{F}) = x(1 + 3x + x^2) + (1 + 4x + 3x^2)(1 + x)$$
$$= 1 + 6x + 10x^2 + 4x^3$$

and so
$$R(x, \mathscr{E}) = 1 + 7x + 16x^2 + 14x^3 + 4x^4 \qquad (iii)$$

Substitution of (*ii*) and (*iii*) in (*i*) yields

$$R(x, \mathscr{C}) = 1 + 8x + 20x^2 + 17x^3 + 4x^4 \qquad (iv)$$

2.90 An executive attending a weeklong business seminar has 5 suits of different colors. On Mondays, she does not wear blue or green; on Tuesdays, red or green; on Wednesdays, blue, white, or yellow; on Fridays, white. How many ways can she dress without repeating a color during the seminar?

The scenario here can be represented by the 5×5 board of Fig. 2-5, in which the rows respectively represent Monday through Friday; the columns respectively represent colors R, B, G, W, Y; and the forbidden subboard represents the constraints on the colors worn. Hence, (*i*) of Problem 2.81 may be immediately applied, with the $r_k(\mathscr{C}'')$ taken from (*iv*) of Problem 2.89 (note that $r_5 = 0$):

$$\text{No. of ways} = 1(5!) - 8(4!) + 20(3!) - 17(2!) + 4(1!) = 18 \qquad (i)$$

2.91 Find the number of permutations of a set of n elements such that exactly j elements ($j = 0, 1, \ldots, n$) are in forbidden positions.

This is a matter of redoing the last part of Problem 2.81, with application of Theorem 2.8 (replace m by n) instead of the simple sieve formula:

$$e_j = s_j - C(j+1, 1)s_{j+1} + C(j+2, 2)s_{j+2} - \cdots + (-1)^{n-j}C(n, n-j)s_n$$

Here, $s_k = r_k(n-k)!$, r_k being the number of ways of placing k nontaking rooks on the forbidden (red) subboard.

2.92 The generating polynomial of the numbers e_j of Problem 2.91,

$$H(x) = e_0 + e_1 x + \cdots + e_n x^n$$

is known as the **hit polynomial** (since an object in a forbidden position constitutes a "hit"). Show that

$$H(x) = \sum_{k=0}^{n} s_k (x-1)^k = \sum_{k=0}^{n} r_k (n-k)!(x-1)^k$$

$H(x)$ is just $E(x)$ renamed; see Problem 2.66.

2.93 Find the hit polynomial for Problem 2.90.

Use the result of Problem 2.91; the s_k may be picked up as the magnitudes of the summands in (*i*) of Problem 2.90.

$$H(x) = 1(5!) + 8(4!)(x-1) + 20(3!)(x-1)^2 + 17(2!)(x-1)^3 + 4(1!)(x-1)^4$$
$$= 120 + 192(x-1) + 120(x^2 - 2x + 1)$$
$$+ 34(x^3 - 3x^2 + 3x - 1) + 4(x^4 - 4x^3 + 6x^2 - 4x + 1)$$
$$= 18 + 38x + 42x^2 + 18x^3 + 4x^4$$

The constant term (no hit) is 18. So there are 18 ways she can "lawfully" wear her suits.

The coefficient of x is 38. This means there are 38 ways she can have 1 hit. In other words, there are 38 ways she can dress lawfully on all but 1 day of the seminar.

2.94 A corporation has to assign 4 executives—P, Q, R, and S—to 4 cities—Atlanta (A), Boston (B), Chicago (C), and Denver (D). Owing to internal politics, executive P is not welcome in D; executive Q is not welcome in C or D; executive R is not welcome in B or C; and executive S is not welcome in B. Show that if the assignment is made at random, at least 1 executive will be welcome.

Drawing the 4×4 board of the problem, we note that the forbidden subboard occupies only 3 columns (cities). Hence, $r_4 = 0$; that is, there is no assignment that makes all 4 executives unwelcome.

2.95 It is evident that the rook polynomial of an arbitrary board is invariant under permutations of rows and/or columns. (In Problem 2.90, for example, nothing changes if we agree to call red "white" and to call white "red.") Use this fact to compute the rook polynomial of the forbidden subboard in Fig. 2-9.

	1	2	3	4	5	6
1		×		×		
2	×					
3			×			
4	×				×	
5		×		×		
6						×

Fig. 2-9

The row- and column-permutations indicated in Fig. 2-10 transform the forbidden subboard into a union of 4, pairwise disjunct boards. Applying Problem 2.88,

$$R(x) = (1 + 4x + 2x^2)(1 + x)(1 + 3x + x^2)(1 + x)$$

$$= 1 + 9x + 30x^2 + 47x^3 + 37x^4 + 14x^5 + 2x^6$$

	2	4	3	1	5	6
1	×	×				
5	×	×				
3			×			
2				×		
4				×	×	
6						×

Fig. 2-10

2.96 Five workers (P, Q, R, S, T) are to be assigned to 4 jobs (a, b, c, d), with the forbidden jobs as in Fig. 2-11. Find the hit polynomial.

	a	b	c	d	δ
P			×		
Q	×	×			
R				×	
S				×	
T			×		

Fig. 2-11

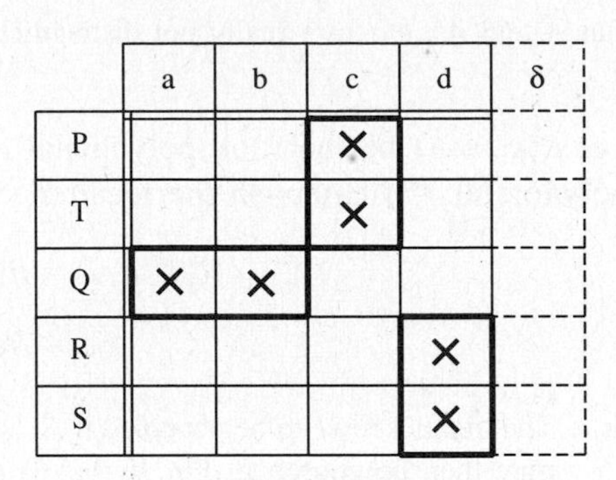

Fig. 2-12

If this 5×4 assignment problem is to be viewed as a problem in restricted permutations (bijective mappings), it is necessary to append a fifth, dummy job (δ) which is open to all 5 workers. Clearly, any hit in the 5×5 problem is a hit in the 5×4 problem, and conversely.

Transforming the 5×5 tableau by a row-c permutation and applying Problem 2.88,

$$R(x) = (1 + 2x)^3 = 1 + 6x + 12x^2 + 8x^3$$

Hence, by Problem 2.92,

$$\begin{aligned} H(x) &= 1(5!) + 6(4!)(x-1) + 12(3!)(x-1)^2 + 8(2!)(x-1)^3 \\ &= 32 + 48x + 24x^2 + 16x^3 \end{aligned}$$

2.97 Find the number of 6-letter words that can be formed using the letters A, B, and C (each twice) in such manner that A does not appear in the first 2 positions, B does not appear in the third position, and C does not appear in the fourth and fifth positions.

Temporarily making the letters distinguishable, we have the board of Fig. 2-13. The rook polynomial of the forbidden subboard is, by Problem 2.88,

$$(1 + 4x + 2x^2)^2(1 + 2x) = 1 + 10x + 36x^2 + 56x^3 + 36x^4 + 8x^5$$

	1	2	3	4	5	6
A	×	×				
A′	×	×				
B			×			
B′			×			
C				×	×	
C′				×	×	

Fig. 2-13

Then (i) of Problem 2.81 yields

$$r_6 = (6!)(1) - (5!)(10) + (4!)(36) - (3!)(56) + (2!)(36) - (1!)(8) = 112$$

But A and A', etc., are really not distinguishable. Thus (see Theorem 2.1) the answer is $112/(2!)(2!)(2!) = 14$.

2.98 Let $R(m, n; x)$ be the rook polynomial of a $m \times n$ chessboard. (*a*) Find the coefficient of x^k in this polynomial. (*b*) Establish the recurrence relation

$$R(m, n; x) = R(m-1, n; x) + nxR(m-1, n-1; x)$$
$$= R(m, n-1; x) + mxR(m-1, n-1; x)$$

(*a*) The distinct rows to be occupied by k nontaking rooks may be chosen in $C(m, k)$ ways. The distinct columns may then be chosen in $P(n, k) = k!\, C(n, k)$ ways. Thus, the required coefficient is

$$r_k(m, n) = k!\, C(m, k)C(n, k)$$

(Note the necessary symmetry in m and n.)

(*b*) The number of ways of placing k nontaking rooks, with no rook in the first row, is $r_k(m-1, n)$. The number of ways of placing k nontaking rooks, with a rook on the jth square of the first row, is $r_{k-1}(m-1, n-1)$. Therefore,

$$r_k(m, n) = r_k(m-1, n) + \sum_{j=1}^{n} r_{k-1}(m-1, n-1)$$
$$= r_k(m-1, n) + nr_{k-1}(m-1, n-1)$$

which implies $R(m, n; x) = R(m-1, n; x) + nxR(m-1, n-1; x)$.

The other relation follows from symmetry.

2.99 Refer to Problem 2.83. Regain the solution to the ménage problem by determining the rook polynomial of the forbidden (shaded) subboard, $\mathscr{C}''$, of Fig. 2-14.

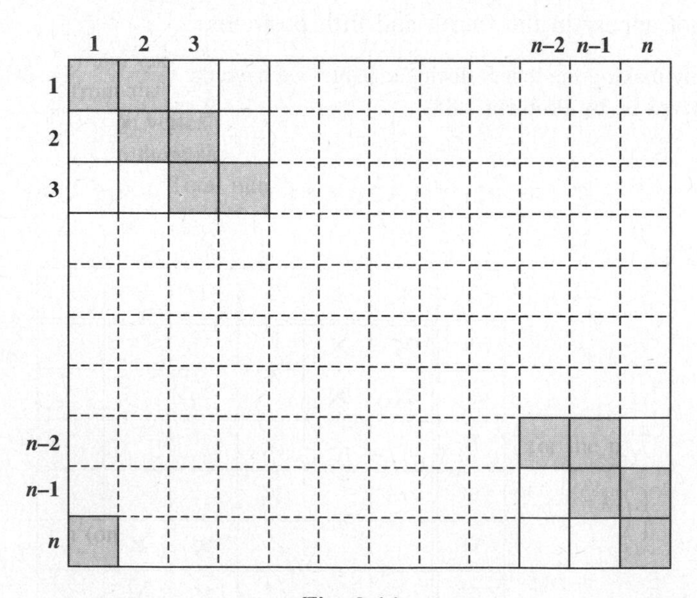

Fig. 2-14

Before applying to $\mathscr{C}''$ the expansion formula of Problem 2.84, we have to look at two geometrically distinct kinds of subboards. One, which shall be denoted $\mathscr{S}_n$, is composed of the squares on the main diagonal of an $n \times n$ board, together with the squares just above the main diagonal. The other type, $\mathscr{T}_{n-1}$, is a subboard of an $(n-1) \times n$ board; it consists of the squares $(1, 1)$ and $(1, 2)$, $(2, 2)$ and $(2, 3)$, $(3, 3)$ and $(3, 4), \ldots,$ $(n-1, n-1)$ and $(n-1, n)$.

Referring to Fig. 2-14, let us temporarily delete square $(n, 1)$ from $\mathscr{C}''$; this leaves $\mathscr{S}_n$. Choosing (n, n) as the special square for $\mathscr{S}_n$, we find:

$$R(x, \mathscr{S}_n) = xR(x, \mathscr{S}_{n-1}) + R(x, \mathscr{T}_{n-1}) \qquad (i)$$

Now decompose $\mathscr{T}_{n-1}$ relative to the special square $(n-1, n)$:

$$R(x, \mathscr{T}_{n-1}) = xR(x, \mathscr{T}_{n-2}) + R(x, \mathscr{S}_{n-1}) \qquad (ii)$$

Next, replace n by $n-1$ in (i):

$$R(x, \mathscr{S}_{n-1}) = xR(x, \mathscr{S}_{n-2}) + R(x, \mathscr{T}_{n-2}) \qquad (iii)$$

Elimination of $R(x, \mathscr{T}_{n-1})$ and $R(x, \mathscr{T}_{n-2})$ among (i), (ii), and (iii) yields the difference equation

$$f_n - (2x+1)f_{n-1} + x^2 f_{n-2} = 0 \qquad (iv)$$

for $f_n \equiv R(x, \mathscr{S}_n)$.

The starting values for (iv) are

$$f_1 = 1 + x = C(2, 0) + C(2-1, 1)x$$

$$f_2 = 1 + 3x + x^2 = C(4, 0) + C(4-1, 1)x + C(4-2, 2)x^2$$

and one obtains

$$f_3 = 1 + 5x + 6x^2 + x^3 = C(6, 0) + C(6-1, 1)x + C(6-2, 2)x^2 + C(6-3, 3)x^3$$

$$f_4 = C(8, 0) + C(8-1, 1)x + C(8-2, 2)x^2 + C(8-3, 3)x^3 + C(8-4, 4)x^4$$

....

It is evident—and easily proved by induction—that

$$f_n = R(x, \mathscr{S}_n) = \sum_{k=0}^{n} C(2n-k, k)x^k \qquad (n = 1, 2, \ldots) \qquad (v)$$

Now back to Fig. 2-14 and the decomposition of $\mathscr{C}''$. Taking $(n, 1)$ as the special square,

$$R(x, \mathscr{C}'') = R(x, \mathscr{S}_n) + xR(x, \mathscr{S}_{n-1}) \qquad (vi)$$

(To get the last term, turn Fig. 2-14 upside down.) Thus, by use of (v)

$$r_k(\mathscr{C}'') = r_k(\mathscr{S}_n) + r_{k-1}(\mathscr{S}_{n-1})$$

$$= C(2n-k, k) + C(2n-k-1, k-1) = \frac{2n}{2n-k} C(2n-k, k)$$

We have recovered the numbers p_k of Problem 2.73. Hence, (i) of Problem 2.81 will yield the same expression for $M_n = r_n(\mathscr{C}')$ as was obtained in Problem 2.73.

SYSTEMS OF DISTINCT REPRESENTATIVES (SDR); MATCHINGS AND COVERINGS IN GRAPHS

2.100 Why is the "marriage theorem" so called?

Let $W_1, W_2, \ldots, W_N$ denote the unmarried women of the town of F______, in which bigamy is forbidden. Let A_i $(i = 1, 2, \ldots, N)$ be the set of single men of F______ acceptable to W_i. Then, if every k of the women among them know at least k of the men, all the women can find acceptable husbands.

2.101 If $r_i > 0$ distinct elements, composing a set B_i, can be chosen from each set A_i of a family $\{A_i : i = 1, 2, \ldots, N\}$ of sets, such that the sets B_i are pairwise disjoint, then the family $\{B_i : i = 1, 2, \ldots, N\}$ is a **generalized SDR** for the family $\{A_i\}$. Find a necessary and sufficient condition for the existence of a generalized SDR.

The given family $\{A_1, A_2, \ldots, A_N\}$ will have a generalized SDR if and only if the expanded family

$$\{\overbrace{A_1, A_1, \ldots, A_1}^{r_1 \text{ times}}, \overbrace{A_2, A_2, \ldots, A_2}^{r_2 \text{ times}}, \ldots, \overbrace{A_N, A_N, \ldots, A_N}^{r_N \text{ times}}\}$$

has an (ordinary) SDR. Consider a k-sample, S_k, of the expanded family in which some set A_j appears fewer than r_j times. Since $A_j \cup A_j = A_j$, it is apparent that S_k will perforce obey the marriage condition if the larger sample obtained by adjoining to S_k the missing A_j's obeys the marriage condition. Thus we need enforce the marriage condition only on those $2^N - 1$ distinct samples whose elements are "completely represented"; our criterion becomes

$$|A_{i_1} \cup A_{i_2} \cup \cdots \cup A_{i_m}| \geq r_{i_1} + r_{i_2} + \cdots + r_{i_m}$$

for every m combination $(i_1, i_2, \ldots, i_m)$ of $\{1, 2, \ldots, N\}$, where $m = 1, 2, \ldots, N$.

2.102 A **graph** $\mathcal{G} = (V, E)$ is a set V of elements called **vertices** and a set E of distinct 2-element subsets of V, called **edges**. Graph $\mathcal{G}$ is **finite** whenever V is finite (*herewith assumed*). If $e = \{x, y\}$ is an edge of $\mathcal{G}$, vertices x and y are both **incident on** e, and edge e is **incident on** both x and y, we also say that e **joins** x and y. A **matching** M in $\mathcal{G}$ is a set of edges such that each vertex of $\mathcal{G}$ is incident on at most one edge in M. If the set V can be partitioned into 2 sets X and Y such that every edge in E is incident on 1 vertex from X and 1 vertex from Y, the graph $\mathcal{G} = (X, Y, E)$ is known as a **bipartite graph**. In $\mathcal{G} = (X, Y, E)$ there exists a **complete matching from X to Y** (also called an **X-saturated matching**) if there exists a matching M such that every vertex in X is incident on exactly 1 edge in M. Obtain a necessary and sufficient condition for a bipartite $\mathcal{G}(X, Y, E)$ to incorporate a complete matching from X to Y.

Let $X = \{1, 2, \ldots, n\}$ and $K_i = \{y \in Y : \text{some } e \in E \text{ joins } i \in X \text{ and } y\}$. Then a complete matching from X to Y exists if and only if the family $\{K_1, K_2, \ldots, K_n\}$ has a system of distinct representatives; that is, if and only if the marriage condition holds for the family.

2.103 The **degree** of a vertex in a graph is the number of edges incident on that vertex. A graph is ***r* regular** if all its vertices have the same degree, r. Show that if $\mathcal{G} = (X, Y, E)$ is an r-regular $(r > 0)$ bipartite graph, then there is a complete matching from X to Y, and vice versa.

Let A be any subset of X and let $f(A)$ be the set of vertices in Y which are joined to at least 1 vertex in A. Let E_1 be the set of edges incident on vertices in A and let E_2 be the set of edges incident on vertices in $f(A)$. Then $|E_1| \leq |E_2|$. But $|E_1| = r\,|A|$ and $|E_2| = r\,|f(A)|$. Thus the marriage condition, $|f(A)| \geq |A|$, holds for every subset A of X, and hence there is a one-to-one mapping from X to a subset of Y.

Now, if the total number of edges is m, then $r\,|X| = r\,|Y| = m$, or $|X| = |Y|$. Hence, the above mapping is also one-to-one from Y to X.

2.104 If $\mathcal{G} = (X, Y, E)$ is a bipartite graph and if the degree of any vertex in X is greater than or equal to the degree of any vertex in Y, then there is a complete matching from X to Y.

Let r be the minimum of all the degrees of vertices in X and let s be the maximum of all the degrees of vertices in Y; by hypothesis, $r \geq s$. Then, with A, $f(A)$, E_1, and E_2 as in Problem 2.103,

$$|E_1| \leq |E_2| \qquad |E_1| \geq r\,|A| \qquad |E_2| \leq s\,|f(A)|$$

which, together with $r \geq s$, imply the marriage condition, $|f(A)| \geq |A|$.

2.105 If C is any set of vectors in a finite-dimensional vector space, the dimension of the **span** of C (the subspace consisting of all linear combinations of vectors in C) is denoted by $d(C)$. Show that

$$d(A \cup B) + d(A \cap B) \leq d(A) + d(B)$$

where A and B are any two subsets of the vector space.

Let P be a basis of the span of $A \cap B$. Then P is a linearly independent subset of A, and so it can be extended to a basis, Q, of the span of A by adjoining vectors from A to B to P. Next, Q can be extended to a basis, R, of the span of $A \cup B$ by adjoining vectors from B to A to Q. By these constructions,

$$d(A \cap B) = |P| \qquad d(A) = |Q| \qquad d(A \cup B) = |R| \tag{i}$$

(These results are schematized in Fig. 2-15.)

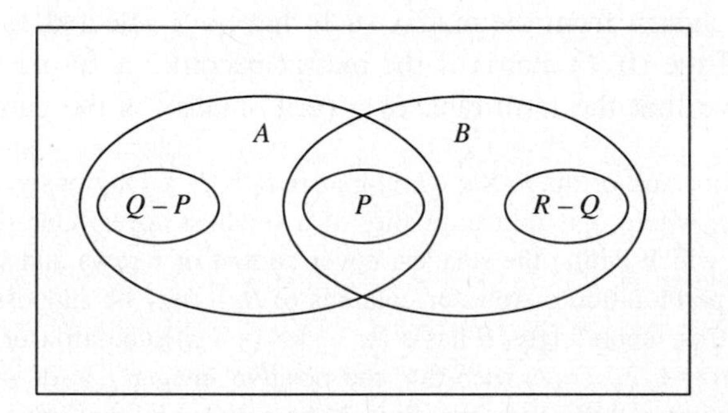

Fig. 2-15

Noting that P and $R - Q$ are disjoint and that $P \cup (R - Q)$ is a *linearly independent* subset of B, we have $|P| + |R - Q| \leq d(B)$, or

$$|P| + |R - Q| + |Q| = |P| + |R| \leq d(B) + |Q| \tag{ii}$$

Substitution of the expressions (i) in the inequality (ii) yields the required result.

2.106 (*Rado's Theorem*) Show that a linearly independent SDR exists for a family $\{B_1, B_2, \ldots, B_n\}$ of subsets of a finite-dimensional vector space if and only if, for every subset K of $\{1, 2, \ldots, n\}$,

$$d\Big(\bigcup_{i \in K} B_i\Big) \geq |K|$$

Here $d(\ \ )$ is defined as in Problem 2.105.

Proof of sufficiency (necessity is obvious) follows the proof of Theorem 2.6: merely replace the previously used equality

$$n(X \cup Y) + n(X \cap Y) = n(X) + n(Y)$$

by the inequality established in Problem 2.105. The SDR composed of the n distinct singleton sets (vectors) ultimately obtained must be linearly independent, because—by hypothesis—these n vectors span a subspace of dimension n or greater.

2.107 An $r \times n$ matrix ($r \leq n$) in which each entry is an element of $N = \{1, 2, \ldots, n\}$ is a **latin rectangle** if (i) each row is a permutation of N and (ii) each column is an r-permutation of N. If $r = n$, the latin rectangle is a latin square (Problem 1.88). Show that any $r \times n$ latin rectangle L, with $r < n$, can be extended into an $(r+1) \times n$ latin rectangle. (This guarantees the existence of latin squares of every order n.)

The proof follows readily from Problem 2.103. Let C_j denote the jth column of L and let S_j be the (unordered) subset of N composed of the $n - r$ elements that do not appear in C_j. Construct the bipartite graph $\mathscr{G} = (N, \{S_j\}, E)$, where an edge joins $q \in N$ with S_k if and only if $q \in S_k$. Each vertex S_j of $\mathscr{G}$ is of degree $|S_j| = n - r > 0$, and the same is true of each vertex $q \in N$. In fact, by condition (i) above, q occurs r times in L (once in each row); and (ii) requires that these r appearances are in r distinct columns of L. Hence, q belongs to exactly $n - r$ of the sets $\{S_j\}$; i.e., the degree of q in $\mathscr{G}$ is $n - r$. As an $(n - r)$-regular bipartite graph, $\mathscr{G}$ incorporates a complete matching from $\{S_j\}$ to N. Under the matching let $S_1 \to q_1, S_2 \to q_2, \ldots, S_n \to q_n$. Then $(q_1 \quad q_2 \quad \cdots \quad q_n)$ is the required $(r+1)$st row of the expanded latin rectangle.

2.108 (*König–Egerváry Theorem*) A matrix in which each entry is either 0 or 1 is called a **(0, 1)-matrix**. A **line** in a matrix is either a row or a column. The **term rank** of a (0, 1)-matrix is the largest number of 1s that can be chosen from the matrix such that no 2 selected 1s lie on the same line. A set S of lines is a **cover** of the (0, 1)-matrix if the matrix becomes a zero matrix after all the lines in S have been deleted. Prove that the term rank of a (0, 1)-matrix is the cardinality of its smallest cover.

Let p be the term rank of the $m \times n$ (0, 1)-matrix $B = [b_{ij}]$. Obviously, any cover of B must contain at least p lines; hence $p \leq q$, where q is the cardinality of a smallest cover. One now shows that $p \geq q$.

Let the q lines which define the smallest cover consist of r rows and s columns ($r + s = q$). Since p and q are unaffected by a permutation of rows or columns of B, it may be supposed that these are the first r rows and the first s columns. Correspondingly, B has a $(m - r) \times (n - s)$ submatrix of zeros at its lower right-hand corner. Define the sets A_i ($i = 1, 2, \ldots, r$) such that the positive integer j is in A_i if and only if $j > s$ and the entry $b_{ij} = 1$. If A_i were empty, all the 1s in row i would be covered by the first s columns; i.e., B would have a cover of fewer than q lines. Hence, each set A_i contains at least 1 element.

More generally, the union of any k sets from the family $\{A_i\}$ has at least k elements. Indeed, if this were not the case, certain k of the first r rows could be replaced in the *minimum* covering by a *smaller* number of columns drawn from among the last $n - s$. Therefore, by Hall's marriage theorem, the family $\{A_i\}$ has an SDR consisting of r elements. This means that the $r \times (n - s)$ submatrix at the upper right-hand corner contains r 1s, no 2 of them on the same line.

Repeating the argument for the $(m - r) \times s$ submatrix at the lower left-hand corner, one shows that this submatrix contains s 1s, no 2 on the same line. But the upper-right and lower-left submatrices are disjunct (Problem 2.88), which implies that B contains at least $r + s$ 1s, no 2 on the same line. Hence, $p \geq r + s = q$.

2.109 (*König's Theorem*) A **covering** (of the edges) in a graph is a set C of vertices such that each edge of the graph is incident on at least 1 vertex in C. Show that in a bipartite graph the cardinality of a maximum matching is equal to the cardinality of a minimum covering.

Let the bipartite graph be $\mathscr{G} = (X, Y, E)$, with $X = \{x_1, x_2, \ldots, x_m\}$ and $Y = \{y_1, y_2, \ldots, y_n\}$. Construct a (0, 1) matrix $B = [b_{ij}]$—called the **adjacency matrix** of $\mathscr{G}$—with m rows and n columns by defining $b_{ij} = 1$ if and only if there is an edge joining the vertex x_i and the vertex y_j. Now, the term rank of B is the cardinality of a maximum matching in $\mathscr{G}$; and the size of a smallest cover of B is the size of a smallest covering of the edges of $\mathscr{G}$. Hence König's theorem is implied by the König–Egerváry theorem. [Conversely, since any (0, 1) matrix can be interpreted as the adjacency matrix of a bipartite graph, König's theorem implies the König–Egerváry theorem. Thus the two theorems are equivalent, and—as will be shown in Problem 2.112—are both equivalent to Hall's marriage theorem.]

2.110 Connect the notions "term rank" and "rook polynomial."

Treat a given $(0, 1)$ matrix as a (rectangular) chessboard, and let the 1s in the matrix define the forbidden subboard. Then, if the rook polynomial of the forbidden subboard has the form $R(x) = 1 + \cdots + \alpha x^p$, the term rank of the matrix is p (and the number of ways of choosing p linewise disjunct 1s is α).

2.111 Use Problem 2.110 to determine the term rank of the matrix

$$\begin{bmatrix} 1 & 1 & 1 & 0 \\ 0 & 0 & 1 & 0 \\ 0 & 0 & 1 & 0 \\ 0 & 0 & 1 & 0 \\ 0 & 0 & 1 & 1 \end{bmatrix}$$

Taking the $(1, 3)$ element as the "special square" (Problem 2.84), we have the expansion

$$\begin{aligned} R(x) &= x(1 + x) + (1 + 2x)(1 + 5x + 3x^2) \\ &= 1 + 8x + 14x^2 + 6x^3 \end{aligned}$$

whence the term rank is 3.

2.112 Prove that König's theorem implies Hall's theorem. (The reverse implication was established in Problems 2.108 and 2.109.)

Let $X = \{1, 2, \ldots, m\}$ and let $\{A_i : i \in X\}$ be a family of m nonempty sets whose union is $Y = \{y_1, y_2, \ldots, y_n\}$. Construct a bipartite graph $\mathcal{G} = (X, Y, E)$ in which $\{i, y_j\}$ is an edge if and only if $y_j \in A_i$.

If K is any subset of X, define $f(K)$ to be the set of vertices in Y such that each vertex in $f(K)$ is joined to at least 1 vertex in K; equivalently, $f(K) = \bigcup_{i \in K} A_i$. It suffices to show that if $|f(K)| \geq |K|$ (the marriage condition) holds for all K, then there exists in $\mathcal{G}$ a complete matching from X to Y.

Now, by König's theorem, a complete matching from X to Y exists if and only if the cardinality of every covering of E is at least m. But, if the vertex subset C is a covering, and if C' denotes the intersection of C and X, then $f(X - C')$ is necessarily a subset of $C - C'$ (*cute!*). Thus,

$$|C| = |C'| + |C - C'| \geq |C'| + |f(X - C')| \geq |C'| + |X - C'| = m$$

2.113 Let $X = \{x_1, x_2, \ldots, x_n\}$ $(n \geq 2)$ and, for each i, $A_i = X - \{x_i\}$. Show that the family $\{A_i\}$ has an SDR. Find the number of distinct SDRs for this family.

Each set in the family $(k = 1)$ contains at least 1 element. For $k = 2, 3, \ldots, n$, the union of any k of the sets is X, and so this union contains $n \geq k$ elements. It is evident that any SDR is a derangement of X; conversely, any derangement of X furnishes an SDR. Hence the number of SDRs is D_n.

2.114 Let $\{A_1, A_2, \ldots, A_n\}$ and $\{B_1, B_2, \ldots, B_n\}$ be 2 families of nonempty subsets of a set E. Show that these 2 families will have a common SDR, or *SCR*, if and only if

$$\left| \left(\bigcup_{i \in I'} A_i \right) \cap \left(\bigcup_{j \in J'} B_j \right) \right| \geq |I'| + |J'| - n \qquad (*)$$

where I' and J' are any subsets of $I = \{1, 2, \ldots, n\}$.

In the special case where $B_j \equiv \bigcup_{i \in I} A_i$, for every j, (*) holds for arbitrary J' if and only if

$$\left| \bigcup_{i \in I'} A_i \right| \geq |I'|$$

Thus the present theorem is a generalization of Hall's theorem.

Assume that $E = \{n+1, n+2, \ldots, n+m\}$, where $m \geq n$. Define a family $\{X_l\}$ of $n+m$ subsets of the set $F = I \cup E = \{1, 2, \ldots, n+m\}$, as follows:

$$X_i = A_i \quad \text{when } i \in I \qquad \text{and} \qquad X_k = \{k\} \cup \{j : k \in B_j\} \quad \text{when } k \in E$$

Because B_j is assumed nonempty for each $j \in I$, the union of all $n+m$ sets in the family $\{X_l\}$ is F itself. So, *if this family has an SDR, that SDR can only be F.*

Suppose, then, that $\{X_l\}$ does have the SDR F. Reindex the subfamily $\{X_i\}$ so that $n+i \in X_i$ for $1 \leq i \leq n$; further, reindex the subfamily $\{X_k\}$ so that $j \in X_{n+j}$ for $1 \leq j \leq n$. But $j \in X_{n+j}$ if and only if $n+j \in B_j$. We thus have $n+j$ as an element in both A_j and B_j. In other words, if the family $\{X_l\}$ has an SDR, then the families $\{A_i\}$ and $\{B_j\}$ have a common SDR. By reversing the argument, it is easy to establish that if the 2 families have a common SDR, then F is the unique SDR of the family $\{X_l\}$.

An arbitrary subset, F', of the set F can be expressed as $F' = I' \cup E'$, where I' and E' are subsets of I and E, respectively. Therefore, the marriage condition for $\{X_l\}$ reads:

$$\left|\bigcup_{l \in F'} X_l\right| = \left|\left(\bigcup_{i \in I'} A_i\right) \cup \left(\bigcup_{k \in E'} X_k\right)\right| \geq |I'| + |E'|$$

Let $\cup_{i \in I'} A_i \equiv V$ and $\cup_{k \in E'} X_k \equiv W$. Then V is a subset of E, and

$$W = E' \cup J \qquad \text{where } J = \{j \in I : B_j \cap E' \neq \emptyset\}$$

Now, $$|V \cup W| = |V| + |W| - |V \cap W| = |V| + |E'| + |J| - |V \cap E'|$$

Thus the marriage condition is equivalent to $|V| - |V \cap E'| + |J| \geq |I'|$, or

$$|V \cap (E - E')| + |J| \geq |I'| \tag{**}$$

All that remains is to show that (**) implies (*), and vice versa.

(1) If I' and J' are any 2 subsets of I, define E' by

$$E - E' \equiv \bigcup_{j \in J'} B_j$$

Then, from (**),

$$\left|\left(\bigcup_{i \in I'} A_i\right) \cap \left(\bigcup_{j \in J'} B_j\right)\right| + |J| \geq |I'|$$

But, for this E', $J = \{j \in I : B_j \cap E' \neq \emptyset\} = I - J'$, whence $|J| = n - |J'|$. Thus (**) implies (*).

(2) Let I' be any subset of I and let E' be any subset of E. Define

$$J' \equiv \{j \in I : B_j \cap E' = \emptyset\} = I - J$$

According to this definition, B_j is a subset of $E - E'$ for each $j \in J'$, whence $\cup_{j \in J'} B_j$ is also a subset of $E - E'$. Consequently, for any set V—in particular, for $V = \cup_{i \in I'} A_i$—$V \cap (\cup_{j \in J'} B_j)$ is a subset of $V \cap (E - E')$. Then (*) gives

$$|V \cap (E - E')| \geq \left|V \cap \left(\bigcup_{j \in J'} B_j\right)\right| \geq |I'| + (n - |J|) - n$$

which is just (**).

2.115 A (0, 1)-matrix (Problem 2.108), P, is a **permutation matrix** if PP^T is the identity matrix, I (here P^T is the transpose of P). It is obvious that a square (0, 1)-matrix is a permutation matrix if and only if it has a single 1 in each row and each column. Suppose that A is an $n \times n$ matrix with nonnegative entries in which each line sum is equal to the positive number M. Show that there exist positive numbers c_i and $n \times n$ permutation matrices P_i, where $i = 1, 2, \ldots, k$, such that

$$A = c_1P_1 + c_2P_2 + \cdots + c_kP_k$$

The notions of term rank and cover may be extended to A by treating each positive entry as a 1. Let a minimum cover of A consist of r rows and s columns. Since the total of all the entries in the matrix is nM,

$$rM + sM \geq nM \qquad \text{or} \qquad r + s \geq n$$

But, by the König–Egerváry theorem, $r + s = p$, the term rank of A. Since $p \leq n$ for an $x \times n$ matrix, we must have

$$n \geq p = r + s \geq n \qquad \text{or} \qquad p = n$$

Let $\{x_1, x_2, \ldots, x_n\}$ be a set of maximum cardinality of linewise disjunct, positive entries of A, and let P_1 be the $n \times n$ permutation matrix derived from A by replacing each x_i with a 1 and replacing all other entries with 0s. Now let $B = A - c_1P_1$, where $c_1 = \min\{x_1, x_2, \ldots, x_n\}$. Then all entries of B are nonnegative and every line sum in B is $M - c_1$. Moreover, the number of zeros in B exceeds the number of zeros in A by at least 1. Repeating this argument, one must eventually reach the zero matrix:

$$0 = A - c_1P_1 - c_2P_2 - \cdots - c_kP_k \qquad \text{or} \qquad A = c_1P_1 + c_2P_2 + \cdots + c_kP_k$$

in which $c_1 + c_2 + \cdots + c_k = M$.

2.115(a) Show that a square (0, 1) matrix in which each line sum is r is the sum of exactly r-permutation matrices.

In the expansion formula of Problem 2.115, each c_i (which is one of the x_j) is 1; and the c_i must sum to r.

SPERNER'S THEOREM AND SYMMETRIC CHAIN DECOMPOSITION

2.116 (*Sperner's Theorem, Part 1*) A collection F of nonempty subsets of a set X is called an **antichain** or **clutter** or **Sperner system**) in X if no set in F is properly contained in another set in F. Prove that if $|X| = n$, the number of sets in any antichain in X cannot exceed $C(n, n')$, where $n' = \lfloor (n+1)/2 \rfloor$. [*More pregnant:* Any $C(n, n') + 1$ subsets of a given n-set must include 2 subsets such that one contains the other.]

The collection of all k subsets of X is obviously an antichain, and the cardinality of this collection is $C(n, k)$. Now, by Problem 1.74, $C(n, k) \leq C(n, n')$ for all k. Thus the number of sets in any antichain in which all sets have the same number of elements is bounded above by $C(n, n')$.

To lift the restriction that all sets in the antichain have the same number of elements, we introduce a new object. A **maximal chain** in the n-set X is a sequence $\langle A_0, A_1, A_2, \ldots, A_n \rangle$ of subsets of X, where A_0 is the empty set and $A_n = X$; $|A_i| = i$, for each i; and A_i is a proper subset of A_{i+1}, for $i = 1, 2, \ldots, n-1$. The number of maximal chains in X is $n!$, since each permutation $\langle x_1, x_2, \ldots, x_n \rangle$ of the set X defines a unique maximal chain

$$\emptyset \subset \{x_1\} \subset \{x_1, x_2\} \subset \cdots \subset \{x_1, x_2, \ldots, x_n\}$$

and conversely. Let $F = \{B_1, B_2, B_3, \ldots, B_t\}$ be an antichain in X, with $|B_i| = b_i$ $(i = 1, 2, \ldots, t)$. By definition, any chain in X—maximal or not—can contain at most 1 of the sets B_i. Now, there are $b_i!$ ways of forming a chain from $\emptyset$ to B_i, and there are $(n - b_i)!$ ways of extending this chain to a maximal chain. Thus the number of maximal chains that contain the set B_i is $b_i!\,(n - b_i)!$; we therefore have the inequality

$$\sum_{i=1}^{t} b_i!\,(n - b_i)! \leq n! \tag{i}$$

Suppose that among the cardinalities $\{b_i\}$ there are p_1 1s, p_2 2s, ..., and p_n n's; here, the p's are nonnegative integers with sum t. Then (i) becomes, after dividing through by $n!$, the **Lubell–Yamamoto–Meschalkin (LYM) inequality**:

$$\sum_{k=1}^{n} \frac{p_k}{C(n, k)} \leq 1 \qquad (ii)$$

From the LYM inequality it follows at once that

$$\frac{t}{C(n, n!)} = \sum_{k=1}^{n} \frac{p_k}{C(n, n')} \leq \sum_{k=1}^{n} \frac{p_k}{C(n, k)} \leq 1$$

or $t \leq C(n, n')$.

2.117 (*Sperner's Theorem, Part 2*) Prove that if $n = |X|$ is even, the only antichain of maximum cardinality is the collection of all $(n/2)$-subsets of X.

Observe first that if $n = 2m$, the greatest cardinality allowed by Problem 2.116 is

$$C(n, n') = C\left(n, \left\lfloor m + \frac{1}{2} \right\rfloor\right) = C(n, m) = C\left(n, \frac{n}{2}\right)$$

which is in fact the cardinality of the given antichain.

By Problem 1.74, $1/C(n, k) > 1/C(n, n/2)$ for all $k \neq n/2$. Hence, for an arbitrary antichain (of cardinality t), the LYM inequality yields

$$0 \leq \sum_{k=1}^{n} p_k \left[\frac{1}{C(n, k)} - \frac{1}{(n, n/2)}\right] \leq 1 - \frac{t}{C(n, n/2)}$$

It follows that, if t assumes the value $C(n, n/2)$, we must have $p_k = 0$ for all $k \neq n/2$ and $p_{n/2} = C(n, n/2)$. This proves uniqueness of the maximal antichain.

2.118 (*Sperner's Theorem, Part 3*) Prove that if $n = |X|$ is odd, there are exactly 2 maximal antichains: the collection of all $[(n-1)/2]$ subsets of X and the collection of all $[(n+1)/2]$ subsets of X.

Let $n = 2m + 1$. Using Problem 1.74 and the LYM inequality as in Problem 2.117, one establishes that in an antichain F of maximum cardinality,

$$t_{\max} = C(n, n') = C(2m + 1, m + 1)$$

each set has either $m = (n-1)/2$ or $m + 1 = (n+1)/2$ elements. Assuming there are k m sets and $t_{\max} - k$ $(m + 1)$ sets in F, we now prove that $k = 0$ or $k = t_{\max}$.

To this end, let $X = \{1, 2, \ldots, 2m + 1\}$; $A = \{1, 2, \ldots, m + 1\}$; and, for some $2 \leq i \leq m + 1$,

$$B = \{i, i + 1, i + 2, \ldots, i + m\}$$

Suppose that the $(m + 1)$-set A belongs to F, but the $(m + 1)$-set B does not belong to F. Then there is a positive integer $j < i$ such that $C = \{j, j + 1, \ldots, j + m\}$ is in F, while $D = \{j + 1, j + 2, \ldots, j + m + 1\}$ is not in F. Let $E = C \cap D = \{j + 1, \ldots, j + m\}$; does E belong to F or does it not?

Notice that E and D are sets in the maximal chain

$$\emptyset \subset \{j + 1\} \subset \{j + 1, j + 2\} \cdots \subset E \subset D \subset \cdots \subset X$$

By Problem 2.116, our *maximal* antichain F must be intersected in exactly 1 element by any maximal chain in X.

For the above maximal chain, that element can only be the m set E or the $(m+1)$ set D (since only these 2 cardinalities are represented in F). But D is not an element of F. Hence, E must belong to F.

On the other hand, E is a proper subset of C, and C belongs to F. Therefore, E cannot belong to F.

The only way out of the contradiction is to admit *either* (*i*) that when one $(m+1)$-subset belongs to F, they all belong to F (i.e., $k=0$); *or* (*ii*) that no $(m+1)$-subset belongs to F (i.e., $k = t_{\max}$).

2.119 A sequence $\langle A_1, A_2, \ldots, A_k \rangle$ of subsets of a set X of cardinality n is called a **symmetric chain** in X if (*i*) $A_i \subset A_{i+1}$ $(i = 1, 2, \ldots, k-1)$; (*ii*) $|A_{i+1}| = |A_i| + 1$ $(i = 1, 2, \ldots, k-1)$; (*iii*) $|A_1| + |A_k| = n$. Show that symmetric chains exist for $k = n+1,\ n-1,\ n-3,\ n-5, \ldots$, and that no other k values are possible.

Maximal chains (which certainly exist) are symmetric chains with $k = n+1$. Delete the first and last sets from a maximal chain, and you have a symmetric chain with $k = n-1$; delete the first two and the last two, and you get $k = n-3$; and so on.

To prove necessity, consider condition (*ii*) above as a difference equation:

$$|A_{i+1}| - |A_i| = 1$$

A particular solution is $|A_i| = i$; the general solution of the homogeneous equation is $|A_i| = c = \text{const}$. Hence, $|A_i| = i + c$. The "boundary condition" (*iii*) determines

$$c = \frac{n-(k+1)}{2}$$

Because c has to be an integer, we infer that *k is opposite in parity to n*. In addition, it is necessary that $|A_1| = 1 + c \geq 0$, which implies $k \leq n+1$. Putting together these two facts about k, we conclude that $n+1$, $n-1$, $n-3, \ldots$ are the only possible k values.

2.120 (*Existence of SCDs*) Given an n set—which can always be notated as $X(n) = \{1, 2, \ldots, n\}$—prove that $2^{X(n)}$ (Problem 1.22) can be partitioned into symmetric chains in $X(n)$. [Such a partition is called a **symmetric chain decomposition (SCD)** of $2^{X(n)}$.]

The theorem is true for $n=1$: the partition of $2^{X(1)}$ consists of the single cell

$$\emptyset \subset \{1\}$$

The inductive proof becomes transparent when it is borne in mind that each element S of $2^{X(n-1)}$ gives rise to 2 elements, $S \cup \emptyset$ and $S \cup \{n\}$, of $2^{X(n)}$; moreover, every element of $2^{X(n)}$ is generated in this way. Thus, if the symmetric chain $\langle A_1, A_2, \ldots, A_k \rangle$ in $X(n-1)$ is a cell in the partition of $2^{X(n-1)}$ (which exists by the induction hypothesis), then the 2 symmetric chains in $X(n)$

$$\langle A_1 \cup \emptyset, A_2 \cup \emptyset, \ldots, A_k \cup \emptyset, A_k \cup \{n\} \rangle$$

and

$$\langle A_1 \cup \{n\}, A_2 \cup \{n\}, \ldots, A_{k-1} \cup \{n\} \rangle$$

will serve as 2 cells in a partition of $2^{X(n)}$. (*Exception:* When the second chain is empty, it is not used as a cell.)

Figure 2-16 shows how the above mapping produces SCDs for the first few values of n. (The vertices of the tree are the symmetric chains, in a condensed notation.)

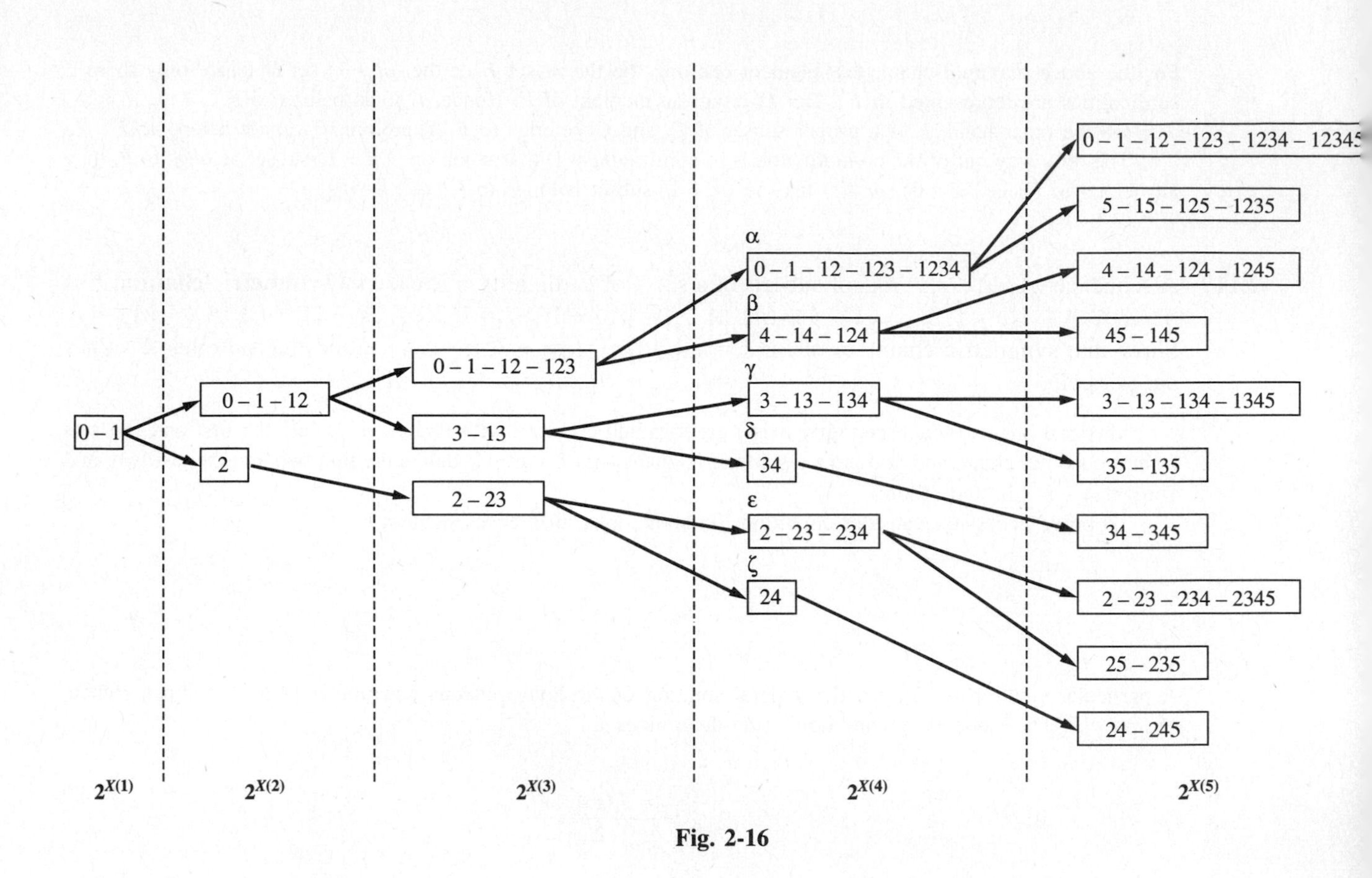

Fig. 2-16

2.121 Show that any SCD of $2^{X(n)}$ involves exactly $C(n, n')$ symmetric chains.

Let s be the number of cells (symmetric chains) into which $2^{X(n)}$ is partitioned. The crucial condition (*iii*) of Problem 2.119 ensures that, in any cell, the size of the smallest subset is less than or equal to $\lfloor (n+1)/2 \rfloor = n'$, while the size of the largest subset is greater than or equal to n'. (Otherwise the two sizes could not sum to n.) Condition (*ii*) then implies that the cell contains exactly 1 subset A_j of size n'. The cells being disjoint, it follows that the number of n' subsets cannot be smaller than the number of cells:

$$C(n, n') \geq s$$

But, since each n' subset must belong to some cell of the partition, and since no 2 can belong to the same cell,

$$s \geq C(n, n')$$

2.122 Use the fact (Problem 2.120) that the class of all subsets of a set of cardinality n has an SCD to establish the first part of Sperner's theorem (Problem 2.116).

Suppose $2^{X(n)}$ is partitioned into symmetric chains in $X(n)$. If F is an antichain in $X(n)$, each set in F must belong to a different cell of the partition. Thus, the cardinality of F cannot exceed the number of cells, which (Problem 2.121) is $C(n, n')$.

2.123 Verify that $2^{X(n)}$ allows at least $n!$ SCDs.

The inductive process of Problem 2.120 and Fig. 2-16, starting with the unique SCD of $2^{X(1)}$, generates a single SCD for $2^{X(n)}$. In this SCD permute the integers $\{1, 2, \ldots, n\}$ arbitrarily, to obtain $n!$ SCDs in all.

2.124 (*G.O.H. Katona, 1972*) A **directed graph** $\mathscr{G}$ (or **digraph**) is a pair (V, E) where E is a set of *ordered* pairs from V. The ordered pair (x, y) in E is *called* the **arc from x to y**. Suppose that the set V of vertices can be partitioned into disjoint subsets $V_0, V_1, \ldots, V_n$ such that every arc (x, y) in E has $x \in V_i$ and $y \in V_{i+1}$, for some $0 \le i \le n-1$. [Note that such a partition is impossible if both (p, q) and (q, p) belong to E. If the partition *is* possible, the digraph must be **acyclic** (no continuously directed cycles).] If v is a vertex in V_i, one defines its **rank** as $r(v) \equiv i$. A **symmetric chain** in $\mathcal{G}$ is the vertex set of a directed path starting from a vertex x and ending in a vertex y where $r(x) + r(y) = n$. $\mathcal{G}$ is a **symmetric chain graph** if there exists a partition of V into symmetric chains.

Restate Sperner's theorem (Problem 2.116) in graph-theoretic terminology.

Just as the general n set is modeled by $X(n) = \{1, 2, \ldots, n\}$, the general *connected* symmetric chain graph (which will be a tree) may be modeled by $\mathcal{G}(n)$, the **subset graph** of $X(n)$, which is constructed as follows. For the vertex set of $\mathcal{G}(n)$ choose $V = 2^{X(n)}$. Let V be partitioned into subsets $V_i = \{A \subset X(n) : |A| = i\}$ $(i = 0, 1, 2, \ldots, n)$. Define the arc set E thus: (A, B) is an arc in E if and only if $A \subset B$ and $|B| = |A| + 1$. That $\mathcal{G}(n)$ is in fact a symmetric chain graph is guaranteed by Problem 2.120. Figure 2-17 is a diagram of $\mathcal{G}(4)$; the Greek letters indicate how the SCD of Fig. 2-16 provides a set of disjoint directed paths (2 of them of length zero) that cover all the vertices of the graph.

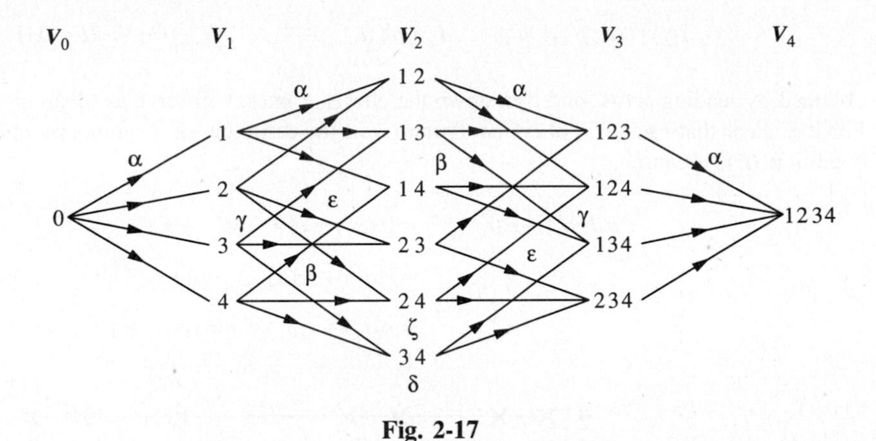

Fig. 2-17

An antichain in $X(n)$ corresponds in $\mathcal{G}(n)$ to a subset, F, of V with the property that no 2 vertices in F are joined by a directed path. Therefore [abandoning the model $\mathcal{G}(n)$] the graph-theoretic version of Sperner's theorem reads:

> Let $\mathcal{G} = (V, E)$ be a symmetric chain graph with a vertex partition $\langle V_0, V_1, \ldots, V_n \rangle$. Then any selection of $|V_{n'}| + 1$ vertices must include a pair of vertices that are joined by a directed path of $\mathcal{G}$. [Here, as usual, $n' = \lfloor (n+1)/2 \rfloor$.]

2.125 (*de Bruijn, Tengbergen, and Kruiswijk, 1952*) The sum of the exponents in the prime factorization of a positive integer m is its **degree**, $r(m)$. A sequence $\langle d_1, d_2, \ldots, d_k \rangle$ of divisors of m is called a **symmetric chain** of m if (*i*) d_{i+1}/d_i is a prime for $1 \le i \le k-1$, and (*ii*) $r(d_1) + r(d_k) = r(m)$. Two divisors of m are **incomparable** if neither divides the other. Prove: (*a*) The set of divisors of m has an SCD. (*b*) If $r(m)$ is even, the largest set of mutually incomparable divisors is the set of all divisors of degree $r(m)/2$. (*c*) If $r(m)$ is odd, there are 2 largest sets of mutually incomparable divisors: the set of all divisors of degree $[r(m) - 1]/2$ and the set of all divisors of degree $[r(m) + 1]/2$.

(*a*) The proof is by induction on the number, n, of distinct prime factors of m. If $n = 1$, then $m = p^t$ and $r(m) = t$. The required SCD involves the single symmetric chain $\langle 1, p, p^2, \ldots, p^t \rangle$. Suppose the theorem holds for $n = k$. Any number m with $k + 1$ distinct prime factors will be of the form $m_1 p^t$, where p is a prime that does not divide m_1 and where the set of divisors of m_1 (which has k distinct prime factors) has an SCD, by the induction hypothesis. From this SCD, one derives an SCD for the set of divisors of m, as follows.

Let $S = \langle d_1, d_2, \ldots, d_h \rangle$ be a symmetric chain in the SCD for m_1; correspondingly let

$$X = \{d_i p^s : 1 \le i \le h; 0 \le s \le t\}$$

The elements of X, all of which are divisors of m, may be displayed in $t+1$ rows and h columns:

	col 1	col 2					col h
row 0	d_1	d_2	d_3	$\cdots$	d_{h-2}	d_{h-1}	d_h
row 1	$d_1 p$	$d_2 p$	$d_3 p$	$\cdots$	$d_{h-2} p$	$d_{h-1} p$	$d_h p$
row 2	$d_1 p^2$	$d_2 p^2$	$d_3 p^2$	$\cdots$	$d_{h-2} p^2$	$d_{h-1} p^2$	$d_h p^2$
....							
row t	$d_1 p^t$	$d_2 p^t$	$d_3 p^t$	$\cdots$	$d_{h-2} p^t$	$d_{h-1} p^t$	$d_h p^t$

Figure 2-18 shows how the tableau of divisors may be covered by a set of disjoint, L-shaped lines, on the assumption that $h \ge t+1$. (As the reader may verify, everything goes through when $h < t+1$.) We claim that the sequences

$$T(i) = \langle d_1 p^i, d_2 p^i, \ldots, d_{h-i} p^i, d_{h-i} p^{i+1}, \ldots, d_{h-i} p^t \rangle \qquad (i = 0, 1, \ldots, t)$$

obtained by reading across and then down the covering lines, will serve as (disjoint) cells of an SCD for m. For it is clear that each $T(i)$ obeys condition (i) of the definition of a symmetric chain. Further, $T(i)$ obeys condition (ii), because

$$\begin{aligned} r(d_1 p^i) + r(d_{h-i} p^t) &= [r(d_1) + i] + [r(d_{h-i}) + t] \\ &= r(d_1) + [r(d_{h-i}) + i] + t \\ &= r(d_1) + r(d_h) + t = r(m_1) + t = r(m) \end{aligned}$$

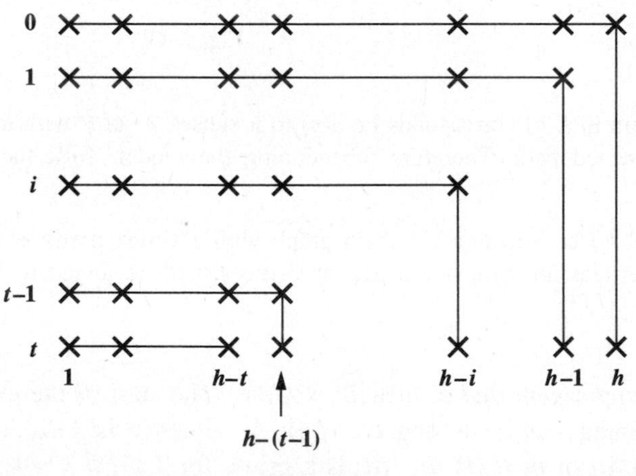

Fig. 2-18

Collecting the symmetric chains T generated by all the symmetric chains S, we obtain an SCD for m. This completes the induction.

(*b*) If $r(m)$ is even, choose the divisor of degree $r(m)/2$ from each symmetric chain in the decomposition.

(*c*) If $r(m)$ is odd, choose the divisor of degree $[r(m) - 1]/2$ from each symmetric chain or choose the divisor of degree $[r(m) + 1]/2$ from each chain.

2.126 Use the inductive process of Problem 2.125 to generate an SCD for the squarefree integer $m = 42 = (2)(3)(7)$.

For squarefree integers—the divisors of which are also squarefree—all tableaux are two-rowed. See Fig. 2-19.

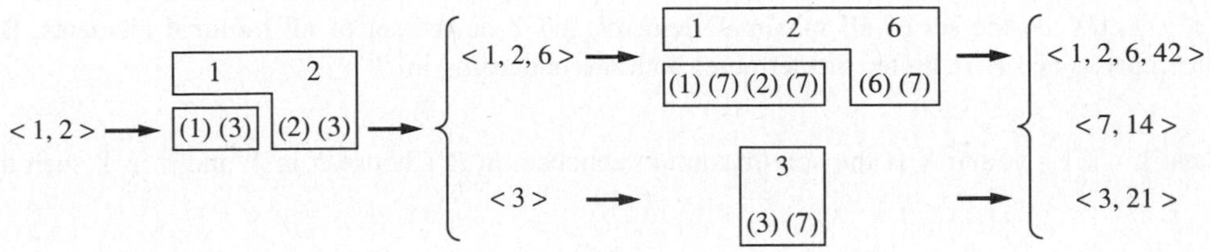

<1, 2> → [1 2 / (1)(3) (2)(3)] → <1, 2, 6> → [1 2 6 / (1)(7) (2)(7) (6)(7)] → <1, 2, 6, 42>, <7, 14>

<3> → [3 / (3)(7)] → <3, 21>

Fig. 2-19

2.127 A set S of divisors of a positive integer m is called **convex** whenever $d_1 \in S$, $d_2 \in S$, $d_1 \mid d_3$, and $d_3 \mid d_2$ together imply $d_3 \in S$. If n is the number of distinct prime factors of m, show that

$$\left| \sum_{d \in S} \mu(d) \right| \le C(n, n')$$

where $n' = \lfloor (n+1)/2 \rfloor$ and μ is the Möbius function (Problem 2.51).

It is enough to consider the case when m is squarefree, since, otherwise, $\mu(d) \equiv 0$ over S. Hence the degree of m is $r(m) = n$. By Problems 2.125(*a*) and 2.121, the set of divisors of m has an SCD $\langle C_1, C_2, \ldots, C_k \rangle$, with $k = C(n, n')$. Thus,

$$\left| \sum_{d \in S} \mu(d) \right| = \left| \sum_{i=1}^{k} \sum_{d \in S \cap C_i} \mu(d) \right|$$

Since S is convex, each $S \cap C_i$ is either empty or consists of a number of consecutive elements of C_i. The inner sum on the right can therefore only be 0, +1, or −1; whence

$$\left| \sum_{d \in S} \mu(d) \right| \le \sum_{i=1}^{k} 1 = k$$

PARTIALLY ORDERED SETS AND DILWORTH'S THEOREM

2.128 A partially ordered set (**poset**) is a set, P, and a binary relation, $\le$, on P such that, for any a, b, and c in P, the following conditions hold:

(*i*) $a \le a$, (*ii*) $a \le b$ and $b \le a$ imply $a = b$, and (*iii*) $a \le b$ and $b \le c$ imply $a \le c$. Two elements x and y in P form a **comparable pair** if $x \le y$ or $y \le x$. A **chain** in P is a subset in which every pair of elements is a comparable pair; an **antichain** in P is a subset in which no pair is a comparable pair. How do these notions generalize the concept of set inclusion?

The power set, 2^X, of an arbitrary set X, together with the binary relation $\subset$, constitutes a poset. The antichains of this particular poset are the same objects F contemplated in Sperner's theorem. However—even in the case of a finite X—there are many chains other than the simple "staircase" chains previously dealt with. Note also that the binary relation $\subset$ is defined for every pair of elements in 2^X, a property not shared by $\le$ in a general poset.

2.129 (*Dilworth's Theorem*) Prove that in a finite poset the cardinality of a maximum antichain is equal to the minimum number of (disjoint) chains into which the poset can be partitioned.

Obviously, if p is the size of an antichain in a poset and if the poset can be partitioned into q chains, then $p \le q$, and therefore the maximum value of p cannot exceed the minimum value of q. So it is enough to show that if m is the size of a maximal antichain, there exists a partition of P into m (disjoint) chains. The proof is by induction on $|P| = n$.

Suppose the theorem is true for any poset with fewer than n elements. An element x in P is a **minimal element** if there is no y in P such that $y \leq x$. Element x is a **maximal element** if there is no y in P such that $x \leq y$. Let X be the set of all maximal elements and Y be the set of all minimal elements. Both X and Y are nonempty, since P is finite; furthermore, both are antichains in P.

Case 1. $|X| = m$ and X is the sole maximum antichain in P. Choose b in X and a in Y such that $a \leq b$. Define

$$X' \equiv X - \{b\} \qquad P' \equiv P - \{a, b\}$$

Clearly, P' is a smaller poset than P, and X' is an antichain in P' of size $m-1$. Now, P' cannot support an antichain of size m (by the uniqueness of X). Hence, X' is maximum in P', and the induction hypothesis guarantees a partition of P' into $m-1$ chains. These chains and the chain I give a partition of P into m chains.

Case 2. $|Y| = m$ and Y is the sole maximum antichain in P. Reverse the relation $\leq$ and this becomes Case 1.

Case 3. P has a maximum antichain, A, that is not a subset of $X \cup Y$. The set

$$A^+ \equiv \{x \in P : \text{there exists } p \text{ in } A \text{ such that } p \leq x\}$$

is nonempty (because $A \neq X$) and smaller than P (because $A \neq Y$). By condition (i) above, $A \subset A^+$; therefore, by the induction hypothesis, A^+ can be partitioned into m chains. Similarly, the set

$$A^- = \{x \in P : \text{there exists } q \text{ in } A \text{ such that } x \leq q\}$$

which also contains A as a subset, can be partitioned into m chains. Now, by the maximality of A, $A^+ \cup A^- = P$. Also, $A^+ \cap A^- = A$; for if $t \in A^+ \cap A^-$, then there exist u and v in A such that $u \leq t \leq v$, which implies $u = t = v$, or $t \in A$. For each $a \in A$, join the chain in A^+ of which the smallest element is a with the chain in A^- of which the largest element is a, thereby obtaining m pairwise disjoint chains that exhaust $A^+ \cup A^- = P$.

2.130 (*Mirsky's Dual of Dilworth's Theorem*) Prove that in a finite poset the cardinality of a maximum chain is equal to the minimum number of disjoint antichains into which the poset can be partitioned.

Here chains and antichains play the roles of antichains and chains in Problem 2.129. Thus we must show that if m is the size of a maximum chain in P, there exists a partition of P into m (disjoint) antichains.

The proof is by induction on m. The assertion holds trivially when $m = 1$. Assume that the theorem is true for $m-1$. Let P be any poset with a maximum chain consisting of m elements. As we know, the set X of all maximum elements of P is an antichain in P. The subposet $P - X$ cannot contain a chain with m elements; so a maximum chain in $P - X$ has at most $m-1$ elements. But, if a maximum chain in $P - X$ has fewer than $m-1$ elements and if M is a maximum chain in P, the antichain X must intersect M in 2 or more elements—an impossibility. So any maximum chain in $P - X$ has exactly $m-1$ elements, whence, by the induction hypothesis, $P - X$ can be partitioned into $m-1$ (disjoint) antichains. Adjoining the antichain X, we have a partition of P into m antichains. The induction is complete.

(A deep theorem in graph theory, due to Lovasz, shows that Mirsky implies Dilworth, and conversely.)

2.131 A finite poset $(P, \leq)$ can be represented by its **Hasse diagram**, which is a digraph $\mathcal{G} = (P, E)$, where (u, v) is an arc in E if and only if v covers u in P (see Problem 2.137). Show how Case 3 of Problem 2.129 is exemplified by the poset whose Hasse diagram is Fig. 2-20.

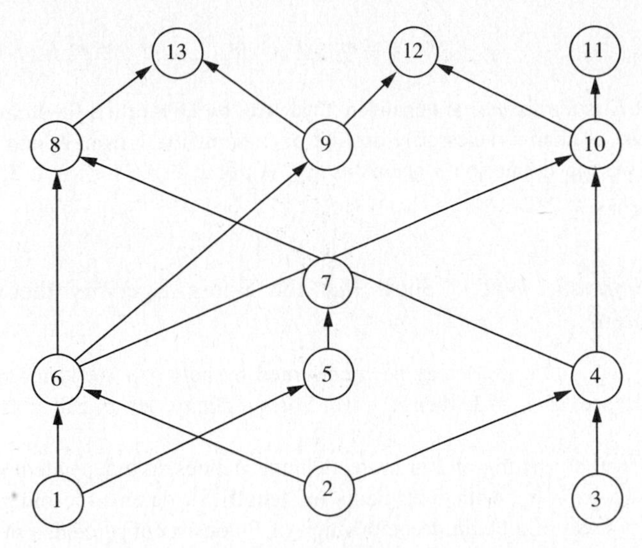

Fig. 2-20

In this poset, corresponding to the maximum antichain $A = \{7, 8, 9, 10\}$ we have

$$A^+ = \{7, 8, 9, 10, 11, 12, 13\} \qquad \text{and} \qquad A^- = \{7, 8, 9, 10, 1, 2, 3, 4, 5, 6\}$$

The 4 disjoint chains in A^- that terminate in the 4 elements of A are

1—5—7   2—6—8   9   3—4—10

The 4 disjoint chains in A^+ that emanate from the 4 elements of A are

7   8—13   9—12   10—11

Thus the promised partition of P has as its 4 cells the chains

1—5—7   2—6—8—13   9—12   3—4—10—11

2.132 Let $\mathscr{G} = (V, E)$ be a finite acyclic digraph (Problem 2.124). Define a partial order $\leq$ on E as follows: For 2 arcs e and f, $e \leq f$ if and only if e precedes f in some directed path of $\mathscr{G}$. Reformulate Dilworth's theorem in terms of directed paths of $\mathscr{G}$.

"The maximum number of arcs in E with the property that no 2 of them belong to a directed path of $\mathscr{G}$ is equal to the minimum number of arc-disjoint directed paths into which E can be partitioned."

2.133 Show that Dilworth's theorem implies Hall's theorem.

Let $I = \{1, 2, \ldots, n\}$ and let $\{A_i : i \in I\}$ be a family of subsets of $E = \{x_1, x_2, \ldots, x_m\}$ satisfying Hall's marriage condition: $J \subset I$ implies $|J| \leq |\cup\{A_i : i \in J\}|$. It is to be proved that the family has a system of distinct representatives.

On the set $X = \{x_1, x_2, \ldots, x_m, A_1, A_2, \ldots, A_n\}$ define a strict partial order, $<$, where $x_i < A_j$ if and only if $x_i \in A_j$. [The partial order is not defined for pairs (x_r, x_s) or pairs (A_r, A_s).] In the poset X, the set E is an antichain of cardinality m. Suppose there is an antichain D in X consisting of p elements from E and q sets from the family; without loss of generality we may write

$$D = \{x_1, x_2, \ldots, x_p, A_1, A_2, \ldots, A_q\}$$

Then none of these p elements x_i from D can belong to the union of these q sets A_j from D. So the union of these q sets can have at most $m - p$ elements. Thus, by the marriage condition,

$$q \leq m - p \qquad \text{or} \qquad p + q \leq m$$

which shows that E is a *maximal* antichain in X. Hence, by Dilworth's theorem, there exists a partition of X into m chains. Of these, n chains necessarily consist of 2 elements, 1 from E and 1 from the family. By a suitable reindexing of X, we can notate the n chains as $x_i < A_i$, or $x_i \in A_i$, for $i = 1, 2, \ldots, n$. In other words the family has the SDR $\langle x_1, x_2, \ldots, x_n \rangle$.

2.134 (*Danzig and Hoffman, 1956*) Show that the König–Egerváry theorem (Problem 2.108) implies Dilworth's theorem.

A poset $(\{x_1, x_2, \ldots, x_n\}, \leq)$ may be represented by an $n \times n$ (0, 1) matrix $A = [a_{ij}]$, where $a_{ij} = 1$ if and only if $x_i < x_j$. (Hence, if $a_{ij} = 1$, then $a_{ji} = 0$.) For simplicity, let us call a set of linewise disjunct 1s in A an **independent set**.

Every chain in P consisting of 2 or more elements defines an independent set. For example, assuming $n \geq 8$, a chain x_2—x_3—x_6—x_7—x_8, with 5 elements (**of length** 5), defines the independent set $\{a_{23}, a_{36}, a_{67}, a_{78}\}$, of cardinality $5 - 1 = 4$. So, if a chain decomposition of P consists of p_k chains of length $q_k > 1$ $(k = 1, 2, \ldots,)$ and q chains of length 1, this *disjoint* decomposition corresponds to an independent set of cardinality $n - q - p$, where $p \equiv \Sigma\, p_k$. In other words,

$$\begin{gathered}(\text{Number of chains in a decomposition of } P)\\ +\,(\text{size of corresponding independent set})\\ = (\text{size of } P)\end{gathered} \qquad (i)$$

It follows from (i) that if there exists a *maximum* independent set of size t (i.e., if the term rank of A is t), there is a *minimum* chain decomposition of P consisting of q chains of length 1 and $n - t - q$ chains of length greater than 1. But, by the König–Egerváry theorem, A has a cover of t (and no fewer) lines. This minimum cover corresponds to a set D of $n - t - q$ elements of P (one from each of the $n - t - q$ nonsingleton chains). The set E of elements which constitute chains of length 1 is of cardinality q. Then $F = D \cup E$ is a set of cardinality $n - t$ the elements of which are pairwise noncomparable. Thus there exist an antichain in P of cardinality $n - t$ and a chain decomposition of P consisting of $n - t$ chains.

2.135 Verify the complete equivalence of the König, König–Egerváry, Dilworth, and Hall theorems.

One circle of implications is:

$$\textbf{K–E} \overset{\text{Prob.}}{\underset{2.134}{\Longrightarrow}} \textbf{D} \overset{\text{Prob.}}{\underset{2.133}{\Longrightarrow}} \textbf{H} \overset{\text{Probs.}}{\underset{2.108,\ 2.109}{\Longrightarrow}} \textbf{K} \overset{\text{Prob.}}{\underset{2.109}{\Longrightarrow}} \textbf{K–E}$$

(The circle could be widened to include two fundamental graph-theoretical results, **Menger's theorem** and the **Ford–Fulkerson theorem**. See Appendix.)

2.136 Prove that a poset P of cardinality $mn + 1$ has either a chain of cardinality $m + 1$ or an antichain of cardinality $n + 1$.

If there is no antichain of cardinality $n + 1$, the cardinality, r, of a maximum antichain satisfies $r \leq n$. By Dilworth's theorem, there exists a decomposition of P ("pigeons") into disjoint chains (pigeonholes) $C_1, C_2, \ldots, C_r$. Theorem 1.4 ensures that some hole contains at least $m + 1$ pigeons.

2.137 If x and y are 2 elements in a poset $Q = (X, \leq)$ then x **covers** y if $y < x$ and $y \leq t \leq x$ imply $y = t$ or $t = x$. A poset P is called a **ranked poset** if there exists a function r defined on P such that $r(x) = 0$ whenever x is a minimal element and $r(x) = r(y) + 1$ whenever x covers y. The set Q_k of all elements of rank k is the ***k*th level set**; obviously, Q_k is an antichain in P. The cardinality of Q_k is called the

kth Whitney number of P. Show that the poset $(2^{X(n)}, \subset)$ is a ranked poset. Give its Whitney numbers.

For each subset A of $X(n)$ define $r(A)$ as the cardinality of A. The kth Whitney number is then $C(n, k)$. Note that for any ranked poset,

$$\text{Largest Whitney number} \leq \text{size of a maximal antichain} \qquad (*)$$

If $(*)$ holds with equality, the ranked poset is termed a **Sperner set**. Clearly, $(2^{X(n)}, \subset)$ is a Sperner set, because, by Sperner's theorem, both members of $(*)$ are equal to $C(n, n')$.

2.138 Obtain a sufficient condition for a ranked poset $P = (X, \leq)$ to be a Sperner set.

If there is a chain decomposition of X such that every constituent chain intersects the largest level set, $Q_{\max}$, then $Q_{\max}$ must be a *maximum* antichain, and $(*)$ of Problem 2.137 will hold with equality.

A more interesting sufficient condition is obtained by generalizing the concept of an SCD (Problems 2.119 and 2.120) along the lines sketched in Problem 2.124. Suppose the level sets of a ranked poset are $\{Q_0, Q_1, \ldots\}$, with corresponding Whitney numbers $\{w_0, w, \ldots\}$. The poset is **symmetric** if there exists an integer m such that $w_i = 0$ for $i > m$, and $w_i = w_{m-i}$ for all other i. (*Caution:* Symmetry by itself does not imply unimodality.) The number m is the **rank** of the symmetric poset. A chain $a_1 < a_2 < \cdots < a_k$ in a symmetric poset of rank m is a **symmetric chain** if $r(a_1) + r(a_k) = m$. *If a symmetric poset P has a decomposition into symmetric chains*, then a consideration of the intersections of the level sets with these chains—cf. Problem 2.121—establishes that the sequence of Whitney numbers is unimodal, with the largest level set intersecting every chain of the decomposition. In other words, if a symmetric poset has an SCD, then it is a Sperner set.

2.139 Let $\mathscr{D}_n$ be the set all divisors of a positive integer n, with the prime factorization

$$n = p_1^{e_1} p_2^{e_2} \cdots p_m^{e_m}$$

The number n can be specified as the m vector $[e_1 \quad e_2 \quad \cdots \quad e_m]$, in which all components are positive integers; any divisor of n is represented by an m vector with nonnegative integer components. If x and y are 2 divisors of n, with respective representations $[f_1 \quad f_2 \quad \cdots \quad f_m]$ and $[g_1 \quad g_2 \quad \cdots \quad g_m]$, then the ordering $x \leq y$ defined by $f_i \leq g_i$ (all i) is a partial order on $\mathscr{D}_n$. Show that the poset $(\mathscr{D}_n, \leq)$ is a ranked poset and find the Whitney numbers.

A ranking function on $\mathscr{D}_n$ may be defined by

$$r(x) \equiv \text{degree of } x \equiv \text{sum of components in representation}$$

(compare Problem 2.125). In the ranked poset the kth level set, Q_k, is the set of all divisors of n of degree k. Consequently, the kth Whitney number, w_k, is the number of solutions in integers of $u_1 + u_2 + \cdots + u_m = k$, subject to $0 \leq u_i \leq e_i$ for $i = 1, 2, \ldots, m$. See Problem 2.21 for a counting method.

2.140 Is the poset of Problem 2.139 a Sperner set?

The answer is yes. First, note that the poset is a symmetric poset. In fact: (*i*) $w_i = 0$ for $i > e_1 + e_2 + \cdots + e_m \equiv q$; and (*ii*) for each $x \in D_n$ of rank (degree) i, there is one and only one $y \in D_n$ (namely, $y = n/x$) of rank $q - i$, whence $w_i = w_{q-i}$. Second, it is clear that whenever y covers x, the order relation $x \leq y$ becomes

$$\frac{y}{x} = \text{a prime}$$

This means that in Problem 2.125 the implicit ordering of $\mathscr{D}_n$ was precisely $\leq$. Hence, by Problem 2.125(*a*), our symmetric poset has an SCD. Then, by Problem 2.138, it is a Sperner set.

[The existence of an SCD for $(\mathscr{D}_n, \leq)$ yields a bonus: Without even counting, one knows that the numbers of integral solutions in Problem 2.139 have a unimodal property.]

2.141 Calculate the Whitney numbers of $(\mathscr{D}_{42}, \leq)$ (*a*) by the method of Problem 2.139, and (*b*) by use of the SCD found in Problem 2.126.

(*a*) For a squarefree n $(42 = 2' \times 3' \times 7')$ the constrained equation of Problem 2.139 has exactly $C(m, k)$ solutions. Here $m = 3$, whence

$$w_0 = C(3,0) = 1 \qquad w_1 = C(3,1) = 3 \qquad w_2 = C(3,2) = 3 \qquad w_3 = C(3,3) = 1$$

(*b*) Reading Fig. 2-19 vertically, we have

$$Q_0 = \{1\} \qquad Q_1 = \{2, 7, 3\} \qquad Q_2 = \{6, 14, 21\} \qquad Q_3 = \{42\}$$

whence $w_0 = 1$, $w_1 = 3$, $w_2 = 3$, $w_3 = 1$.

2.142 Let Π_n be the set of all partitions of a set X of cardinality n. [By Problem 1.161, $|\Pi_n| = B_n$ (the nth Bell number).] If $\pi = \{A_1, A_2, \ldots, A_r\}$ and $\pi' = \{B_1, B_2, \ldots, B_s\}$ are 2 partitions of X, write $\pi \leq \pi'$ if and only if π' is obtained from π by partitioning each of the sets A_i into one or more (nonempty) sets. Then $(\Pi_n, \leq)$ is a poset. Show that it is a ranked poset and obtain the corresponding Whitney numbers.

Let the rank of a partition be 1 less than the number of cells in the partition. Then, *by definition* (Problem 1.146), the kth Whitney number of the ranked poset is $S(n, k+1)$.

It is interesting to observe that this poset has a unimodal sequence of Whitney numbers (Problem 1.155), even though the poset is not symmetric (see Table 1-2).

Supplementary Problems

2.143 Find the number of ways in which the complete collection of letters that appear in the word MISSISSIPPI can be arranged in a row, if (*a*) there is no restriction on the locations of the letters; (*b*) all the S's must stay together; and (*c*) no two S's may be adjacent.
Ans. (*a*) $P(11; 4, 4, 2, 1)$; (*b*) $P(8; 1, 4, 2, 1)$; (*c*) $P(7; 4, 2, 1)C(8, 4)$

2.144 Find the number of ways in which the complete collection of letters that appear in TALLAHASSEE can be arranged in a row so that (*a*) T appears at the beginning and E appears at the end, (*b*) there are no adjacent A's.
Ans. (*a*) $P(9; 3, 2, 2, 1, 1)$; (*b*) $P(8; 2, 2, 2, 1, 1)C(9, 3)$

2.145 Find the number of ways of (*a*) assigning 10 students to 12 single rooms; (*b*) installing 10 identical telephones in 12 rooms; and (*c*) installing 10 color telephones (4 red, 3 white, 3 green) in 12 rooms.
Ans. (*a*) $P(12, 10)$; (*b*) $C(12, 10)$; (*c*) $C(12; 4, 3, 3)$

2.146 There are 20 students in a class. Find the number of ways of: (*a*) allocating them to 4 distinct dormitories so that the first dormitory gets 3 students, the second dormitory gets 5 students, and the third dormitory gets 4 students; (*b*) dividing them into 4 groups of 3, 5, 4, and 8; (*c*) allocating them to 4 distinct dormitories so that each dormitory gets 5 students; and (*d*) dividing them into 4 equal groups.
Ans. (*a*) $C(20; 3, 5, 4, 8)$; (*b*) $C(20; 3, 5, 4, 8)$; (*c*) $C(20; 5, 5, 5, 5)$; (*d*) $(1/4!)C(20; 5, 5, 5, 5)$

2.147 Find the coefficient of $p^2q^3r^3s^4$ in the expansion of $(2p - 3q + 2r - s)^{12}$. *Ans.* $(-864)C(12; 2, 3, 3, 4)$

2.148 Find the number of ways of distributing 10 distinct books to 4 students so that each gets at least 2 books.
Ans. $C(4, 1)C(10; 4, 2, 2, 2) + C(4, 2)C(10; 3, 3, 2, 2)$

2.149 From an abundant supply of red, yellow, blue, and green marbles, how many rows of 10 marbles can be made if each row must contain at least 2 marbles of each color? (*Hint*: This is the dual of Problem 2.148.)
Ans. $C(4, 1)C(10; 4, 2, 2, 2) + C(4, 2)C(10; 3, 3, 2, 2)$

2.150 Find the coefficient of x^5 in $(a + bx + cx^2)^8$.
Ans. $a^3b^5C(8; 3, 5) + a^4b^3cC(8; 4, 3, 1) + a^5bc^2C(8; 5, 1, 2)$

2.151 Disregarding order within a box, find the number of ways of packing: (*a*) 12 distinct books in 3 distinct boxes so that 3 books are in box 1, 4 books are in box 2, and 5 books are in box 3; (*b*) 12 distinct books in 3 distinct boxes so that 1 of them has 3 books, another has 4 books, and yet another has 5 books; (*c*) 12 distinct books in 3 identical boxes, if they are to contain 3, 4, and 5 books; (*d*) 12 distinct books in 3 distinct boxes, each to contain 4 books; and (*e*) 12 distinct books in 3 identical boxes, each to contain 4 books.
Ans. (*a*) $C(12; 3, 4, 5)$; (*b*) $(3!)C(12; 3, 4, 5)$; (*c*) $C(12; 3, 4, 5)$; (*d*) $C(12; 4, 4, 4)$; (*e*) $(1/3!)C(12; 4, 4, 4)$

2.152 Find the sum of all 4-digit integers formed by permuting the digits 1, 2, 3, and 4. (*Hint*: In each column of the sum each digit must appear 3! times.) *Ans.* $(60)[1 + (10)^1 + (10)^2 + (10)^3] = 66{,}660$

2.153 Find the sum of all 4-digit integers formed by permuting 1, 2, 2, and 5.
Ans. $(30)[1 + (10)^1 + (10)^2 + (10)^3 = 33{,}330$

2.154 Find the number of ways of choosing 3 distinct integers from the set $X = \{1, 2, \ldots, 100\}$ so that their sum will be divisible by 3. (*Hint*: First partition X into residue classes modulo 3.)
Ans. $C(33, 3) + C(33, 1)C(34, 1)C(33, 1) + [C(34, 3) + C(33, 3)]$

2.155 Find the number of ways of giving $3n$ different toys to Maddy, Jimmy, and Tommy so that Maddy and Jimmy together get $2n$ toys. *Ans.* $C(3n, n)2^{2n}$

2.156 Find the number of r sequences that can be formed using the first 7 letters of the English alphabet, if (*a*) $r = 4$ and no letter repeats; (*b*) $r = 4$; and (*c*) $r = 9$.
Ans. (*a*) $P(7, 4) = 940$; (*b*) $7^4 = 2401$; (*c*) $7^9 = 40{,}353{,}607$

2.157 Repeat Problem 2.156 for r selections.
Ans. (*a*) $C(7, 4) = 35$; (*b*) $C(4 + 7 - 1, 7 - 1) = 210$; (*c*) $C(9 + 7 - 1, 7 - 1) = 5005$

2.158 (*a*) Find the number of terms in the multinomial expansion of $(p + q + r + s)^{25}$. (*Hint*: See Problem 1.139.) (*b*) Find the number of terms in which the exponents of p, q, r, and s are at least 1, 2, 3, and 4, respectively. (*Hint*: See Problem 1.142.) *Ans.* (*a*) $C(25 + 4 - 1, 4 - 1)$; (*b*) $C(25 - 10 + 4 - 1, 4 - 1)$

2.159 Find the number of ways of distributing 7 identical pens and 7 identical pencils to 5 students so that each gets at least 1 pen and at least 1 pencil. (*Hint*: Two independent distributions.)
Ans. $C(7 - 5 + 4, 4)C(7 - 5 + 4, 4) = 225$

2.160 Determine the number of positive integers which do not exceed 100 and which are not divisible by 2, 3, or 5.
Ans. 26

2.161 Find the number of solutions in nonnegative integers of the equation $a + b + c + d + e + f = 20$ in which no variable is greater than 8. *Ans.* $C(25, 20) - C(6, 1)C(16, 11) + C(6, 2)C(7, 2)$

2.162 Find the number of positive integers smaller than 291,060 and relatively prime to it. *Ans.* 60,480

2.163 Find the number of ways of arranging the 26 letters of the alphabet so that no one of the sequences *ABC*, *EFG*, *PQRS*, and *XYZ* appears. *Ans.* $26! - [3(24!) + 23!] + [3(22!) + 3(21!)] - [20! + 3(19!)] + 17!$

2.164 Find the number of 4-digit integers involving the first 6 positive digits in which each of the first 3 positive digits appears at least once. *Ans.* $6^4 - 3(5^4) + 3(4^4) - 3^4$

2.165 Find the number of permutations of the 9 positive digits in which (*a*) the blocks 12, 34, and 567 do not appear; (*b*) the blocks 415, 12, and 23 do not appear.
Ans. (*a*) $9! - (8! + 8! + 7!) + (7! + 6! + 6!) - 5!$; (*b*) $9! - (8! + 8! + 7!) + (7! + 0 + 6!)$

2.166 Find the number of ways of assigning 6 students (A, B, C, D, E, F) to 6 dormitories (1, 2, 3, 4, 5, 6), with the following restrictions: A does not go to 1 or 2, B does not go to 4, C does not go to 1 or 5, D does not go to 2, E does not go to 4, and F does not go to 4 or 6. (*Hint*: The rook polynomial may be used.) *Ans.* 132

2.167 From Problems 2.30 and 2.31 conclude that n and D_n are always of opposite parity.
Ans. If n is odd, D_n is even by Problem 2.30. If n is even, D_n is the sum of an even number and an odd number by the next problem.

2.168 From Problems 2.32 and 2.167 conclude that T_n is always odd.
Ans. T_n is the sum of an even number and an odd number for any n.

2.169 In a 6×6 chessboard the forbidden squares are (1, 2), (2, 1), (2, 5), (3, 4), (4, 1), (4, 5), and (6, 6). Determine the rook polynomial of the forbidden subboard. (*Hint*: Permute rows and columns.)
Ans. $(1 + 4x + 2x^2)(1 + x)^3 = 1 + 7x + 17x^2 + 19x^3 + 10x^4 + 2x^5$

2.170 A final exam is given each day of the week (Monday through Friday), and each exam should have a different professor in charge. Unfortunately, only 4 professors are available, under the following constraints: Professor A is not free Mondays and Tuesdays; Professor B is not free Tuesdays; Professor C is not free Wednesdays and Thursdays; and Professor D is not free Thursdays and Fridays. In how many ways can 4 of the exams be covered? (*Hint*: Add a dummy Professor E who is never free.)
Ans. $5! - 7(4!) + 16(3!) - 13(2!) + 3(1!) = 25$

2.171 Count the positive integers smaller than 1 million that (*a*) include all the digits 2, 4, 6, and 8; (*b*) include only the digits 2, 4, 6, and 8.
Ans. (*a*) $10^6 - 4(9^6) + 6(8^6) - 4(7^6) + 1(6^6) = 23{,}160$; (*b*) $4 + 4^2 + \cdots + 4^6 = 5460$

2.172 Find the number of permutations of 11223344 such that no 2 adjacent positions are occupied by the same digit.
Ans. $C(8; 2, 2, 2, 2) - 4C(7; 2, 2, 2, 1) + 6C(6; 2, 2, 1, 1) - 4C(5; 2, 1, 1, 1) + C(4; 1, 1, 1) = 864$

2.173 Find the number of ways of assigning r distinguishable objects to n girls and p boys so that each girl receives at least 1 object. (*Hint*: Use inclusion-exclusion.)
Ans. $(n + p)^r - C(n, 1)(n + p - 1)^r + C(n, 2)(n + p - 2)^r - \cdots + (-1)^n C(n, n)(n + p - n)^r$

2.174 An elevator starts with 9 people at the first floor. At each floor at least 1 person leaves the elevator and nobody enters the elevator. It becomes empty at the fifth floor where it stays till it is activated again. Find the number of ways of unloading the people. *Ans.* $5^9 - C(5, 1)4^9 + C(5, 2)3^9 - C(5, 3)2^9 + C(5, 4)$

2.175 Use Theorem 2.9 (Problem 2.78) to compute the permanents of

$$A = \begin{bmatrix} 1 & 2 & 3 \\ 2 & 3 & 4 \\ 3 & 4 & 5 \end{bmatrix} \qquad B = \begin{bmatrix} 1 & 1 & 1 & 1 \\ 1 & 2 & 3 & 4 \\ 1 & 1 & 1 & 1 \end{bmatrix}$$

Ans. per $(A) = 126$, per $(B) = 60$

2.176 Use Problem 2.77 and Theorem 2.9 to establish the following identities:

(*a*) $n! = \sum_{r=0}^{n-1} (-1)^r C(n, r)(n-r)^n$ (*b*) $D_n = \sum_{r=0}^{n-2} (-1)^r C(n, r)(n-r)^r(n-r-1)^{n-r}$

(*Hint*: $n! = \text{Per } I_n$ and $D_n = \text{Per } (J_n - I_n)$.)

2.177 Verify that the Ménage numbers (Problem 2.73) satisfy the difference equation ($n \geq 3$)

$$(n-2)M_n - n(n-2)M_{n-1} - nM_{n-2} = 4(-1)^{n+1}$$

(*Hint*: M_n is the sum of $(n+1)$ terms.) Let $M_n = x_1 + x_2 + x_3 + x_4 + \cdots + x_n + X_{n+1}$, $M_{n-1} = y_1 + y_2 + y_3 + y_4 + \cdots + y_n + 0$, and $M_{n-2} = 0 + 0 + 0 + z_1 + \cdots + z_{n-2} + z_{n-1}$. Then $(n-2)(x_1 + x_2) = n(n-1)(y_1 + y_2)$, $(n-2)x_{n+1} = nz_{n-1} + 4(-1)^{n+1}$ and $(n-2)x_r = n(n-2)y_r + nz_{r-2}$ $(3 \leq r \leq n-2)$.

2.178 In the special case of Problem 2.114 where $\{A_i\}$ and $\{B_j\}$ are 2 *partitions* of E, show that an SCR exists if and only if, for $k = 2, \ldots, n$, no k of the A_i are contained in fewer than k of the B_j. (*Hint*: Consider the family of subsets of I defined by $X_i \equiv \{j : A_i \cap B_j \neq \emptyset\}$.)

2.179 (*Birkhoff–von Neumann Theorem*) A square matrix of nonnegative real numbers is **doubly stochastic** if each line sum is 1. Show that a matrix is doubly stochastic if and only if it is a convex combination of permutation matrices. *Ans.* Put $r = 1$ in Problem 2.115*a*.

2.180 Exhibit simple bipartite graphs that allow (*a*) both an X-saturated and a Y-saturated matching; (*b*) neither an X-saturated nor a Y-saturated matching.

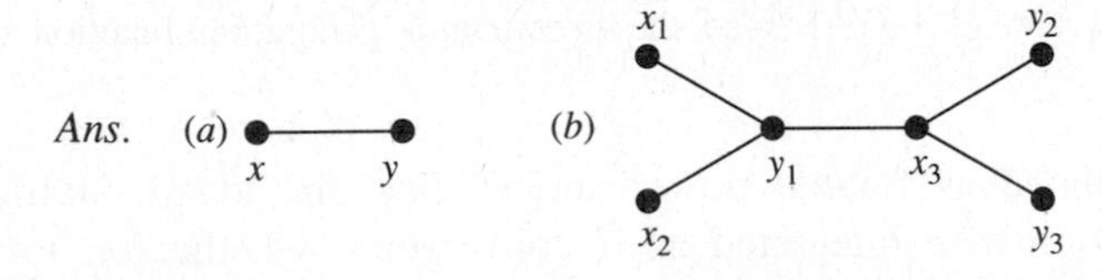

2.181 (*König, 1914*) Prove that the edges of an r-regular bipartite graph, $\mathscr{G}(X, Y, E)$, can be painted with r colors in such a manner that all colors are represented at every vertex. [*Hint*: If M is a complete matching in the graph, then $\mathscr{G}(X, Y, E - M)$ is an $(r-1)$-regular bipartite graph.]

2.182 How about interchanging "chain" and "antichain" in the proof of Problem 2.130, thereby obtaining a proof of Dilworth's theorem that is much simpler than that given in Problem 2.129?
Ans. This cannot be done: the subposet $P - X$ can contain a maximum *antichain*.

2.183 By suitably orienting the edges of a pentagon, demonstrate that not every poset is a ranked poset.
Ans. Suppose the 5 corners (represented as the vertices of a directed graph) are labeled clockwise, A, B, C, D, E with orientation A——→B——→C——→D——→E——→A. Then $r(\text{A}) = 0$ implies $r(\text{C}) = 2$ or 3.

Chapter 3

Generating Functions and Recurrence Relations

3.1 ORDINARY AND EXPONENTIAL GENERATING FUNCTIONS

Given a sequence of real numbers $\langle a_0, a_1, a_2, \ldots \rangle$ and a dummy variable x, one makes the

Definition: The (**ordinary**) generating function of the sequence is

$$g(x) = a_0 + a_1 x + a_2 x^2 + \cdots$$

The **exponential generating function** of the same sequence is

$$G(x) = a_0 + a_1 \frac{x}{1!} + a_2 \frac{x^2}{2!} + \cdots$$

Example 1. For the sequence of combination symbols $\langle C(n,0), C(n,1), C(n,2), \ldots, C(n,n) \rangle$, the generating function is

$$\sum_{k=0}^{n} C(n,k)x^k = (1+x)^n$$

Since $C(n,k) = P(n,k)/k!$ (Theorem 1.1), one infers that $(1+x)^n$ is also the *exponential* generating function of the sequence of permutation symbols.

Generating functions of either sort are defined as **formal** power series: they are added, multiplied, scalar-multiplied, termwise differentiated, and termwise integrated as if convergent—whether or not they actually are so. Some general properties of generating functions are listed as

Theorem 3.1. (*i*) If $g(x)$ is the ordinary generating function of $\langle a_r \rangle$, then $(1-x)g(x)$ is the ordinary generating function of $\langle a_r - a_{r-1} \rangle$.

(*ii*) If $g(x)$ is the ordinary generating function of $\langle a_r \rangle$, then $(1-x)^{-1}g(x)$ is the ordinary generating function of $\langle a_0 + a_1 + a_2 + \cdots + a_r \rangle$.

(*iii*) If $g(x)[G(x)]$ is the ordinary (exponential) generating function of $\langle a_r \rangle$, then $xg'(x)[xG'(x)]$ is the ordinary (exponential) generating function of $\langle ra_r \rangle$.

(*iv*) If $g(x)[G(x)]$ is the ordinary (exponential) generating function of $\langle a_r \rangle$, and $h(x)[H(x)]$ is the ordinary (exponential) generating function of $\langle b_r \rangle$, then $g(x)h(x)[G(x)H(x)]$ is the ordinary (exponential) generating function of the **convolution** (**binomial convolution**)

$$\left\langle \sum_{i=0}^{r} a_i b_{r-i} \right\rangle \qquad \left[\left\langle \sum_{i=0}^{r} C(r,i) a_i b_{r-i} \right\rangle \right]$$

(*v*) If $S(x)$ is the (ordinary or exponential) generating function of $\langle a_r \rangle$, and $T(x)$ is the generating function (same type) of $\langle b_r \rangle$, then $pS(x) + qT(x)$ generates $\langle pa_r + qb_r \rangle$, for any real p and q.

Example 2. The **reciprocal**, $h(x) = 1/g(x)$, of the generating function $g(x) = a_0 + a_1 x + \cdots$ may be determined by means of the multiplication rule, provided $a_0 \neq 0$. Indeed, if $h(x) = b_0 + b_1 x + \cdots$, then

$$g(x)h(x) = (a_0 + a_1 x + a_2 x^2 + \cdots)(b_0 + b_1 x + b_2 x^2 + \cdots) \equiv 1$$

implies the triangular system

$$\begin{aligned} a_0 b_0 &= 1 \\ a_1 b_0 + a_0 b_1 &= 0 \\ a_2 b_0 + a_1 b_1 + a_0 b_2 &= 0 \\ &\cdots\cdots\cdots \end{aligned}$$

which has a unique solution for the b_k.

If by combinatorial reasoning or other means a generating function $g(x)$ has been found, then the particular sequence of which $g(x)$ is the generating function may be recovered via the formula

$$a_k = \frac{1}{k!} \frac{d^k g}{dx^k}\bigg|_{x=0} \qquad (k = 0, 1, 2, \ldots)$$

Similarly, if the exponential generating function $G(x)$ is known,

$$a_k = \frac{d^k G}{dx^k}\bigg|_{x=0} \qquad (k = 0, 1, 2, \ldots)$$

Example 3. For each fixed m, the Stirling numbers of the second bind, $S(n, m)$, may be defined via an exponential generating function. Consider the identity

$$(e^x - 1)^m = \left(x + \frac{x^2}{2!} + \cdots + \frac{x^{n_1}}{n_1!} + \cdots\right)\left(x + \frac{x^2}{2!} + \cdots + \frac{x^{n_2}}{n_2!} + \cdots\right)$$
$$\cdots\left(x + \frac{x^2}{2!} + \cdots + \frac{x^{n_m}}{n_m!} + \cdots\right) \qquad (i)$$

When the product on the right of (i) is expanded using the multiplication rule, the coefficient of x^n is found as

$$\sum_{\substack{n_1+n_2+\cdots+n_m=n \\ n_i \geq 1}} \left(\frac{1}{n_1!}\right)\left(\frac{1}{n_2!}\right)\cdots\left(\frac{1}{n_m!}\right) = \frac{1}{n!} \sum_{\substack{n_1+n_2+\cdots+n_m=n \\ n_i \geq 1}} \frac{n!}{n_1! n_2! \cdots n_m!} \qquad (ii)$$

By the proof of Theorem 2.2, the quantity $n!/n_1! \cdots n_m!$ represents the number of ordered partitions of an n set X into cells of respective sizes $n_1, n_2, \ldots, n_m$. Therefore, the right-hand side of (ii) equals $1/n!$ times the number of ordered partitions of X into m cells; or, $m!/n!$ times the number of (unordered) partitions of X into m cells; or, finally, $(m!/n!)S(n, m)$. In short, $(e^x - 1)^m$ is the exponential generating function of the sequence $\langle m!\, S(n, m)\rangle_{n \geq 0}$; therefore

$$m!\, S(n, m) = \frac{d^n}{dx^n}(e^x - 1)^m\bigg|_{x=0} \qquad (n = 0, 1, 2, \ldots) \qquad (iii)$$

Observe that (iii) properly makes $S(n, m) = 0$ for $n < m$.

3.2 PARTITIONS OF A POSITIVE INTEGER

Generating functions play an essential role in the theory of partitions. Recall from Problem 2.24 that a partition of a positive integer r is a collection of positive integers (parts) with sum r. Some useful notations are:

$p(r) \equiv$ number of distinct partitions of r

$p_n(r) \equiv$ number of partitions of r into parts at most equal to n

$\equiv$ number of solutions in nonnegative integers of

$$1u_1 + 2u_2 + 3u_3 + \cdots + nu_n = r$$

$q_n(r) \equiv$ number of partitions of r into at most n parts
$\equiv$ number of distributions of r identical objects (1s) among n identical places, empty places being permitted

Example 4. (*a*) The partitions of 5 are

$$5 \quad 4+1 \quad 3+2 \quad 3+1+1 \quad 2+2+1 \quad 2+1+1+1 \quad 1+1+1+1+1$$

So $p(5)=7$. (*b*) The partitions of 5 having no part greater than 2 are

$$2+2+1 \quad 2+1+1+1 \quad 1+1+1+1+1$$

Thus $p_2(5)=3$. (*c*) The partitions of 5 into 3 or fewer parts are

$$5 \quad 4+1 \quad 3+2 \quad 3+1+1 \quad 2+2+1$$

So $q_3(5)=5$.

Theorem 3.2. For all r and n, $p_n(r)=q_n(r)$.

Proof. Any partition Π of a positive integer r may be represented in a **star diagram** composed of r asterisks that are arranged in rows corresponding to the parts. The rows are nonincreasing in length as one moves from top to bottom of the diagram (see Fig. 3-1). If the star diagram of Π is read by columns, a **conjugate** partition Π^* of r is obtained. The relation between Π and Π^* is clearly reciprocal, with the number of parts of the one equal to the the largest part of the other. Thus, for every Π counted in $p_n(r)$ there is a unique Π^* counted in $q_n(r)$; i.e., $p_n(r)=q_n(r)$.

```
*   *   *   *   *

*   *   *   *   *

*   *   *   *

*

*

*
```

Fig. 3-1 The conjugate partitions: $5+5+4+1+1+1=17=6+3+3+3+2$

The second defining expression for $p_n(r)$ arises from the fact that a partition is uniquely determined if we give the number of 1s, the number of 2s, etc., among the parts. From it, a generating function for $\langle p_n(0)\equiv 1, p_n(1), p_n(2), p_n(3), \ldots\rangle$ is simply obtained. Consider the product of n geometric series

$$\begin{aligned}
\frac{1}{1-x}\,\frac{1}{1-x^2}\,\frac{1}{1-x^3}\cdots\frac{1}{1-x^n}
&= [1+x^1+(x^1)^2+\cdots+(x^1)^{u_1}+\cdots] \\
&\quad \times [1+x^2+(x^2)^2+\cdots+(x^2)^{u_2}+\cdots] \\
&\quad \times [1+x^3+(x^3)^2+\cdots+(x^3)^{u_3}+\cdots] \\
&\quad \cdots\cdots\cdots\cdots\cdots\cdots\cdots\cdots \\
&\quad \times [1+x^n+(x^n)^2+\cdots+(x^n)^{u_n}+\cdots]
\end{aligned}$$

It is seen that the coefficient of x^r on the right will be

$$\sum_{\substack{1u_1+2u_2+3u_3+\cdots+nu_n=r \\ u_i \geq 0}} \equiv p_n(r)$$

That is to say:

Theorem 3.3. The ordinary generating function of $\langle p_n(r)\rangle = \langle q_n(r)\rangle$ is

$$g_n(x) = \frac{1}{(1-x)(1-x^2)(1-x^3)\cdots(1-x^n)}$$

As a matter of definition, $p_n(r) = p(r)$ for all $n > r$. Consequently, as $n \to \infty$, $p_n(r) \to p(r)$ for all r; and Theorem 3.3 yields in the limit (remember that we are dealing with *formal* power series):

Theorem 3.4 (*Euler*). The ordinary generating function of $\langle p(r)\rangle$ is

$$g(x) = \frac{1}{(1-x)(1-x^2)(1-x^3)\cdots}$$

While in principle they solve the counting problem for partitions, Theorems 3.3 and 3.4 are, by themselves, of small practical use. Noncombinatorial techniques from the theory of analytic functions must be employed to obtain, e.g., asymptotic estimates of $p(r)$.

3.3 RECURRENCE RELATIONS

If $\langle a_0, a_1, \ldots, a_k, \ldots\rangle$ is a sequence of real numbers such that there is an equation relating the term a_n (for any $n \geq n_0$) to one or more of its predecessors in the sequence, then this equation is a **recurrence relation** obeyed by the sequence.

Example 5. The sequence $\langle 0!, 1!, 2!, \ldots\rangle$ satisfies the recurrence relation

$$a_n = na_{n-1} \qquad (n \geq 1)$$

Conversely, given this relation and the **initial condition** $a_0 = 1$, one can recover the entire sequence by iteration:

$$a_n = n[(n-1)a_{n-2}] = n(n-1)[(n-2)a_{n-3}]$$
$$= \cdots = n(n-1)\cdots(1) = n!$$

Example 6. The idea of a recurrence relation extends to sequences that depend on 2 or more indices. Thus Pascal's identity (Problem 1.37) is a recurrence relation for the binomial coefficients.

As illustrated in Chapters 1 and 2 (see especially Problems 1.149 to 1.150 and Problems 2.30 to 2.31), a powerful method for solving counting problems consists in first obtaining, by combinatorial reasoning, a recurrence relation for the counting numbers, and then solving that relation, subject to appropriate initial conditions, for those numbers. A complete solution theory is available when the recurrence relation falls under a certain broad category. This will be outlined in the next section.

3.4 ALGEBRAIC SOLUTION OF LINEAR RECURRENCE RELATIONS WITH CONSTANT COEFFICIENTS

Definition: The recurrence relation

$$a_n = c_1 a_{n-1} + c_2 a_{n-2} + \cdots + c_r a_{n-r} + f(n) \qquad (*)$$

in which c_i $(i = 1, 2, \ldots, r)$ are constants, with $c_r \neq 0$, is called a **linear recurrence relation with constant coefficients**, of **order** r.

From linear algebra we have a fundamental theorem, *which holds even if the coefficients c_i are functions of n.*

Theorem 3.5. The general solution of $(*)$ is given by

$$a_n = h(n) + p(n)$$

where $p(n)$ is any particular solution of $(*)$ and $h(n)$ is the general solution—a solution that linearly involves r arbitrary constants—of the homogeneous relation $[(*)$ with $f(n) \equiv 0]$.

If a combinatorial or other problem has $(*)$ as its recurrence relation and if there are r initial conditions attached to the problem, there are two possibilities:

1. The initial conditions are **consecutive**, which means that $a_0, a_1, \ldots, a_{r-1}$ are prescribed. Then the arbitrary constants C_i in $h(n)$ are uniquely fixed, so that the problem has a unique solution.
2. The initial (or boundary) conditions are not consecutive. In this case the problem may have a single solution, more than one solution, or no solution at all.

Example 7. The general solution of $a_n = 4a_{n-2}$ (a homogeneous relation) is obviously $a_n = C_1 2^n + C_2(-2)^n$.

(*i*) Prescribing $a_0 = x$ and $a_1 = y$, we get

$$\begin{aligned} C_1 + C_2 &= x \\ 2C_1 - 2C_2 &= y \end{aligned} \quad \text{or} \quad C_1 = \frac{2x+y}{4}, \quad C_2 = \frac{2x-y}{4}$$

(unique solution).

(*ii*) Prescribing $a_0 = 1$ and $a_2 = 4$ (nonconsecutive),

$$\begin{aligned} C_1 + C_2 &= 1 \\ 4C_1 + 4C_2 &= 4 \end{aligned} \quad \text{or} \quad C_1 = \text{arbitrary}, \quad C_2 = 1 - C_1$$

(a single infinity of solutions).

(*iii*) Prescribing $a_0 = 1$ and $a_2 = 5$,

$$\begin{aligned} C_1 + C_2 &= 1 \\ 4C_1 + 4C_2 &= 5 \end{aligned}$$

(no solution).

(*iv*) Prescribing $a_0 = 2$ and $a_3 = 0$,

$$\begin{aligned} C_1 + C_2 &= 2 \\ 8C_1 - 8C_2 &= 0 \end{aligned} \quad \text{or} \quad C_1 = C_2 = 1$$

(unique solution).

Taking $f(n) \equiv 0$ in $(*)$, we see that the homogeneous relation will have a solution of the form $a_n = t^n$ $(t \neq 0)$ provided

$$t^r - c_1 t^{r-1} - c_2 t^{r-2} - \cdots - c_r = 0$$

This polynomial equation, of degree r, is called the **characteristic equation** of (*), and its r roots (counted according to multiplicity) are the **characteristic roots** of (*). Note that, even when all coefficients c_i are real, complex charcteristic roots are possible; these must occur in conjugate pairs.

Theorem 3.6. Let the characteristic roots of (*) be $t_1, t_2, \ldots, t_s$, of respective multiplicities $m_1, m_2, \ldots, m_s$ ($\Sigma\, m_i = r$). Then the general homogeneous solution of (*) may be written as $h(n) = \Sigma\, h_i(n)$, where

$$h_i(n) = (C_{i0} + C_{i1} + C_{i2}n^2 + \cdots + C_{i,m_i-1}n^{m_i-1})t_i^n \qquad (i = 1, 2, \ldots, s)$$

Corollary. If all the characteristic roots are simple,

$$h(n) = C_1 t_1^n + C_2 t_2^n + \cdots + C_r t_r^n$$

Example 8. (*i*) For the recurrence relation $a_n = 6a_{n-1} - 8a_{n-2} + 3^n$, the characteristic roots are 2 and 4. By the corollary to Theorem 3.6, $h(n) = C_1 2^n + C_2 4^n$; and it is known that $p(n) = (-9)(3^n)$ is a particular solution. Thus we have the general solution $a_n = C_1 2^n + C_2 4^n - 9(3^n)$. (*ii*) For the recurrence relation

$$a_n = 6a_{n-1} - 12a_{n-2} + 8a_{n-3} + 3^n$$

the characteristic roots are 2, 2, and 2. By Theorem 3.6, $h(n) = (C_0 + C_1 n + C_2 n^2)(2^n)$. It is known that $p(n) = 27(3^n)$ is a particular solution. Thus the general solution is

$$a_n = (C_0 + C_1 n + C_2 n^2)(2^n) + 27(3^n)$$

If a particular solution of (*) cannot be guessed, it can, in theory, be constructed from $f(n)$ and the homogeneous solutions $h_i(n)$. Fortunately, this usually tedious procedure can be bypassed in two important special cases.

Theorem 3.7. (*i*) If $f(n)$ is a polynomial in n of degree d, and if 1 is not a characteristic root of (*), then (*) has the particular solution $A_0 + A_1 n + A_2 n^2 + \cdots + A_d n^d$ [where the constants A_i are evaluated by substitution of $p(n)$ in (*)]. If 1 is a characteristic root, of multiplicity m,

$$p(n) = A_0 n^m + A_1 n^{m+1} + \cdots + A_d n^{m+d}$$

(*ii*) If $f(n) = b^n$ (an exponential in n), and if b is not a characteristic root, then $p(n) = Ab^n$. If b is a characteristic root, of multiplicity m, then $p(n) = An^m b^n$. Again the constant A is evaluated by substitution.

Example 9. A student, who has forgotten the formula for the sum of the first n squares, sets

$$1^2 + 2^2 + 3^2 + \cdots + n^2 \equiv a_n$$

and obtains the linear recurrence relation

$$a_n = a_{n-1} + n^2 \qquad (a_0 = 0)$$

The characteristic equation, $t - 1 = 0$, has the single root $t = 1$ ($m = 1$). Hence $h(n) = C$ and, by Theorem 3.7(*i*),

$$p(n) = A_0 n + A_1 n^2 + A_2 n^3$$

Upon substituting $p(n)$ in the recurrence relation, she finds:

$$A_0n + A_1n^2 + A_2n^3 = (-A_0 + A_1 - A_2) + (A_0 - 2A_1 + 3A_2)n$$
$$+ (A_1 - 3A_2 + 1)n^2 + A_2n^3$$

Balancing the equation from the top down, she successively gets: nothing, $A_2 = \frac{1}{3}$, $A_1 = \frac{1}{2}$, $A_0 = \frac{1}{6}$. The initial condition forces $C = 0$; so,

$$a_n = \frac{1}{6}n + \frac{1}{2}n^2 + \frac{1}{3}n^3 = \frac{n(n+1)(2n+1)}{6}$$

3.5 SOLUTION OF RECURRENCE RELATIONS USING GENERATING FUNCTIONS

In many cases it is possible to transform a recurrence relation for a sequence into an algebraic or differential equation for the generating function of that sequence. If this equation is solvable, the sequence can be recovered by differentiation, as previously described, or by other means.

Example 10. Rework Example 9 by use of an ordinary generating function.

Define $g(x) = \sum_{n=0}^{\infty} a_n x^n$. Now multiply the recurrence relation through by x^n and sum from $n = 0$ to $n = \infty$, taking into account the initial conditions $a_{-1} = a_0 = 0$:

$$\sum_{n=0}^{\infty} a_n x^n = x \sum_{n=1}^{\infty} a_{n-1} x^{n-1} + \sum_{n=0}^{\infty} n^2 x^n$$
$$g(x) = xg(x) \qquad + \frac{x(1+x)}{(1-x)^3}$$

[To evaluate $\sum n^2 x^n$, write

$$\frac{1}{1-x} = \sum x^n \qquad \frac{x}{(1-x)^2} = \sum nx^n \qquad \frac{2x^2}{(1-x)^3} = \sum n(n-1)x^n$$

and add the last two series.] Thus

$$(1-x)g(x) = \frac{x(1+x)}{(1-x)^3} \qquad \text{or} \qquad g(x) = \frac{x(1+x)}{(1-x)^4}$$

Now see Problem 3.82.

The importance of the generating-function method lies not in its applications to linear recurrence relations with constant coefficients (where, of course, it must produce the same results as the characteristic-equation method) but in its applications to linear recurrence relations with variable coefficients and to certain nonlinear recurrence relations.

Example 11. As was shown in Problem 1.131, the Catalan numbers satisfy the nonlinear recurrence relation

$$a_n = \sum_{i=0}^{n} a_i a_{n-i} \qquad (n \geq 2)$$

with starting values $a_0 \equiv 0$ and $a_1 = 1$. Then

$$g(x) \equiv \sum_{n=0}^{\infty} a_n x^n = 0 + x + \sum_{n=2}^{\infty} a_n x^n = x + \sum_{n=2}^{\infty} \left[\sum_{i=0}^{n} a_i a_{n-i} \right] x^n$$
$$= x + \sum_{n=0}^{\infty} \left(\sum_{i=0}^{n} a_i a_{n-i} \right) x^n = x + [g(x)]^2$$

or $[g(x)]^2 - g(x) + x = 0$, of which the solution that vanishes at $x = 0$ is

$$g(x) = \tfrac{1}{2} - \sqrt{\tfrac{1}{4} - x}$$

To recover the Catalan numbers, differentiate $g(x)$ n times at $x = 0$:

$$g^{(n)}(0) = 1 \cdot 3 \cdot 5 \cdots (2n-3)2^{n-1}$$

But $2 \cdot 4 \cdot 6 \cdots (2n-2) = 2^{n-1}(n-1)!$, so that

$$g^{(n)}(0) = \frac{[1 \cdot 3 \cdot 5 \cdots (2n-3)][2 \cdot 4 \cdot 6 \cdots (2n-2)]}{(n-1)!}$$

and

$$a_n = \frac{1}{n!} g^{(n)}(0) = \frac{(2n-2)!}{n(n-1)!\,(n-1)!} = \frac{1}{n} C(2n-2, n-1) \equiv C_n$$

Solved Problems

ORDINARY GENERATING FUNCTIONS

3.1 Find the ordinary generating functions for the following sequences:

(*a*) $\langle 1, 1, 1, 1, \ldots \rangle$ (*c*) $\langle 1, 2, 3, 4, \ldots \rangle$ (*e*) $\langle 0, 1, 2, 3, \ldots \rangle$

(*b*) $\langle 1, -1, 1, -1, \ldots \rangle$ (*d*) $\langle 1, -2, 3, -4, \ldots \rangle$

(*a*) $$1 + x + x^2 + x^3 + \cdots = (1-x)^{-1} \equiv f(x)$$

(*b*) $$1 - x + x^2 - x^3 + \cdots = f(-x) = (1+x)^{-1}$$

(*c*) $$1 + 2x + 3x^2 + \cdots = f'(x) = (1-x)^{-2}$$

(*d*) $$1 - 2x + 3x^2 - \cdots = f'(-x) = (1+x)^{-2}$$

(*e*) $$0 + x + 2x^2 + 3x^3 + \cdots = xf'(x) = x(1-x)^{-2}$$

3.2. Using the generating function of Example 1, establish Pascal's identity,

$$C(n+1, r) = C(n, r) + C(n, r-1)$$

The coefficient of x^r in $(1+x)^{n+1}$ is $C(n+1, r)$. But $(1+x)^{n+1} = (1+x)^n$, and the coefficient of x^r in the right-hand side is $C(n, r) + C(n, r-1)$.

3.3 Give the ordinary generating function of the sequence $\langle n(3+5n) \rangle$.

We have $n(3+5n) = 3n + 5n^2$. By Example 10, the respective generating functions of $\langle n \rangle$ and $\langle n^2 \rangle$ are

$$\frac{x}{(1-x)^2} \quad \text{and} \quad \frac{x(1+x)}{(1-x)^3}$$

Hence, by Theorem 3.1(v), the answer is

$$\frac{3x}{(1-x)^2}+\frac{5x(1+x)}{(1-x)^3}=\frac{8x+2x^2}{(1-x)^3}$$

3.4 Find the ordinary generating function of the sequence $\langle C(r+n-1, n-1)\rangle_{r\geq 0}$ (*a*) by a combinatorial argument, and (*b*) by differentiation of the infinite geometric series.

(*a*) From Problem 1.139 it is known that $C(r+n-1, n-1)$ counts the nonnegative integral solutions of

$$u_1+u_2+\cdots+u_n=r$$

Therefore, one can write

$$\frac{1}{(1-x)^n}=\left(\frac{1}{1-x}\right)\left(\frac{1}{1-x}\right)\cdots\left(\frac{1}{1-x}\right)$$

$$=(1+x+x^2+\cdots+x^{u_1}+\cdots)(1+x+x^2+\cdots+x^{u_2}+\cdots)$$

$$\cdots(1+x+x^2+\cdots+x^{u_n}+\cdots)$$

$$=\sum_{r=0}^{\infty}\left(\sum_{\substack{u_1+u_2+\cdots+u_n=r\\ u_i\geq 0}}1\right)x^r=\sum_{r=0}^{\infty}C(r+n-1, n-1)x^r$$

(*b*) Differentiate

$$\frac{1}{1-x}=\sum_{k=0}^{\infty}x^k$$

$n-1$ times to obtain

$$\frac{(n-1)!}{(1-x)^n}=\sum_{k=n-1}^{\infty}k(k-1)\cdots(k-n+2)x^{k-n+1}$$

$$=\sum_{r=0}^{\infty}(r+n-1)(r+n-2)\cdots(r+1)x^r=\sum_{r=0}^{\infty}\frac{(r+n-1)!}{r}x^r$$

Division of both sides by $(n-1)!$ reproduces the result of (*a*).

3.5 Find the sequences corresponding to the ordinary generating functions (*a*) $(3+x)^3$, (*b*) $3x^3+e^{2x}$, and (*c*) $2x^2(1-x)^{-1}$.

(*a*) $(3+x)^3=27+27x+9x^2+x^3$; the sequence is $\langle 27, 27, 9, 1, 0, 0, 0, \ldots\rangle$.

(*b*) $$3x^3+e^{2x}=1+2x+\frac{2^2}{2!}x^2+\left(3+\frac{2^3}{3!}\right)x^3+\frac{2^4}{4!}x^4+\frac{2^5}{5!}x^5+\cdots$$

The sequence is $\langle 1, 2, 2^2/2!, 2^3/3!+3, 2^4/4!, \ldots\rangle$.

(*c*) $2x^2(1-x)^{-1}=2x^2(1+x+x^2+x^3+\cdots)$; the sequence is $\langle 0, 0, 2, 2, 2, \ldots\rangle$.

3.6 Find the coefficients of x^{27} in (*a*) $(x^4+x^5+x^6+\cdots)^5$ and (*b*) $(x^4+2x^5+3x^6+\cdots)^5$.

(*a*) Since $(x^4+x^5+\cdots)^5=x^{20}(1-x)^{-5}$, what is required is the coefficient of x^7 in $(1-x)^{-5}$. By Problem 3.4, this is $C(11, 4)$.

(*b*) Since $(x^4+2x^5+3x^6+\cdots)^5=x^{20}[(1-x)^{-2}]^5=x^{20}(1-x)^{-10}$, we require the coefficient of x^7 in $(1-x)^{-10}$, which is $C(16, 9)$.

3.7 Repeat Problem 3.4 for the sequence $\langle C(r-1, n-1)\rangle$ $(r \geq 0)$. (The first n terms are 0s.)

(*a*) By Problem 1.142, $C(r-1, n-1)$ counts the solutions in *positive* integers of

$$u_1 + u_2 + \cdots + u_n = r$$

Therefore, one can write

$$\left(\frac{x}{1-x}\right)^n = \left(\frac{x}{1-x}\right)\left(\frac{x}{1-x}\right)\cdots\left(\frac{x}{1-x}\right)$$

$$= (x + x^2 + \cdots + x^{u_1} + \cdots)(x + x^2 + \cdots + x^{u_2} + \cdots)$$

$$\cdots(x + x^2 + \cdots + x^{u_n} + \cdots)$$

$$= \sum_{r=0}^{\infty}\left(\sum_{\substack{u_1+u_2+\cdots+u_n=r \\ u_i \geq 1}} 1\right)x^r = \sum_{r=0}^{\infty} C(r-1, n-1)x^r$$

(*b*) From Problem 3.4(*b*),

$$\left(\frac{x}{1-x}\right)^n = x^n \frac{1}{(1-x)^n} = x^n \sum_{s=0}^{\infty} C(s+n-1, n-1)x^s$$

$$= \sum_{s=0}^{\infty} C(s+n-1, n-1)x^{s+n} = \sum_{r=n}^{\infty} C(r-1, n-1)x^r$$

3.8 (*Restricted Partitions*) Given a collection K of n distinct positive integers, $\alpha_1 < \alpha_2 < \cdots < \alpha_n$, and an arbitrary positive integer r, let

$f_n(r) \equiv$ number of partitions of r into parts selected (with replacement) from K

Determine the ordinary generating function of $\langle f_n(0) \equiv 1, f_n(1), f_n(2), \ldots\rangle$.

Here we want to count the solutions in nonnegative integers of

$$\alpha_1 u_1 + \alpha_2 u_2 + \cdots + \alpha_n u_n = r$$

and so we write

$$\prod_{i=1}^{n} \frac{1}{1-x^{\alpha_i}} = [1 + x^{\alpha_1} + (x^{\alpha_1})^2 + \cdots + (x^{\alpha_1})^{u_1} + \cdots$$

$$\times [1 + x^{\alpha_2} + (x^{\alpha_2})^2 + \cdots + (x^{\alpha_2})^{u_2} + \cdots]$$

$$\times \cdots\cdots\cdots\cdots\cdots\cdots\cdots\cdots\cdots\cdots\cdots$$

$$\times [1 + x^{\alpha_n} + (x^{\alpha_n})^2 + \cdots + (x^{\alpha_n})^{u_n} + \cdots]$$

$$= \sum_{r=0}^{\infty} f_n(r)x^r$$

3.9 If throwing a die 5 times constitutes a *trial*, with the 5 throws considered distinguishable, find the number of trials that produce a total of 12 or fewer dots.

Let

$a_r \equiv$ number of trials that produce r dots

$A_r \equiv$ number of trials that produce at most r dots

Clearly, $\langle a_r\rangle$ has the generating function

$$(x^1 + x^2 + \cdots + x^6)^5 = \left(\frac{x - x^7}{1 - x}\right)^5$$

Hence, by Theorem 3.1(*ii*), $\langle A_r \rangle$ has the generating function

$$(1-x)^{-1}\left(\frac{x - x^7}{1 - x}\right)^5 = (x - x^7)^5(1-x)^{-6}$$

$$= (x^5 - 5x^{11} + 10x^{17} - \cdots) \sum_{r=0}^{\infty} C(r+5, 5)x^r$$

The coefficient of x^{12} on the right is

$$A_{12} = 1 \cdot C(12, 5) - 5 \cdot C(6, 5) = 762$$

3.10 In an *experiment*, 4 differently colored dice are thrown simultaneously, and the numbers are added. Find the numbers of distinct experiments such that (*a*) the total is 18 and (*b*) the total is 18 and the green die shows an even number.

(*a*) The answer is the coefficient of x^{18} in the generating function

$$(x^1 + x^2 + \cdots + x^6)^4 = (x - x^7)^4(1 - x^{-4})$$

$$= (x^4 - 4x^{10} + 6x^{16} - \cdots) \sum_{r=0}^{\infty} C(r+3, 3)x^r$$

which is seen to be

$$1 \cdot C(17, 3) - 4 \cdot C(11, 3) + 6 \cdot C(5, 3) = 80$$

(*b*) Now the generating function is

$$(x^2 + x^4 + x^6)(x^1 + x^2 + \cdots + x^6)^3 = (x^2 + x^4 + x^6)(x^3 - 3x^9 + 3x^{15} - x^{21}) \sum_{r=0}^{\infty} C(r+2, 2)x$$

$$= (x^5 + x^7 + x^9 - 3x^{11} - 3x^{13} - 3x^{15} + 3x^{17} + \cdots) \sum_{r=0}^{\infty} C(r+2, 2)x$$

in which the coefficient of x^{18} is

$$1 \cdot C(15, 2) + 1 \cdot C(13, 2) + 1 \cdot C(11, 2) - 3 \cdot C(9, 2) - 3 \cdot C(7, 2) - 3 \cdot C(5, 2) + 3 \cdot C(3, 2)$$

or 46.

3.11 Use the generating-function method (*a*) to count the distinct binary solutions of

$$u_1 + u_2 + \cdots + u_n = r$$

(*b*) to establish the pigeonhole principle (Section 1.3).

(*a*) The generating function (on r) is

$$(1+x)^n = \sum_{r=0}^{n} C(n, r)x^r$$

Thus there are $C(n, r)$ solutions ($= 0$ for $r > n$).

(*b*) The function $(1+x)^n$ is also the generating function corresponding to the problem of distributing r identical pigeons among n distinct pigeonholes so that each hole receives fewer than 2 pigeons. The coefficient of x^{n+1} in the generating function is zero. Hence, when $n + 1$ pigeons are distributed, some hole receives at least 2 pigeons.

3.12 A box contains many identical red, blue, white, and green marbles. Find the ordinary generating function corresponding to the problem of finding the number of ways of choosing r marbles from the box such that the sample does not have more than 2 red, more than 3 blue, more than 4 white, and more than 5 green.

The generating function is

$$(1 + x + x^2)(1 + x + x^2 + x^3)(1 + x + \cdots + x^4)(1 + x + \cdots + x^5)$$
$$= (1 - x^3)(1 - x^4)(1 - x^5)(1 - x^6)(1 - x)^{-4}$$

3.13 Find the number of ways of forming a committee of 9 people drawn from 3 different parties so that no party has an absolute majority in the committee.

If any party is excluded, one of the other parties will have an absolute majority. So there must be at least 1 person from each party. And no party can have more than 4 representatives in the committee. Thus the generating function is

$$f(x) = (x + x^2 + x^3 + x^4)^3 = (x^3 - 3x^7 + 3x^{11} - x^{15}) \sum_{r=0}^{\infty} C(r + 2, 2)x^r$$

The answer is the coefficient of x^9 in $f(x)$, which is

$$1 \cdot C(8, 2) - 3 \cdot C(4, 2) = 10$$

3.14 Let $f(x) = (1 + x + \cdots + x^n)^3$ and $g(x) = (1 + x + \cdots + x^{n-1})^3$. Use a combinatorial argument to show that the coefficient of x^{2n+1} in $f(x)$ is equal to the coefficient of x^{2n-2} in $g(x)$.

Consider the equation $a + b + c = 2n + 1$, where the 3 variables are nonnegative integers at most equal to n. The number of solutions is the coefficient of x^{2n+1} in $f(x)$. But no variable in this equation is 0, for then 1 of the remaining 2 variables would have to exceed n. So the number of solutions is also equal to the coefficient of x^{2n+1} in

$$(x + x^2 + \cdots + x^n)^3 = x^3 g(x)$$

which is the coefficient of x^{2n-2} in $g(x)$.

3.15 Find the number of ways of changing a dollar bill into coins (pennies, nickels, dimes, quarters, and half dollars).

Required is the number of partitions of 100 when the parts are restricted to the numbers $\alpha_1 = 1$, $\alpha_2 = 5$, $\alpha_3 = 10$, $\alpha_4 = 25$, and $\alpha_5 = 50$. By Problem 3.8 the generating function is

$$(1 - x^1)^{-1}(1 - x^5)^{-1}(1 - x^{10})^{-1}(1 - x^{25})^{-1}(1 - x^{50})^{-1}$$

After a tedious computation one finds that the coefficient of x^{100} is 292.

3.16 For a fixed real number k, find the ordinary generating functions of the sequences (indexed by n) (*a*) $\langle k^n \rangle$; (*b*) $\langle nk^n \rangle$; (*c*) $\langle c_n \rangle$, where $c_n = 1 + k + 2k^2 + 3k^3 + \cdots + nk^n$; and (*d*) $\langle k^n/n! \rangle$.

(*a*) $f(x) \equiv 1/(1 - kx)$. (*b*) $xf'(x)$, by Theorem 3.1(*iii*). (*c*) $(1 - x)^{-1}xf'(x)$, by Theorem 3.1(*ii*). (*d*) e^{kx}.

3.17 Find an ordinary generating function that solves the problem of finding the number of positive 5-digit integers with digit sum r.

The leading digit is at least 1 and most 9; the other 4 digits are nonnegative and at most 9. Hence the generating function is

$$(x+x^2+\cdots+x^9)(1+x+x^2+\cdots+x^9)^4 = x(1-x^9)(1-x^{10})(1-x)^{-2}$$

3.18 Given a positive integer k, use the generating-function method to find the number of solutions in nonnegative integers of $u_1+u_2+\cdots+u_k=n$, when (*a*) the first 2 variables are at most 2; and (*b*) the sum of the first two variables is at most 2.

(*a*)
$$f(x) = (1+x+x^2)^2(1+x+x^2+\cdots)^{k-2} = (1-x^3)^2(1-x)^{-k}$$

(*b*) **Case 1.** Both variables are 0, and the sum is 0.

Case 2. One of them is 1 and the other is 0, and the sum is 1.

Case 3. Both are 1 or one of them is 2 and the other is 0, and the sum is 2.

These cases occur in 1, 2, and 3 ways, respectively. So the generating function is

$$(1+2x+3x^2)(1+x+x^2+\cdots)^{k-2} = (1+2x+3x^2)(1-x)^{-k+2}$$

3.19 Rework Problem 1.144 by means of a generating function with respect to n.

For fixed r,

$$a_n \equiv \text{number of } r\text{-combinations of } \{1,2,3,\ldots,n\} \text{ with the desired property}$$

Then, from the constrained equation derived in Problem 1.144,

$$(1+x+x^2+\cdots)(x+x^2+\cdots)^{r-1}(1+x+x^2+\cdots) = \sum_{n-0}^{\infty} a_n x^{n-r}$$

or

$$x^{r-1}(1-x)^{-(r+1)} = x^{-r}\sum_{n=0}^{\infty} a_n x^n$$

or, finally,

$$\sum_{n=0}^{\infty} a_n x^n = x^{2r-1}(1-x)^{-(r+1)} \qquad (i)$$

3.20 Verify that the correct generating function was obtained in Problem 3.19.

We know that

$$(1-x)^{-(r+1)} = \sum_{j=0}^{\infty} C(j+r,r)x^j \qquad (*)$$

Hence (*i*) of Problem 3.19 gives [in (*) take $j=n-2r+1$]

$$a_n = C(n-r+1,r)$$

in agreement with Problem 1.144.

3.21 (*Uniqueness of Base-b Representation*) If b is an integer greater than 1, prove by means of a generating function that an arbitrary positive integer r can be written as

$$r = r_0b^0 + r_1b^1 + r_2b^2 + \cdots \qquad (0 \le r_i \le b-1) \qquad (*)$$

in 1 and only 1 way.

The generating function for the number of solution vectors $(r_0, r_1, r_2, \ldots)$ of (*) is obviously

$$[1 + x^{1(1)} + x^{2(1)} + \cdots + x^{(b-1)(1)}][1 + x^{1(b)} + x^{2(b)} + \cdots + x^{(b-1)(b)}]$$

$$\times [1 + x^{1(b^2)} + x^{2(b^2)} + \cdots + x^{(b-1)(b^2)}] \cdots$$

$$= \frac{1 - x^b}{1 - x} \frac{1 - x^{b^2}}{1 - x^b} \frac{1 - x^{b^3}}{1 - x^{b^2}} \cdots$$

$$= \frac{1}{1 - x} = 1 + x + x^2 + x^3 \cdots$$

Each coefficient in the generating function is 1; i.e., any r has a unique base-b representation.

PARTITIONS OF INTEGERS AND THEIR GENERATING FUNCTIONS

3.22 Prove Theorem 3.2 without drawing the star diagram.

The system

$$1u_1 + 2u_2 + 3u_3 + \cdots + nu_n = r \qquad (u_i \text{ a nonnegative integer})$$

which by definition has precisely $p_n(r)$ solutions, is taken by the substitution

$$\begin{aligned} u_n &= w_1 \\ u_{n-1} &= w_2 - w_1 \\ u_{n-2} &= w_3 - w_2 \\ &\cdots\cdots\cdots \\ u_1 &= w_n - w_{n-1} \end{aligned} \qquad (*)$$

into the system

$$w_1 + w_2 + w_3 + \cdots + w_n = r$$

$$w_1 \le w_2 \le w_3 \le \cdots \le w_n$$

$$(w_i \text{ a nonnegative integer})$$

This latter system has precisely $q_n(r)$ solutions. But the mapping (*) is obviously bijective, so that $p_n(r) = q_n(r)$.

3.23 Establish the recurrence relation

$$q_n(r) = q_{n-1}(r) + q_n(r - n)$$

[also satisfied by $p_n(r)$] (*a*) by solving a distribution problem; and (*b*) by use of Theorem 3.3.

(*a*) Imagine you are given a heap of r identical 1s and a row of n identical boxes. Partitioning r into *exactly n* parts—which, by definition, can be accomplished in $q_n(r) - q_{n-1}(r)$ ways—is tantamount to first putting a 1 into each box (*1 way*) and then arbitrarily distributing the remaining $(r - n)$ 1s among the n boxes [$q_n(r - n)$ *ways*]. Thus, by the product rule,

$$q_n(r) - q_{n-1}(r) = 1 \cdot q_n(r - n)$$

(*b*) From Theorem 3.3, $(1 - x^n)g_n(x) = g_{n-1}(x)$. Equating coefficients of x^r,

$$q_n(r) - q_n(r-n) = q_{n-1}(r)$$

3.24 Let

$$p(r, n) \equiv \text{number of partitions of } r \text{ with largest part } n$$
$$q(r, n) = \text{number of partitions of } r \text{ into exactly } n \text{ parts}$$

(*a*) Prove that, for all r and n, $p(r, n) = q(r, n)$. (*b*) Determine the ordinary generating function (on r) of either sequence.

(*a*) $$p^*(r, n) = p_n(r) - p_{n-1}(r) = q_n(r) - q_{n-1}(r) = q(r, n)$$

(*b*) In the notation of Theorem 3.3,

$$f_n(x) \equiv \sum_{r=0}^{\infty} p(r, n)x^r = \sum_{r=0}^{\infty} p_n(r)x^r - \sum_{r=0}^{\infty} p_{n-1}(r)x^r$$
$$= g_n(x) - g_{n-1}(x) = x^n g_n(x)$$

3.25 Use generating functions to establish (*a*) $p(r, n) = p_n(r-n)$; (*b*) $p(r, n) = p(r-n)$, for $n \geq \lfloor r/2 \rfloor$; and (*c*) $p(2n, n) = p(n)$.

(*a*) By Problem 3.24(*b*), the coefficient of x^r in $f_n(x)$ equals the coefficient of x^{r-n} in $g_n(x)$.

(*b*) By Theorem 3.4,

$$p(r-n) = \text{coefficient of } x^{r-n} \text{ in}$$
$$g(x) = g_n(x)(1 + x^{n+1} + x^{2n+2} + \cdots)(1 + x^{n+2} + x^{2n+4} + \cdots)\cdots$$
$$= (\text{coefficient of } x^{r-n} \text{ in } g_n(x))$$
$$+ (\text{coefficient of } x^{(r-n)-(n+1)} \text{ in } g_n(x))$$
$$+ \cdots$$

Now $n \geq \lfloor r/2 \rfloor$ implies $r - 2n - 1 < 0$, so that the second and all succeeding coefficients on the right-hand side are 0. The first coefficient is just $p_n(r-n) = p(r, n)$.

(*c*) Set $r = 2n$ in (*b*).

3.26 Show that $p(r+n, n) = p(r, 1) + p(r, 2) + \cdots + p(r, n)$.

The left-hand side is $p_n(r)$, by Problem 3.25(*a*). The right-hand side is $p_n(r)$, by definition.

3.27 Let $p^{\#}(r)$ be the number of partitions of r into unequal parts. Obtain the ordinary generating function of $\langle p^{\#}(r) \rangle$.

A given positive integer i appears either 0 times or 1 time among the parts of r; hence

$$(1 + x^1)(1 + x^2)(1 + x^3)\cdots(1 + x^i)\cdots = \sum_{r=0}^{\infty} p^{\#}(r)x^r$$

It is evident that for a particular value of r—say, $r = s$—only the first s factors of the infinite product need be retained.

3.28 Let $p^{\#}(r, \text{EVEN})$ be the number of partitions of r into distinct even parts and let $p^{\#}(r, \text{ODD})$ be the number of partitions of r into distinct odd parts. Find the corresponding generating functions.

$$(1+x^2)(1+x^4)(1+x^6)\cdots \quad \text{and} \quad (1+x)(1+x^3)(1+x^5)\cdots$$

3.29 (*Euler's Theorem*) Let $p(r, \text{ODD})$ be the number of ways of partitioning r into (possibly repeated) odd parts. Show that, for every r, $p(r, \text{ODD}) = p^{\#}(r)$.

The ordinary generating function of $\langle p(r, \text{ODD})\rangle$ is given by

$$\begin{aligned}\frac{1}{(1-x^2)(1-x^3)(1-x^5)\cdots} &= \left(\frac{1-x^2}{1-x^1}\right)\left(\frac{1-x^4}{1-x^2}\right)\left(\frac{1-x^6}{1-x^3}\right)\left(\frac{1-x^8}{1-x^4}\right)\cdots \\ &= (1+x^1)(1+x^2)(1+x^3)(1+x^4)\cdots \\ &= \text{generating function of } \langle p^{\#}(r)\rangle\end{aligned}$$

where the last step uses Problem 3.27.

3.30 Establish the upper bound $p(r) < \exp(3\sqrt{r})$.

If $g(x)$ is the generating function of Theorem 3.4, then for any r and all $0 < x < 1$,

$$p(r)x^r < g(x) \quad \text{or} \quad \log p(r) < \log g(x) - r\log x$$

From the well-known expansion

$$\log\frac{1}{1-u} = u + \frac{u^2}{2} + \frac{u^3}{3} + \cdots \qquad (0 \le u < 1)$$

it follows that

$$\begin{aligned}\log g(x) &= \left(x + \frac{x^2}{2} + \frac{x^3}{3} + \cdots\right) + \left(x^2 + \frac{x^4}{2} + \frac{x^6}{3} + \cdots\right) \\ &\quad + \left(x^3 + \frac{x^6}{2} + \frac{x^9}{3} + \cdots\right) + \cdots \\ &= (x + x^2 + x^3 + \cdots) + \tfrac{1}{2}(x^2 + x^4 + x^6 + \cdots) \\ &\quad + \tfrac{1}{3}(x^3 + x^6 + x^9 + \cdots) + \cdots \\ &= \frac{x}{1-x} + \frac{1}{2}\frac{x^2}{1-x^2} + \frac{1}{3}\frac{x^3}{1-x^3} + \cdots\end{aligned}$$

Now, for $0 < x < 1$ and $k = 1, 2, 3, \ldots,$

$$\begin{aligned}\frac{x^k}{1-x^k} &= \frac{x}{1-x}\,\frac{x^{k-1}}{1+x+x^2+\cdots+x^{k-1}} < \frac{x}{1-x}\,\frac{x^{k-1}}{x^{k-1}+x^{k-1}+x^{k-1}+\cdots+x^{k-1}} \\ &= \frac{x}{1-x}\,\frac{1}{k}\end{aligned}$$

whence
$$\log g(x) < \frac{x}{1-x}\left[1 + \left(\frac{1}{2}\right)^2 + \left(\frac{1}{3}\right)^2 + \cdots\right] = \frac{x}{1-x}\,\frac{\pi^2}{6}$$

Further,
$$-\log x = \log\frac{1}{x} = \int_1^{1/x}\frac{dt}{t} < \int_1^{1/x} dt = \frac{1-x}{x}$$

Therefore we have

$$\log p(r) < \frac{\pi^2}{6}\frac{x}{1-x} + r\frac{1-x}{x} \tag{i}$$

Using standard calculus to minimize the right-hand side of (i) over $0 < x < 1$, one obtains

$$\log p(r) < \frac{2\pi}{\sqrt{6}}\sqrt{r} < 3\sqrt{r}$$

and the proof is complete.

3.31 Establish the lower bound $p(r) \geq 2^q$ for $r \geq 2$, where $q \equiv \lfloor \sqrt{r} \rfloor$.

The bound may be established by inspection for $q = 1, 2$; so assume $q \geq 3$. It is asserted that each nonempty subset S of $X = \{1, 2, 3, \ldots, q\}$ generates a partition of r. In fact, if

$$\sigma(S) \equiv \text{sum of the elements of } S$$

then
$$\sigma(S) \leq \sigma(X) = \frac{q^2+q}{2} \leq \frac{r+\sqrt{r}}{2} < \frac{r+r}{2} = r$$

so that $S \cup \{r - \sigma(S)\}$ is the desired partition of r.

Furthermore, distinct subsets generate distinct partitions. To see that this is so, let

$$S_1 \cup \{r - \sigma(S_1)\} \qquad \text{and} \qquad S_2 \cup \{r - \sigma(S_2)\} \tag{i}$$

be the partitions generated by the distinct k subsets $(1 \leq k \leq q-1)$ S_1 and S_2. For $i = 1, 2$, we have

$$r - \sigma(S_i) \geq q^2 - [\sigma(X) - 1] = q^2 - \left[\frac{q^2+q-2}{2}\right]$$
$$= \frac{q^2 - q + 2}{2} = q + \frac{(q-2)(q-1)}{2}$$

Consequently, for $q \geq 3$,

$$r - \sigma(S_i) > q \tag{ii}$$

If the 2 partitions (i) coincided, and if $\sigma(S_1) = \sigma(S_2)$, then S_1 must coincide with S_2, which is contrary to the hypothesis. On the other hand, if the 2 partitions (i) coincided, and if $\sigma(S_1) \neq \sigma(S_2)$, then $r - \sigma(S_1)$ would have to be an element of S_2, which is ruled out by (ii).

The conclusion is that the number of partitions of r must exceed the number of nonnull subsets of X:

$$p(r) > 2^q - 1 \geq 2^q$$

3.32 The number of partitions of r into n distinct (unequal) parts is denoted by $q^{\#}(r, n)$. Prove that

$$q^{\#}(r, n) = q(r - C(n, 2), n)$$

The following proof is similar to that of Problem 3.22. By definition the system

$$\begin{aligned} u_1 + u_2 + \cdots + u_n &= r \\ 0 < u_1 < u_2 < \cdots &< u_n \end{aligned} \tag{*}$$

has precisely $q^{\#}(r, n)$ solutions in integers u_i. Under the bijective transformation

$$\begin{aligned}
u_1 &= w_1 \\
u_2 &= w_2 + 1 \\
u_3 &= w_3 + 2 \\
&\cdots\cdots\cdots\cdots \\
u_n &= w_n + (n-1)
\end{aligned}$$

(*) goes over into

$$\begin{gathered}
w_1 + w_2 + \cdots + w_n = r - C(n, 2) \\
0 < w_1 \leq w_2 \leq w_3 \leq \cdots \leq w_n
\end{gathered} \tag{**}$$

But, again by definitions, (**) has exactly $q(r - C(n, 2), n)$ solutions in integers w_i.

3.33 A partition that is its own conjugate (see the proof of Theorem 3.2) is called **self-conjugate**. Show that the number of self-conjugate partitions of r is equal to $p^{\#}(r, \text{ODD})$ (Problem 3.28).

A partition is self-conjugate if and only if its star diagram is symmetric about a diagonal of a square [see Fig. 3-2(*a*)]. But then the diagram may be read as nested L shapes, or elbows [see Fig. 3-2(*b*)], giving a partition into distinct odd parts. This procedure is obviously reversible; hence the correspondence is one-to-one.

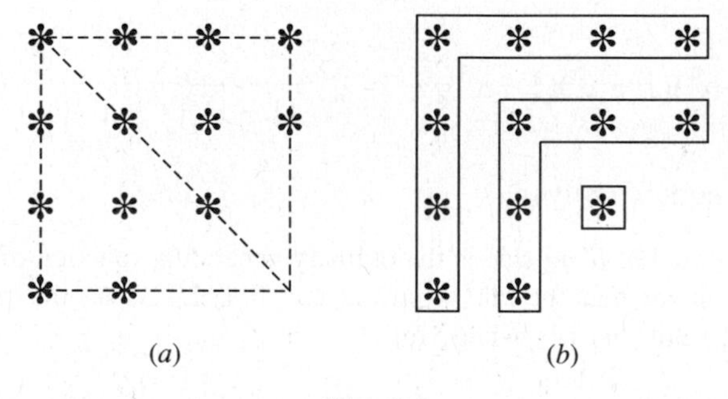

Fig. 3-2

3.34 Show that

$$p^{\#}(r, \text{ODD}) = \sum_{k=1}^{\lfloor\sqrt{r}\rfloor} p_{2k}(r - k^2, \text{EVEN}) = \sum_{k=1}^{\lfloor\sqrt{r}\rfloor} q_k(r - k^2, \text{EVEN})$$

As in Fig. 3-2(*b*), represent a partition of r into distinct odd parts by nested elbows. Let k be the number of parts. Then k is the largest integer such that the diagram contains a $k \times k$ square (called the DURFEE square) having as one corner the asterisk in the first row and first column. Clearly, $1 \leq k \leq \lfloor\sqrt{r}\rfloor$; in Fig. 3-3, which diagrams $23 = 11 + 9 + 3$, one has $k = 3$.

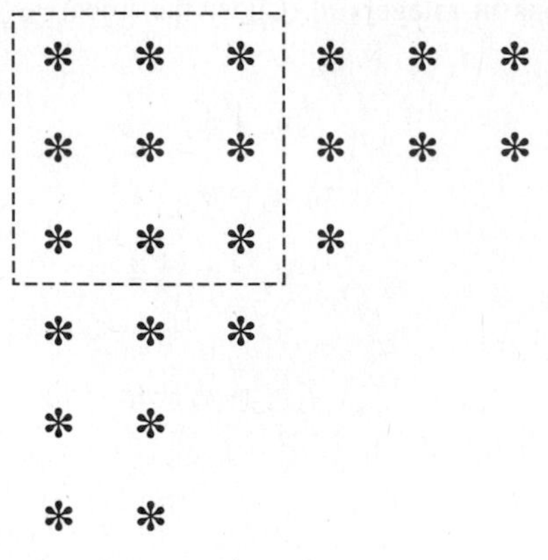

Fig. 3-3

The remaining $r-k^2$ asterisks can be assembled into a partition of $r-k^2$ with all parts even, in two different ways:

(*i*) If the ith part in the partition is the total number of asterisks in the $(k+i)$th row and the $(k+i)$th column, then all parts are less than or equal to $2k$.

(*ii*) If the jth part in the partition is the number of asterisks in the jth **elbow** which lie outside the square, then there are at most k parts.

Since the geometric argument is reversible, we conclude that the number of partitions of r into k distinct odd parts is given by either $p_{2k}(r-k^2, \text{EVEN})$ or $q_k(r-k^2, \text{EVEN})$. Hence, a summation on k yields the required results. Note that, as usual, the sums may be extended from $k=0$ to $k=\infty$, since only null summands are thereby introduced.

3.35 (*Euler's First Identity*) Derive

$$(1+x^1)(1+x^3)(1+x^5)\cdots=\sum_{k=0}^{\infty}\frac{x^{k^2}}{(1-x^2)(1-x^4)(1-x^6)\cdots(1-x^{2k})}$$

(an empty product equals unity).

By Problem 3.28 the left-hand side is the ordinary generating function of $\langle p^{\#}(r, \text{ODD})\rangle$. In view of Problem 3.34 it is enough to prove that the kth summand $(k=0,1,2,\ldots)$ on the right is the generating function of $\langle p_{2k}(r-k^2, \text{EVEN})\rangle_{r\geq 0}$. But this is obvious; for

$$\frac{x^{k^2}}{(1-x^2)(1-x^4)\cdots(1-x^{2k})}=x^{k^2}(1+x^2+x^4+\cdots)(1+x^4+x^8+\cdots)\cdots(1+x^{2k}+x^{4k}+\cdots)$$

$$=x^{k^2}\sum_{s=0}^{\infty}p_{2k}(s, \text{EVEN})x^s=\sum_{s=0}^{\infty}p_{2k}(s, \text{EVEN})x^{s+k^2}$$

$$=\sum_{r=0}^{\infty}p_{2k}(r-k^2, \text{EVEN})x^r$$

3.36 (*Euler's Second Identity*) Show that

$$(1+x^2)(1+x^4)(1+x^6)\cdots=\sum_{k=0}^{\infty}\frac{x^{k(k+1)}}{(1-x^2)(1-x^4)(1-x^6)\cdots(1-x^{2k})}$$

Suppose that we are given a partition of r into k distinct *even* parts. Then subtraction of 1 from each part yields a unique partition of $r - k$ into k distinct *odd* parts. Conversely, addition of 1 ⋯ yields a unique ⋯. By this one-to-one correspondence, and by the result of Problem 3.34,

$$\begin{pmatrix}\text{No. of partitions of } r \text{ into} \\ k \text{ distinct even parts}\end{pmatrix} = \begin{pmatrix}\text{No. of partitions of } r-k \\ \text{into } k \text{ distinct odd parts}\end{pmatrix}$$

$$= p_{2k}((r-k) - k^2, \text{EVEN})$$

$$= p_{2k}(r - k(k+1), \text{EVEN})$$

Therefore, the proof of Euler's second identity reduces to establishing that the ordinary generating function of $\langle p_{2k}(r - k(k+1), \text{EVEN})\rangle_{r\geq 0}$ is just

$$\frac{x^{k(k+1)}}{(1-x^2)(1-x^4)(1-x^6)\cdots(1-x^{2k})}$$

This may be carried out by inspection (compare Problem 3.35).

3.37 Define

$$q^{\#}(r, \text{E}) = \text{number of partitions of } r \text{ into an } \textit{even number} \text{ of unequal parts}$$
$$q^{\#}(r, \text{O}) = \text{number of partitions of } r \text{ into an } \textit{odd number} \text{ of unequal parts}$$

Prove that

$$(1-x)(1-x^2)(1-x^3)\cdots = \sum_{r=0}^{\infty} [q^{\#}(r, \text{E}) - q^{\#}(r, \text{O})]x^r$$

Because

$$(1-x)(1-x^2)(1-x^3)\cdots = [1 + (-1)x][1 + (-1)x^2][1 + (-1)x^3]\cdots$$

any partition of r into an even number, e, of unequal parts will contribute $(-1)^e = +1$ to the coefficient of x^r in the infinite product. Analogously, any partition of r into an odd number, o, of unequal parts will contribute $(-1)^o = -1$. Therefore, the coefficient of x^r is

$$q^{\#}(r, \text{E})(+1) + q^{\#}(r, \text{O})(-1) = q^{\#}(r, \text{E}) - q^{\#}(r, \text{O})$$

as asserted.

EXPONENTIAL GENERATING FUNCTIONS

3.38 Find the exponential generating functions of

(*a*) $\langle 1, 2, 3, 0, 0, 0, \ldots\rangle$ (*c*) $\langle 1, a, a^2, a^3, \ldots\rangle$

(*b*) $\langle 0, 0, 1, 1, 1, \ldots\rangle$ (*d*) $\langle 0, 1, 2a, 3a^2, 4a^3, \ldots\rangle$

(*a*) $1 + 2x + 3\dfrac{x^2}{2!}$ (*c*) $1 + ax + a^2\dfrac{x^2}{2!} + a^3\dfrac{x^3}{3!} + \cdots = e^{ax}$

(*b*) $\dfrac{x^2}{2!} + \dfrac{x^3}{3!} + \cdots = e^x - 1 - x$ (*d*) $x + 2a\dfrac{x^2}{2!} + 3a^2\dfrac{x^3}{3!} + \cdots = xe^{ax}$

3.39 Find the exponential generating function of $\langle a_r\rangle$, where a_r is the number of r sequences of the set $E = \{e_1, e_2, \ldots, e_n\}$.

Looking at the product

$$G(x) \equiv \left(1 + x + \frac{x^2}{2!} + \cdots + \frac{x^{i_1}}{i_1!} + \cdots\right)\left(1 + x + \frac{x^2}{2!} + \cdots + \frac{x^{i_2}}{i_2!} + \cdots\right)\cdots\left(1 + x + \frac{x^2}{2!} + \cdots + \frac{x^{i_n}}{i_n!} + \cdots\right)$$

one sees that the r *sample* of E consisting of i_1 e_1's, i_2 e_2's, ..., i_n e_n's—where $i_1, i_2, \ldots, i_n$ are nonnegative integers with sum r—contributes

$$\frac{r!}{i_1!\,i_2!\cdots i_n!} = P(r; i_1, i_2, \ldots, i_n)$$

to the coefficient of $x^r/r!$ in $G(x)$. But this contribution is, by Theorem 2.1, precisely the number of r *sequences* of E generated by permutation of the given r sample. Hence the total coefficient of $x^r/r!$ is just a_r. That is, $G(x) = (e^x)^n$ is the desired exponential generating function.

3.40 For any positive integer n, prove combinatorially that $(e^x)^n = e^{nx}$.

In Problem 3.39 the exponential generating function of the integers a_r was found to be $(e^x)^n$. On the other hand, Section 2.2 shows that $a_r = n^r$; and so the exponential generating function of the a_r is

$$\sum_{r=0}^{\infty} n^r \frac{x^r}{r!} = e^{nx}$$

3.41 If a leading digit of 0 is permitted, find the numbers of r-digit binary numbers that can be formed using (*a*) an even number of 0s and an even number of 1s; (*b*) an odd number of 0s and an odd number of 1s.

Here we are counting r sequences of the set $\{0, 1\}$ that obey certain restrictions.

(*a*) By analogy with Problem 3.39, the exponential generating function is

$$F_e(x) = \left(1 + \frac{x^2}{2!} + \frac{x^4}{4!} + \cdots\right)^2 = \left(\frac{e^x + e^{-x}}{2}\right)^2 = \frac{1}{4}(e^{2x} + e^{-2x} + 2)$$

The coefficient of $x^r/r!$ in $F_e(x)$ is 2^{r-1} if r is even, and (of course) 0 if r is odd.

(*b*)
$$F_o(x) = \left(x + \frac{x^3}{3!} + \frac{x^5}{5!} + \cdots\right)^2 = \left(\frac{e^x - e^{-x}}{2}\right)^2 = F_e(x) - 1$$

Thus the answer is the same as in (*a*).

3.42 Find the number of r-letter sequences that can be formed using the letters P, Q, R, and S such that in each sequence there are an odd number of P's and an even number of Q's.

The answer is the coefficient of $x^r/r!$ in

$$\left(x + \frac{x^3}{3!} + \frac{x^5}{5!} + \cdots\right)\left(1 + \frac{x^2}{2!} + \frac{x^4}{4!} + \cdots\right)(e^x)(e^x)$$
$$= \left(\frac{e^x - e^{-x}}{2}\right)\left(\frac{e^x + e^{-x}}{2}\right)e^{2x} = \frac{1}{4}(e^{4x} - 1)$$

This coefficient is 4^{r-1}.

3.43 Obtain the result of Problem 1.146 via an exponential generating function.

There exists a one-to-one correspondence between the n sequences of an m set, in each of which every element of the m set appears at least once, and the onto mapings from an n set (of positions in an n sequence) to the above m set. Now, the exponential generating function for these n sequences is clearly

$$\left(x + \frac{x^2}{2!} + \frac{x^3}{3!} + \cdots\right)^m = (e^x - 1)^m$$

which function is also, by Example 3, the exponential generating function of the integers $m!\,S(n, m)$. Hence there are exactly $m!\,S(n, m)$ onto mappings from an n set to an m set.

3.44 The exponential generating function for the Bell numbers was found in Problems 2.9 and 2.10 to be

$$e^{e^x - 1}$$

Check this result by use of the exponential generating function for the Stirling numbers of the second kind.

By the definitions of the two kinds of numbers (see Problems 1.146 and 1.161),

$$B_n = \sum_{m=0}^{\infty} S(n, m) \qquad [B_0 = S(0, 0) = 1]$$

(If $n \geq 1$, only n of the summands are nonzero.) Because the (exponential) generating function (g.f.) of a sum is the sum of the generating functions, Example 3 gives:

$$\text{Exponential g.f. of } \langle B_n \rangle = \sum_{m=0}^{\infty} \frac{1}{m!}(e^x - 1)^m = e^{e^x - 1}$$

3.45 The sequence of **Bernoulli numbers**, $\langle b_n \rangle_{n \geq 0}$, has the exponential generating function

$$\frac{x}{e^x - 1}$$

Show that (*a*) $b_3 = b_5 = b_7 = \cdots = 0$; and (*b*)

$$b_n = \sum_{m=0}^{\infty} \frac{(-1)^m}{m + 1} m!\,S(n, m)$$

(*a*)

$$b_1 \frac{x^1}{1!} + b_3 \frac{x^3}{3!} + b_5 \frac{x^5}{5!} + \cdots = \frac{1}{2}\left(\frac{x}{e^x - 1} - \frac{-x}{e^{-x} - 1}\right) = -\frac{1}{2}x$$

whence $b_1 = -\frac{1}{2}$ and $b_3 = b_5 = \cdots = 0$.

(*b*) Two sequences are identical if and only if they have the same exponential generating function. By Example 3, the exponential g.f. of the numbers

$$\sum_{m=0}^{\infty} \frac{(-1)^m}{m + 1} m!\,S(n, m)$$

is given by

$$\begin{aligned}
\sum_{m=0}^{\infty} \frac{(-1)^m}{m + 1}(e^x - 1)^m &= \frac{1}{1 - e^x} \sum_{m=0}^{\infty} \frac{(1 - e^x)^{m+1}}{m + 1} \\
&= \frac{1}{1 - e^x} \int_0^{1 - e^x} \frac{1}{1 - u}\,du \\
&= \frac{1}{1 - e^x}(-x) = \frac{x}{e^x - 1}
\end{aligned}$$

3.46 Find the number of ways of distributing 10 distinguishable books among 4 distinguishable shelves so that each shelf gets at least 2 and at most 7 books.

This is a problem in restricted sequences, like Problems 3.41 and 3.42. Here we want to count the 10-sequences of a 4 set (the ith term of a sequence is the shelf to which the ith book is assigned) that can exist under the given limitations. The appropriate exponential generating function is

$$G(x) = \left(\frac{x^2}{2!} + \frac{x^3}{3!} + \cdots + \frac{x^7}{7!}\right)^4 = x^8\left(\frac{1}{2!} + \frac{x}{3!} + \frac{x^2}{4!} + \cdots\right)^4$$

and the answer is the coefficient of $x^{10}/10!$ in $G(x)$, which is $10!/16 = 226{,}800$.

3.47 Derive the linear recursion relation for the Bell numbers [see Problem 1.161(*b*)] from the exponential generating function.

Differentiation of

$$e^{e^x - 1} = \sum_{r=0}^{\infty} B_r \frac{x^r}{r!}$$

gives

$$(e^{e^x - 1})(e^x) = \sum_{r=0}^{\infty} B_r \frac{x^{r-1}}{(r-1)!}$$

On the right the coefficient of $x^r/r!$ is B_{r+1}; on the left it is—by Theorem 3.1(*iv*)—

$$\sum_{i=0}^{r} C(r, i)B_i$$

Therefore,

$$B_{r+1} = \sum_{i=0}^{r} C(r, i)B_i \qquad \text{or} \qquad B_r = \sum_{i=0}^{r-1} C(r-1, i)B_i$$

3.48 (*Dobinski's Equality*) Prove that

$$B_n = \begin{cases} 1 & n = 0 \\ e^{-1} \sum_{k=1}^{\infty} k^n/k! & n = 1, 2, 3, \ldots \end{cases}$$

The exponential generating function of the numbers on the right of the asserted equality is

$$\begin{aligned} 1 + e^{-1} \sum_{n=1}^{\infty} \left(\sum_{k=1}^{\infty} \frac{k^n}{k!}\right) \frac{x^n}{n!} &= 1 + e^{-1} \sum_{k=1}^{\infty} \frac{1}{k!} \sum_{n=1}^{\infty} \frac{(kx)^n}{n!} \\ &= 1 + e^{-1} \sum_{k=1}^{\infty} \frac{1}{k!} [(e^x)^k - 1] \\ &= 1 + e^{-1}(e^{e^x} - e) = e^{e^x - 1} \end{aligned}$$

which is just the exponential generating function of $\langle B_n \rangle$.

3.49 Obtain a linear recurrence relation for the Bernoulli numbers (Problem 3.45).

By Problem 3.45,

$$x = (e^x - 1) \sum_{n=0}^{\infty} b_n \frac{x^n}{n!}$$

The left-hand side is the exponential generating function of $\langle 0, 1, 0, 0, 0, \ldots \rangle$. Since $e^x - 1$ generates $\langle 0, 1, 1, 1, \ldots \rangle$, Theorem 3.1(*iv*) tells us that the right-hand side generates

$$\left\langle \sum_{i=1}^{n} C(n, i) b_{n-i} \right\rangle$$

Consequently, $b_0 = 1$ and, for $n \geq 2$,

$$\sum_{i=1}^{n} C(n, i) b_{n-i} = 0 \qquad \text{or} \qquad b_{n-1} = -\frac{1}{n} \sum_{i=2}^{n} C(n, i) b_{n-i}$$

3.50 For every positive integer n let

$$s_0(n) = n \qquad \text{and} \qquad s_m(n) = 1^m + 2^m + 3^m + \cdots + (n-1)^m$$

where m is a positive integer. Obtain the exponential generating function of the sequence $\langle s_m(n) \rangle_{m \geq 0}$.

By linearity,

$$\begin{aligned} \text{g.f. of } \langle s_m(k+1) \rangle_{m \geq 0} - \text{g.f. of } \langle s_m(k) \rangle_{m \geq 0} &= \text{g.f. of } \langle s_m(k+1) - s_m(k) \rangle_{m \geq c} \\ &= \text{g.f. of } \langle k^m \rangle_{m \geq 0} = e^{kx} = (e^x) \end{aligned}$$

This simple recurrence relation is solved by summation $[s_m(0) \equiv 0]$:

$$\text{g.f. of } \langle s_m(n) \rangle_{m \geq 0} = \sum_{k=0}^{n-1} (e^x)^k = \frac{e^{nx} - 1}{e^x - 1}$$

3.51 The sum of the mth powers of the first $n - 1$ positive integers may be expressed in terms of the first $m + 1$ powers of n, as follows:

$$s_m(n) = \frac{1}{m+1} \sum_{i=0}^{m} C(m+1, i) b_i n^{m+1-i}$$

Prove this.

By Problem 3.50 the exponential generating function of $\langle s_m(n) \rangle_{m \geq 0}$ is

$$\frac{e^{nx} - 1}{e^x - 1} = \left(\frac{x}{e^x - 1} \right) \left(\frac{e^{nx} - 1}{x} \right) \tag{*}$$

THe first factor on the right of (*) generates $\langle b_m \rangle_{m \geq 0}$; the second factor generates $\langle n^{m+1}/(m+1) \rangle_{m \geq 0}$. Hence, by Theorem 3.1(*iv*),

$$\begin{aligned} s_m(n) &= \sum_{i=0}^{m} C(m, i) b_i \frac{n^{m+1-i}}{m+1-i} \\ &= \frac{1}{m+1} \sum_{i=0}^{m} C(m+1, i) b_i n^{m+1-i} \end{aligned}$$

3.52 Use the formula of Problem 3.51 to check the result of Example 9.

The recurrence relation of Problem 3.49 gives $b_0 = 1$,

$$b_1 = -\tfrac{1}{2} b_0 = -\tfrac{1}{2} \qquad \text{and} \qquad b_2 = -\tfrac{1}{3}(3b_1 + b_0) = \tfrac{1}{6}$$

so that

$$\begin{aligned} s_2(n+1) &= \tfrac{1}{3} \sum_{i=0}^{2} C(3,i)b_i(n+1)^{3-i} \\ &= \tfrac{1}{3}\left[(n+1)^3 - \tfrac{3}{2}(n+1)^2 + \tfrac{1}{2}(n+1)\right] \\ &= \frac{(n+1)n(2n+1)}{6} \end{aligned}$$

3.53 The **Bernoulli polynomial** of degree m is defined by

$$\mathscr{B}_m(t) = \sum_{i=0}^{m} C(m,i)b_i t^{m-i}$$

(*a*) Show that $s_{m-1}(n) = [\mathscr{B}_m(n) - \mathscr{B}_m(0)]/m$.

(*b*) Obtain the exponential generating function (in the variable x) of the sequence $\langle \mathscr{B}_m(t)\rangle_{m\geq 0}$.

(*a*) By Problem 3.51,

$$\begin{aligned} s_{m-1}(n) &= \frac{1}{m}\left[\sum_{i=0}^{m} C(m,i)b_i n^{m-i} - C(m,m)b_m n^{m-m}\right] \\ &= \frac{1}{m}[\mathscr{B}_m(n) - \mathscr{B}_m(0)] \end{aligned}$$

(*b*) The sequences $\langle b_m\rangle_{m\geq 0}$ and $\langle t^m\rangle_{m\geq 0}$ have respective exponential generating functions

$$\frac{x}{e^x - 1} \qquad \text{and} \qquad e^{tx}$$

Then, by Theorem 3.1(*iv*), the binomial convolution

$$\left\langle \sum_{i=0}^{m} C(m,i)b_i t^{m-i} \right\rangle_{m\geq 0} = \langle \mathscr{B}_m(t)\rangle_{m\geq 0}$$

must have the product $xe^{tx}/(e^x - 1)$ as its exponential generating function.

RECURRENCE RELATIONS AND ASSOCIATED GENERATING FUNCTIONS

3.54 A bank offers r percent interest. If a_n is the amount in deposit for a customer at the end of n years, find recurrence relations for $\langle a_n\rangle$ if (*a*) the interest is simple, and (*b*) the interest is compounded annually.

(*a*) Suppose a_0 is the initial deposit. At the end of each year ra_0 is added to the balance at the beginning of the year. Thus $a_{n+1} = a_n + ra_0$ is the recurrence relation.

(*b*) Now ra_n is added to a_n, so that $a_{n+1} = (1 + r)a_n$.

3.55 There are n lines drawn in a plane such that no 2 lines are parallel and no 3 lines are concurrent. If the plane is thereby divided into $f(n)$ regions, find a recurrence relation for $\langle f(n)\rangle$.

The number of regions formed by $n-1$ lines is $f(n-1)$, and by hypothesis the nth line meets the other $n-1$ lines in $n-1$ distinct points. So these $n-1$ lines divide the nth line into n parts. Each such part divides a former region into 2 subregions. Thus the nth line creates n more regions, and one has

$$f(n) = f(n-1) + n \qquad \text{with } f(1) = 2$$

3.56 There are n circles drawn in a plane such that each circle intersects every other circle in exactly 2 points and no 3 circles meet in the same point. If these n circles create $f(n)$ regions, find a recurrence relation for $\langle f(n) \rangle$.

The nth circle meets the other $n-1$ circles in $2(n-1)$ points, creating $2(n-1)$ more regions. Thus the relation is

$$f(n) = f(n-1) + 2(n-1) \qquad \text{with} \qquad f(1) = 2$$

3.57 (*The Tower of Hanoi*) There are n circular disks of decreasing radii, each with a hole at the center, and 3 pegs (marked A, B, and C) fixed vertically on a table so that the distance between the feet of any 2 of them is greater than the diameter of the largest disk. Initially these disks are slipped onto peg A with the largest disk at the bottom and the others on top of this, in decreasing order of size. A *legal move* is defined as the transfer of the top disk from 1 of the 3 pegs to the top of the stack on 1 of the other 2 pegs where it rests on a larger disk. Let $f(n)$ be the number of legal moves needed to transfer all n disks from peg A to another peg. Find a recurrence relation for $\langle f(n) \rangle$ and solve it.

Suppose the largest disk is fixed (for the time being) at the bottom of peg A. The number of legal moves needed to transfer the other $n-1$ disks to peg C is $f(n-1)$. Now transfer the largest disk from peg A to peg B—in 1 legal move. Then transfer the $n-1$ disks from peg C to peg B, making $f(n-1)$ legal moves. Thus the relation is

$$f(n) = f(n-1) + 1 + f(n-1) = 2f(n-1) + 1$$

with the initial condition $f(1) = 1$.

A particular solution of the linear recurrence relation (*) is $p(n) = -1$; the general homogeneous solution is $h(n) = c2^n$. Therefore,

$$f(n) = c2^n - 1 = 2^n - 1$$

where the initial condition was used to evaluate c.

3.58 Derive Theorem 3.3 from the recurrence relation

$$q_n(r) - q_{n-1}(r) = q_n(r-n)$$

obtained in Problem 3.23(*a*).

Multiply through by x^r, and sum from $r = 0$ to $r = \infty$:

$$g_n(x) - g_{n-1}(x) = x^n g_n(x) \qquad \text{or} \qquad g_n(x) = \frac{1}{1-x^n} g_{n-1}(x)$$

Iteration of this last relation, with use of the starting value

$$g_1(x) = \sum_{r=0}^{\infty} 1x^r = \frac{1}{1-x}$$

yields

$$g_n(x) = \frac{1}{(1-x)(1-x^2)(1-x^3)\cdots(1-x^n)}$$

3.59 Let a_n be the number of n-letter sequences that can be formed using the letters A, B, and C such that any nonterminal A has to be immediately followed by a B. Find the recurrence relation for $\langle a_n \rangle$ and the corresponding generating function.

Either the first letter of a sequence is A or it is not. In the former case, the second letter is B, and so there are

a_{n-2} sequences in this category. In the latter case there are 2 choices for the first letter, and therefore $2a_{n-1}$ possible sequences. Hence,

$$a_n = 2a_{n-1} + a_{n-2}$$

with starting values $a_1 = 3$, $a_2 = 7$.

To obtain $g(x) = \Sigma_{n=1}^{\infty} a_n x^n$, multiply (*) by x^n and sum from $n = 3$ to $n = \infty$:

$$g(x) - 3x - 7x^2 = 2x[g(x) - 3x] + x^2 g(x) \qquad \text{or} \qquad g(x) = \frac{3x + x^2}{1 - 2x - x^2}$$

3.60 The famous **Fibonacci sequence**, $\langle 1, 1, 2, 3, 5, 8, \ldots \rangle$, is defined by the recurrence relation

$$f(n) = f(n-1) + f(n-2) \tag{i}$$

with initial conditions $f(0) = f(1) = 1$. Show that, for $n = 0, 1, 2, \ldots$,

$$f(n) = \frac{1}{\sqrt{5}}\left[\left(\frac{1+\sqrt{5}}{2}\right)^{n+1} - \left(\frac{1-\sqrt{5}}{2}\right)^{n+1}\right] \tag{ii}$$

Use the theory of Section 3.4. The roots of the characteristic equation, $t^2 - t - 1 = 0$, are

$$t_{1,2} = \frac{1 \pm \sqrt{5}}{2}$$

so that

$$f(n) = C_1\left(\frac{1+\sqrt{5}}{2}\right)^n + C_2\left(\frac{1-\sqrt{5}}{2}\right)^n$$

The initial conditions determine the constants as

$$C_1 = \frac{1+\sqrt{5}}{2\sqrt{5}} \qquad C_2 = -\frac{1-\sqrt{5}}{2\sqrt{5}}$$

and (*ii*) (known as the Binet formula) follows.

3.61 Obtain the ordinary generating function, $\phi(x)$, of the Fibonacci sequence.

Treat (*i*) of Problem 3.60 in the usual manner:

$$\sum_{n=2}^{\infty} f(n)x^n = x \sum_{n=2}^{\infty} f(n-1)x^{n-1} + x^2 \sum_{n=2}^{\infty} f(n-2)x^{n-2}$$

$$\phi(x) - 1 - x = x[\phi(x) - 1] \qquad + x^2\phi(x)$$

$$\phi(x) = \frac{1}{1 - x - x^2}$$

3.62 Retrieve (*ii*) of Problem 3.60 from the generating function found in Problem 3.61.

Observe that $\phi(x)$ is a rational function the numerator of which is determined by the initial conditions on (*i*) of Problem 3.60 and the denominator of which has as its roots the *reciprocals* of the characteristic roots. (This kind of thing happens with every linear recurrence having constant coefficients.) Thus there will exist a partial-fractions decomposition of the form

$$\phi(x) = \frac{A_1}{x - 1/t_1} + \frac{A_2}{x - 1/t_2}$$

The conditions on A_1 and A_2 are found to be

$$A_1 + A_2 = 0$$

$$\frac{1}{t_2}A_1 + \frac{1}{t_1}A_2 = 1$$

which yield

$$A_1 = -A_2 = \frac{t_1 t_2}{t_1 - t_2}$$

and give

$$\phi(x) = \frac{t_1 t_2}{t_1 - t_2}\left[\frac{1}{x - 1/t_1} - \frac{1}{x - 1/t_2}\right]$$

Then, by differentiation,

$$f(n) = \frac{1}{n!}\phi^{(n)}(0) = \frac{1}{n!}\frac{t_1 t_2}{t_1 - t_2}\left[\frac{(-1)^n n!}{(-1/t_1)^{n+1}} - \frac{(-1)^n n!}{(-1/t_2)^{n+1}}\right]$$

$$= \frac{t_1 t_2}{t_2 - t_1}(t_1^{n+1} - t_2^{n+1})$$

When numerical values are substituted, this last expression is seen to agree exactly with (*ii*) of Problem 3.60.

3.63 Prove the following formula for the Fibonacci numbers:

$$f(n) = C(n, 0) + C(n-1, 1) + C(n-2, 2) + \cdots$$

(a terminating series).

From Problem 3.61,

$$\phi(x) = \frac{1}{1 - x(1+x)} = 1 + x(1+x) + x^2(1+x)^2 + \cdots$$

$$+ x^{n-1}(1+x)^{n-1} + x^n(1+x)^n + \cdots$$

If the first $n+1$ terms on the right are examined in reverse order, it is seen that the coefficient of x^n in $\phi(x)$ is

$$1 + C(n-1, 1) + C(n-2, 2) + \cdots$$

as asserted.

3.64 (*Combinatorial Definition of the Fibonacci Numbers*) Prove that the set $I_n = \{1, 2, 3, \ldots, n\}$ has exactly $f(n+1)$ subsets (including the null set) that contain no 2 consecutive integers.

Let $a_n \equiv$ (number of subsets of I_n containing no 2 consecutive integers). The subsets counted by a_n fall into two categories:

<u>n is not an element</u>. There are obviously a_{n-1} of these subsets.

<u>n is an element</u>. If A is such a subset, $A - \{n\}$ is a (possibly null) subset of I_{n-2} that contains no 2 consecutive integers. Thus there are a_{n-2} subsets in this category.

Consequently,

$$a_n = a_{n-1} + a_{n-2} \qquad (a_1 = 2, a_2 = 3)$$

and comparison with the Fibonacci recurrence shows that $a_n = f(n+1)$.

3.65 The set I_n of Problem 3.64 is bent around a circle so that elements 1 and n become consecutive. Show that there are now

$$L_n = f(n) + f(n-2) \qquad (n = 3, 4, 5, \ldots)$$

subsets without consecutive integers. (The numbers L_n are called **Lucas numbers.**)

Assume that $n \geq 3$. An acceptable subset that does not contain n is an acceptable subset, in the sense of Problem 3.64, of I_{n-1}. Consequently, there are $f(n)$ of these. An acceptable subset that contains n must be the union of $\{n\}$ and an acceptable subset (per Problem 3.64) of $\{2, 3, \ldots, n-2\}$. So there are $f(n-2)$ of these.

Observe that, by linearity, the Lucas numbers obey the Fibonacci recurrence relation, but with different starting values; namely, $L_1 = 1$ (only the null subset qualifies) and $L_2 = 3$.

3.66 Let

$$A = \begin{bmatrix} 1 & 1 \\ 1 & 0 \end{bmatrix} \quad \text{and} \quad F_{n+1} = \begin{bmatrix} f(n+1) & f(n) \\ f(n) & f(n-1) \end{bmatrix}$$

where the $f(n)$ are the Fibonacci numbers. Show that $A^{n+1} = F_{n+1}$.

The proof is by induction on n. The claim is valid when $n = 1$. Suppose it is true for $n+1$; then

$$A^{n+2} = AF_{n+1} = \begin{bmatrix} f(n+1) + f(n) & f(n+1) \\ f(n) + f(n-1) & f(n) \end{bmatrix}$$

$$= \begin{bmatrix} f(n+2) & f(n+1) \\ f(n+1) & f(n) \end{bmatrix} = F_{n+2}$$

Thus the claim is true for $n+2$ as well.

3.67 (*Zeckendorf's Theorem*) Show that every positive integer can be expressed as a sum of distinct Fibonacci numbers no 2 of which are consecutive in $\langle f(n) \rangle_{n \geq 1}$.

Constructive Proof. For any positive integer n, there is always a positive integer m such that

$$f(m) \leq n < f(m+1)$$

If $n = f(m)$, we are done. Otherwise

$$0 < n - f(m) < f(m+1) - f(m) = f(m-1)$$

Since $n - f(m)$ is positive, there exists a positive integer p such that

$$f(p) \leq n - f(m) < f(p+1)$$

Now, $f(p) \leq n - f(m) < f(m-1)$ implies $p \leq m-2$; i.e., $f(p)$ and $f(m)$ are not consecutive Fibonacci numbers. If $n - f(m) = f(p)$, we have $n = f(m) + f(p)$ and we are done. Otherwise, there exists a positive integer $q \leq p - 2$ such that

$$f(q) \leq n - f(m) - f(p) < f(q+1)$$

and the process continues. Ultimately we must reach the point where the partial sum equals a Fibonacci number—say, $f(t)$—and thereby obtain the desired representation

$$n = f(m) + f(p) + f(q) + \cdots + f(t)$$

3.68 Prove that the *only* partition of a positive integer into distinct, pairwise nonconsecutive Fibonacci numbers is that produced by the Zeckendorf algorithm (Problem 3.67).

Suppose that the positive integer N enjoyed, besides the Zeckendorf representation, another one that satisfied the stated conditions. Let $f(m) \leq N < f(m+1)$, which implies $N \leq f(m+1) - 1$. Under the Zeckendorf algorithm each element of the set $I \equiv \{1, 2, 3, \ldots, f(m+1) - 1\}$ is mapped one-to-one into a nonempty subset of $J \equiv \{f(1), f(2), \ldots, f(m)\}$, which nonempty subset contains no 2 consecutive Fibonacci numbers (a *red* subset). In addition, the element $N \in I$ corresponds to a red subset not already counted. Thus J must possess at least $f(m+1) - 1 + 1 = f(m+1)$ red subsets. But, by Problem 3.64, J possesses exactly $f(m+1) - 1$ red subsets. With this contradiction the proof is achieved.

3.69 Prove that the Fibonacci number $f(n)$ is even if and only if $n = 3k + 2$ for some nonnegative integer k.

The set of nonnegative integers can be partitioned into 3 sets, as follows:

$$A \equiv \{1, 4, 7, 10, 13, \ldots\} = \{3k + 1 : k = 0, 1, 2, \ldots\}$$

$$B \equiv \{2, 5, 8, 11, 14, \ldots\} = \{3k + 2 : k = 0, 1, 2, \ldots\}$$

$$C \equiv \{0, 3, 6, 9, 12, \ldots\} = \{3k : k = 0, 1, 2, 3, \ldots\}$$

We prove by induction that $f(n)$ is even if n is in B, and odd if n is in A or in C. The claim is true when $k = 0$. Suppose it is true for k; i.e., $f(3k+1)$ and $f(3k)$ are odd, $f(3k+2)$ is even. Then:

$$\begin{aligned} f(3(k+1)+1) &= f(3k+4) = f(3k+3) + f(3k+2) \\ &= f(3k+1) + 2f(3k+2) = \text{ODD} + \text{EVEN} = \text{ODD} \\ f(3(k+1)) &= f(3k+2) + f(3k+1) = \text{EVEN} + \text{ODD} = \text{ODD} \\ f(3(k+1)+2) &= f(3k+5) = f(3k+4) + f(3k+3) \\ &= f(3k) + f(3k+1) + 2f(3k+3) = \text{ODD} + \text{ODD} + \text{EVEN} = \text{EVEN} \end{aligned}$$

So it is true for $k+1$ also.

3.70 Prove that every fifth Fibonacci number is a multiple of 5.

In our notation, $f(n-1)$ is the nth Fibonacci number; so we have to show that $f(5k-1)$ $(k = 1, 2, 3, \ldots)$ is divisible by 5. Now, $f(4) = 5$ and, from the recurrence relation,

$$\begin{aligned} f(5(k+1)-1) &= f(5k+4) = f(5k+3) + f(5k+2) \\ &= 3f(5k+1) + 2f(5k) \\ &= 5f(5k) + 3f(5k-1) \end{aligned}$$

An obvious induction secures the result.

3.71 The **Vandermonde determinant** in n variables is

$$V_n(x_1, x_2, \ldots, x_n) \equiv \begin{vmatrix} 1 & 1 & \cdots & 1 \\ x_1 & x_2 & \cdots & x_n \\ x_1^2 & x_2^2 & \cdots & x_n^2 \\ \cdots & \cdots & \cdots & \cdots \\ x_1^{n-1} & x_2^{n-1} & \cdots & x_n^{n-1} \end{vmatrix}$$

By solving the appropriate recurrence relation show that

$$V_n(x_1, x_2, \ldots, x_n) = \prod_{n \geq i > j \geq 1} (x_i - x_j) \qquad (*)$$

Considered as a function of x_n, $V_n(x_1, x_2, \ldots, x_n)$ is a polynomial of degree $n-1$ the roots y which are $x_1, x_2, \ldots, x_{n-1}$ and the leading coefficient of which is $V_{n-1}(x_1, x_2, \ldots, x_{n-1})$. Thus

$$V_n(x_1, x_2, \ldots, x_n) = V_{n-1}(x_1, x_2, \ldots, x_{n-1})[(x_n - x_1)(x_n - x_2) \cdots (x_n - x_{n-1})]$$

Solving this recurrence relation by iteration, one obtains (*).

3.72 Solve

$$a_n = a_{n-2} + 4n \qquad \text{with } a_0 = 3,\ a_1 = 2$$

(*a*) by use of the characteristic equation; and (*b*) through a generating function.

(*a*) The characteristic roots are $t_1 = +1$ and $t_2 = -1$, so the general homogeneous solution is

$$h(n) = C_1 + C_2(-1)^n$$

By inspection, a particular solution is $p(n) = (n+1)^2$. Hence,

$$a_n = C_1 + C_2(-1)^n + (n+1)^2$$

The initial conditions require $C_1 = 0$, $C_2 = 2$; whence

$$a_n = 2(-1)^n + (n+1)^2$$

(*b*) Let $g(x) \equiv \sum_{n=0}^{\infty} a_n x^n$. We have

$$\sum_{n=2}^{\infty} a_n x^n = x^2 \sum_{n=2}^{\infty} a_{n-2} x^{n-2} + 4x \sum_{n=2}^{\infty} n x^{n-1}$$

$$g(x) - 3 - 2x = x^2 g(x) \qquad + 4x\left[\frac{1}{(1-x)^2} - 1\right]$$

$$g(x) = \frac{3-2x}{1-x^2} + \frac{4x}{(1-x^2)(1-x)^2}$$

A partial-fractions expansion yields

$$g(x) = \frac{2}{1+x} - \frac{1}{(1-x)^2} + \frac{2}{(1-x)^3}$$

$$= 2\sum_{n=0}^{\infty} (-1)^n x^n - \sum_{n=0}^{\infty} n x^{n-1} + \sum_{n=0}^{\infty} n(n-1)x^{n-2}$$

whence $$a_n = 2(-1)^n - (n+1) + (n+2)(n+1) = 2(-1)^n + (n+1)^2$$

3.73 (*Right and Wrong Generating Functions*) For the sequence of derangement numbers, $\langle D_n \rangle$, Problem 2.30 gives the recurrence relation

$$D_n = (n-1)D_{n-1} + (n-1)D_{n-2}$$

which holds for $n \geq 2$ if the starting values are $D_0 = 1$ (an arbitrary definition) and $D_1 = 0$. Attempt to solve this relation by use of (*a*) an ordinary and (*b*) an exponential generating function.

(*a*) If $g(x) = \Sigma_{n=0}^{\infty} D_n x^n$, you may verify that

$$x^2 g'(x) = \sum_{n=2}^{\infty} (n-1)D_{n-1}x^n \qquad \text{and} \qquad x^2(xg(x))' = \sum_{n=2}^{\infty} (n-1)D_{n-2}x^n$$

The recurrence relation therefore gives

$$g(x) - 1 = x^2 g'(x) + x^2(xg(x))' \qquad \text{or} \qquad x^2 g' - (1-x)g = -\frac{1}{1+x}$$

as a first-order, linear differential equation for $g(x)$. Nasty to solve, owing to the right-hand side.

(*b*) If $G(x) = \Sigma_{n=0}^{\infty} D_n x^n / n!$, then

$$\int xG'(x)\,dx = \sum_{n=2}^{\infty} (n-1)D_{n-1}\frac{x^n}{n!} \qquad \text{and} \qquad \int xG(x)\,dx = \sum_{n=2}^{\infty} (n-1)D_{n-2}\frac{x^n}{n!}$$

and the recurrence relation yields

$$G(x) - 1 = \int xG'(x)\,dx + \int xG(x)\,dx$$

or, by differentiation,

$$(1-x)G' - xG = 0$$

This separable equation [any solution of which automatically satisfies $G'(0) = 0 = D_1$] is readily integrated. Applying the boundary condition $G(0) = D_0 = 1$, we find:

$$G(x) = \frac{e^{-x}}{1-x} \tag{*}$$

By Leibniz's rule,

$$D_n = G^{(n)}(0) = \sum_{r=0}^{n} C(n,r)\,(e^{-x})^{(r)}\big|_{x=0}\left(\frac{1}{1-x}\right)^{(n-r)}\Bigg|_{x=0}$$

$$= \sum_{r=0}^{n} \frac{n!}{r!(n-r)!}(-1)^r(n-r)!$$

$$= n!\sum_{r=0}^{n} \frac{(-1)^r}{r!}$$

as in Example 10 of Chapter 2.

[For a one-line derivation of Eq. (*) see Problem 3.113.]

3.74 Denote by $V_n(1)$ the volume of an n-dimensional sphere of radius 1. (*a*) Establish a recurrence relation for $\langle V_n(1)\rangle_{n\geq 1}$. (*b*) Solve the relation.

(*a*) All that is needed are two geometrical facts:

(*i*) The volume of an n-sphere is proportional to the nth power of its radius; that is, $V_n(r) = r^n V_n(1)$.

(*ii*) The intersection of an n-sphere and a hyperplane is an $(n-1)$-sphere.

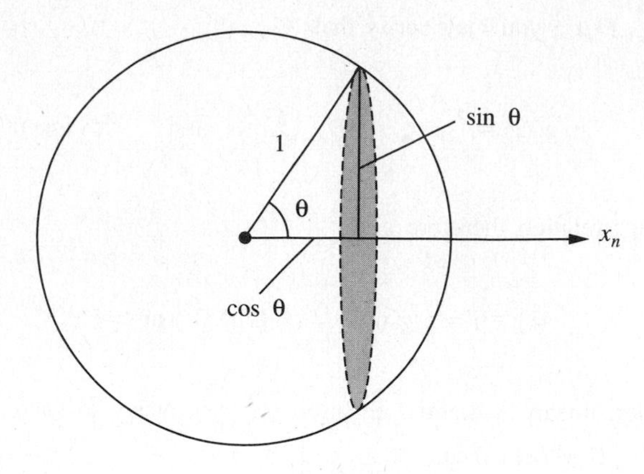

Fig. 3-4

Let us calculate $V_n(1)$ by the integration suggested in Fig. 3-4. The unit n sphere is intersected by the hyperplane $x_n = \cos\theta$ in an $(n-1)$ sphere of radius $\sin\theta$; therefore,

$$V_n(1) = 2\int_{\pi/2}^{0} V_{n-1}(\sin\theta)\,d(\cos\theta) = 2V_{n-1}(1)\int_0^{\pi/2} \sin^n\theta\,d\theta \qquad (*)$$

It would be wrong to stop here, for (*) can be greatly simplified. An integration by parts yields

$$\int_0^{\pi/2} \sin^n\theta\,d\theta = \int_{\pi/2}^{0} \sin^{n-1}\theta\,d(\cos\theta) = (n-1)\int_0^{\pi/2} \cos^2\theta\,\sin^{n-2}\theta\,d\theta$$

$$= (n-1)\int_0^{\pi/2} \sin^{n-2}\theta\,d\theta \;-\; (n-1)\int_0^{\pi/2} \sin^n\theta\,d\theta$$

or

$$\int_0^{\pi/2} \sin^n\theta\,d\theta = \frac{n-1}{n}\int_0^{\pi/2} \sin^{n-2}\theta\,d\theta$$

or, from (*), $V_n(1)/2V_{n-1}(1) = [(n-1)/n]V_{n-2}(1)/2V_{n-3}(1)$; or,

$$\frac{V_n(1)}{V_{n-2}(1)} = \left(\frac{n-1}{n}\right)\frac{V_{n-1}(1)}{V_{n-3}(1)}$$

$$= \left(\frac{n-1}{n}\right)\left(\frac{n-2}{n-1}\right)\left(\frac{n-3}{n-2}\right)\cdots\left(\frac{3}{4}\right)\frac{V_3(1)}{V_1(1)}$$

$$= \left(\frac{3}{n}\right)\frac{4\pi/3}{2} = \frac{2\pi}{n}$$

The desired recurrence relation is thus

$$V_n(1) = \frac{2\pi}{n}V_{n-2}(1) \qquad (n \geq 3) \qquad (**)$$

with starting values $V_1(1) = 2$ and $V_2(1) = \pi$.

(*b*) The solution of (**), by iteration, is

$$V_n(1) = \begin{cases} \dfrac{2(2\pi)^{(n-1)/2}}{1\cdot 3\cdot 5\cdots n} & (n \text{ odd}) \\ \pi^{n/2}/(n/2)! & (n \text{ even}) \end{cases}$$

3.75 Let A be the $m \times m$ matrix in which all diagonal entries are 0 and all nondiagonal entries are 1. Then the diagonal entries of A^n are all $f(n)$, and the nondiagonal entries of A^n are all $g(n)$. Calculate $f(n)$ and $g(n)$.

$$(\text{Row 1 of } A^n)\cdot(\text{column 1 of } A) = (m-1)g(n)$$

which implies

$$f(n+1) = (m-1)g(n) \quad \text{or} \quad f(n) = (m-1)g(n-1) \qquad (i)$$

Also
$$(\text{Row 2 of } A^n)\cdot(\text{column 1 of } A) = f(n) + (m-2)g(n)$$

which implies

$$g(n+1) = f(n) + (m-2)g(n) \qquad (ii)$$

Eliminate $f(n)$ between (i) and (ii) to obtain

$$g(n+1) = (m-1)g(n-1) + (m-2)g(n-2) \qquad (iii)$$

The characteristic roots of (iii) are -1 and $m-1$; thus $g(n) = A(-1)^n + B(m-1)^n$. The initial values $g(1) = 1$ and $g(2) = m-2$ determine $A = -1/m$ and $B = 1/m$. Thus

$$g(n) = \frac{(m-1)^n - (-1)^n}{m} \qquad f(n) = \frac{m-1}{m}[(m-1)^{n-1} - (-)^{n-1}]$$

3.76 A recurrence relation which expresses $f(n)$ in terms of $f(n/b)$, where $b \geq 2$ is an integer, indicates that a problem of size n can be scaled down to size n/b by dividing it into subproblems. Recurrence relations of this kind are usually known as **divide-and-conquer relations**. Solve

$$f(n) = af\left(\frac{n}{b}\right) + cn \qquad \text{with } f(1) = d \qquad (*)$$

It is generally impossible to solve a divide-and-conquer relation for all n. However, a partial solution—for $n = b^{\lambda+1}, b^{\lambda+2}, b^{\lambda+3}, \ldots$—is possible *if* $f(b^\lambda)$ is given as a starting value. Thus, with reference to (*), for which $\lambda = 0$, make the change of variables $n = b^{0+k}$ and $f(b^k) = \phi(k)$, to obtain

$$\phi(k) = a\phi(k-1) + cb^k \qquad \text{with } \phi(0) = d \qquad (**)$$

It is simple to solve (**) either by the characteristic-equation or the generating-function approach. Let us choose the latter, defining $g(x) \equiv \sum_{k=0}^{\infty} \phi(k)x^k$:

$$\sum_{k=1}^{\infty} \phi(k)x^k = ax\sum_{k=1}^{\infty} \phi(k-1)x^{k-1} + c\sum_{k=1}^{\infty} (bx)^k$$

$$g(x) - d = axg(x) \qquad + \frac{cbx}{1-bx}$$

$$g(x) = \frac{d + b(c-d)x}{(1-ax)(1-bx)}$$

Case 1. $a \neq b$. A partial-fractions decomposition gives

$$g(x) = \frac{P}{1-ax} + \frac{Q}{1-bx}$$

in which

$$P = \frac{b(d-c) - ad}{b-a} \qquad Q = \frac{bc}{b-a}$$

By inspection the coefficient x^k in $g(x)$ is

$$\phi(k) = f(b^k) = Pa^k + Qb^k \qquad (k = 0, 1, 2, \ldots)$$

Case 2. $a = b$.

$$g(x) = \left[\frac{d}{b} + (c - d)x\right]\left(\frac{1}{1 - bx}\right)'$$
$$= \left[\frac{d}{b} + (c - d)x\right] \sum_{k=0}^{\infty} kb^k x^{k-1}$$

and the coefficient of x^k is

$$\phi(k) = f(b^k) = \frac{d}{b}(k + 1)b^{k+1} + (c - d)kb^k = (d + ck)b^k$$

3.77 (*a*) Obtain a divide-and-conquer relation for the number, $f(n)$, of single-digit multiplications needed to compute the product of 2 n-digit numbers. (*b*) Solve the relation.

(*a*) The ordinary algorithm for the product of 2 n-digit numbers x and y involves n^2 single-digit multiplications. Now, if n is presumed even, we can write

$$x = x_1 10^{n/2} + x_2 \qquad y = y_1 10^{n/2} + y_2$$

where x_1, x_2, y_1, and y_2 are definite $(n/2)$-digit numbers. Then

$$xy = (x_1 y_1)10^n + (x_1 y_2 + x_2 y_1)10^{n/2} + x_2 y_2$$

Now $x_1 y_2 + x_2 y_1 = (x_1 + x_2)(y_1 + y_2) - x_1 y_1 - x_2 y_2$ involving 3 (not 4) multiplications. Thus

$$f(n) = 3f\left(\frac{n}{2}\right) \qquad \text{with } f(1) = 1$$

(*b*) One is back in Problem 3.76 (Case 1), with $b = 2$, $a = 4$, $c = 0$, and $d = 1$; $P = 1$, and $Q = 0$. Hence,

$$f(2^k) = 3^k < 4^k = n^2$$

3.78 Repeat Problem 3.77 if now $f(n)$ is the number of comparisons needed to find the minimum and maximum entries in a list of $n \geq 3$ numbers.

(*a*) In the ordinary procedure $n - 1$ comparisons are needed to identify the maximum, followed by $n - 2$ comparisons to identify the minimum—for a total of $2n - 3$. In the recursive procedure one supposes n even and divides the list into 2 equal parts. It takes $2f(n/2)$ comparisons to find the maxima and minima of the 2 sublists and then 2 comparisons (max with max and min with min) to find the overall maximum and overall minimum. Hence, for $n \geq 4$,

$$f(n) = 2f\left(\frac{n}{2}\right) + 2 \qquad \text{with } f(2) = 1$$

(*b*) As in Problem 3.76 make the transformation $n = 2^k$, $f(2^k) = \phi(k)$, to obtain, for $k = 2, 3, \ldots,$

$$\phi(k) = 2\phi(k - 1) + 2 \qquad \text{with } \phi(1) = 1$$

The solution is

$$\phi(k) = f(2^k) = \frac{3}{2}(2^k) - 2$$

Thus, when n is a large power of 2, the recursive method requires $\approx \frac{3}{2}n$ comparisons, as against $\approx 2n$ for the ordinary method.

3.79 To multiply two 2×2 matrices, we usually perform eight $(8 = 2^3)$ multiplications. Show that it is possible to find the product by performing 7 multiplications.

Let $\mathbf{A} = [a(i, j)]$, $\mathbf{B} = [b(i, j)]$, $\mathbf{AB} = \mathbf{C} = [c(i, j)]$. Define x_i $(i = 1, 2, \ldots, 7)$ as follows:

$$x_1 = [a(1,1) + a(2,2)][b(1,1) + b(2,2)] \qquad x_5 = [a(1,1) + a(1,2)]b(2,2)$$

$$x_2 = [a(2,1) + a(2,2)]b(1,1) \qquad x_6 = [a(2,1) - a(1,1)][b(1,1) + b(1,2)]$$

$$x_3 = [b(1,2) - b(2,2)]a(1,1) \qquad x_7 = [a(1,2) - a(2,2)][b(2,1) + b(2,2)]$$

$$x_4 = [b(2,1) - a(2,1)]a(2,2)$$

Then the 4 elements of $\mathbf{C}$ can be expressed additively as

$$c(1,1) = x_1 + x_4 - x_5 + x_7 \qquad c(1,2) = x_3 + x_5$$

$$c(2,1) = x_2 + x_4 \qquad c(2,2) = x_1 + x_3 - x_2 + x_6$$

In all there are 7 multiplications.

3.80 (*Strassen's Fast Matrix Multiplication*) Show that the product of two $2^k \times 2^k$ matrices, which ordinarily involves $(2^k)^3 = 8^k$ multiplications, can be computed with $\phi(k) = 7^k$ multiplications.

Partition each factor into four $2^{k-1} \times 2^{k-1}$ submatrices:

$$\mathbf{A} = \left[\begin{array}{c:c} \mathbf{A}(1,1) & \mathbf{A}(1,2) \\ \hdashline \mathbf{A}(2,1) & \mathbf{A}(2,2) \end{array}\right] \qquad \mathbf{B} = \left[\begin{array}{c:c} \mathbf{B}(1,1) & \mathbf{B}(1,2) \\ \hdashline \mathbf{B}(2,1) & \mathbf{B}(2,2) \end{array}\right]$$

This puts us back in Problem 3.79, with $\mathbf{A}(1,1)$ replacing $a(1,1)$, etc., and with matrix arithmetic replacing ordinary arithmetic. The 7 matrix multiplications required to compute $\mathbf{AB}$ will involve $7\phi(k-1)$ (scalar) multiplications; thus

$$\phi(k) = 7\phi(k-1) \qquad \text{with } \phi(1) = 7 \qquad \text{whence } \phi(k) = 7^k$$

H-TABLEAUX AND YOUNG TABLEAUX

3.81 Consider a partition of n as $n = a_1 + a_2 + \cdots + a_m$, where $a_1 \geq a_2 \geq \cdots \geq a_m$. Corresponding to this partition, we can construct an array of n cells with m rows and a_1 columns such that (*i*) the ith row has a_i cells for each i and (*ii*) the jth cell in each row is in the jth column for each j. For each cell (i, j) we define a number $h_{ij} = 1 +$ the number of cells to the right of the (i, j) cell in the ith row $+$ the number of cells below the (i, j) cell in the jth column. The array of cells filled with the numbers h_{ij} is called the **H-tableau** of the partition. The ordered tuple $[a_1\ a_2\ \cdots\ a_m]$ is called the **shape** of the tableau. Construct the H-tableau for the partition $23 = 6 + 5 + 4 + 3 + 3 + 2$.

Here $m = 6$ (the number of summands), and $a_1 = 6$ (the largest summand). The shape is [6 5 4 3 3 2]. There are 6 rows and 6 columns in the tableau. The numbers h_{ij} are computed according to the definition. For example, $h_{11} = 1 + 5 + 5 = 11$, $h_{12} = 1 + 4 + 5 = 10$, etc. The H-tableau is as follows:

$$\begin{array}{llllll} 11 & 10 & 8 & 5 & 3 & 1 \\ 9 & 8 & 6 & 3 & 1 & \\ 7 & 6 & 4 & 1 & & \\ 5 & 4 & 2 & & & \\ 4 & 3 & 1 & & & \\ 2 & 1 & & & & \end{array}$$

3.82 Consider a partition of n and the H-tableau associated with this partition. For each cell (i, j) we have a positive integer h_{ij} and a set $H_{ij} = \{1, 2, \ldots, h_{ij}\}$. We define an ordered tuple S_{ij} as follows: The first part of S_{ij} consists of the number h_{ij} and all the numbers in the tableau in the ith row on the right side of h_{ij} in the order they appear in the row. Then we subtract each number in the jth column below h_{ij} starting with the number $h_{i+1,j}$ in the same order as they appear in the column from the number h_{ij}. Then the front part of S_{ij} is a decreasing tuple, the back part is a decreasing tuple, and S_{ij} has h_{ij} numbers. Construct H_{12} and S_{12} for the partition $23 = 6 + 5 + 4 + 3 + 3 + 2$.

See the previous problem (Problem 3.81) for the H-tableau of this partition. Here $h_{12} = 10$; consequently $H_{12} = \{1, 2, 3, \ldots, 10\}$. The front part of S_{12} is [10 8 5 3 1]. The back part is

$$[10 - 8 \quad 10 - 6 \quad 10 - 4 \quad 10 - 3 \quad 10 - 1],$$

which is equal to [2 4 6 7 9]. Thus $S_{ij} = [10\ 8\ 5\ 3\ 1\ 2\ 4\ 6\ 7\ 9]$.

3.83 Consider a partition of n with m summands in which the summands are in nonincreasing order, and define H_{ij} and S_{ij} as in Problem 3.82. Show that S_{ij} is a permutation of the set H_{ij}.

The number of elements in the sequence S_{ij} is obviously h_{ij} by the definitions of h_{ij} and S_{ij}. It remains to be shown that the numbers in the sequence are distinct. The numbers in S_{ij} that come from h_{ij} and its right side form a decreasing sequence (in the previous problem we have the numbers 10, 8, 5, 3, and 1), and the numbers in S_{ij} that come from the column below h_{ij} when subtracted from h_{ij} form an increasing sequence (the numbers 2, 4, 5, 7, and 9 in the previous problem). Thus it is sufficient to show that there cannot be positive integers s and t such that $h_{i,j+t} = h_{ij} - h_{i+s,j}$. Let the number of integers between h_{ij} and $h_{i,j+t}$ be a. Let the number of integers on the right of $(i, j+t)$ be b. Let the number of integers between h_{ij} and $h_{i+s,j}$ be c. Let the number of integers below $h_{i,j+t}$ be d. Let the number of integers on the right of $h_{i+s,j}$ be e. Finally, let the number of integers below $(i+s, j)$ be f. See the accompanying illustration.

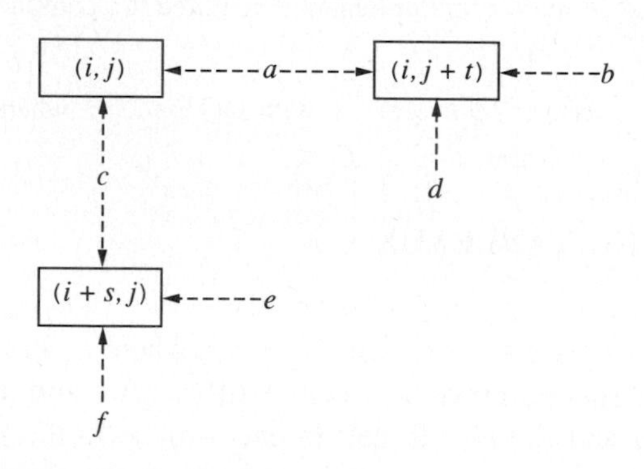

Then

$$h_{ij} = 3 + a + b + c + f$$

$$h_{i,j+t} = 1 + b + d$$

$$h_{i+s,j} = 1 + e + f$$

If $h_{i,j+t} = h_{ij} - h_{i+s,j}$, we have

$$1 + b + d = 3 + a + b + c + f - 1 - e - f = 2 + a + b + c - e$$

Thus $a + c + 1 = d + e$.

Now either $e \leq a$ or $e > a$. In the former case $e \leq a$ and $a + c + 1 = d + e$ imply $a + 1 \leq e$. In the latter case $e > a$ and $a + c + 1 = d + e$ imply $a + 1 > e$. This contradiction establishes the fact that the numbers in the sequence S_{ij} must all be distinct.

3.84 Let $H = (h_{ij})$ be the H-tableau corresponding to a partition of n with m summands the largest of which is a_1. Define $t = (h_{11}!)(h_{21}!)\cdots(h_{m1}!)$ and $t_1 = (h_{11} - h_{21})(h_{11} - h_{31})\cdots(h_{11} - h_{m1})$, $t_2 = (t_{21} - h_{31})(h_{21} - h_{41})\cdots(h_{21} - h_{m1})$, etc., $\ldots$, $t_{m-1} = h_{m-1,1} - h_{m1}$ and $t_m = h_m$. (Notice that these numbers are defined using only the numbers from the first column of H). Show that the product of all the h_{ij} in the H-tableau is $t/[(t_1)(t_2)\cdots(t_m)]$.

The product of the h_{ij} can be computed by finding the products row by row. Using the fact that S_{ij} is a permutation of the set H_{ij} (as proved in the last problem), we see that the product of the h_{ij} in the ith row is $(h_{i1}!)/t_i$.

Example: Suppose the H-tableau is

$$\begin{matrix} 8 & 6 & 5 & 3 & 1 \\ 6 & 4 & 3 & 1 & \\ 4 & 2 & 1 & & \\ 1 & & & & \end{matrix}$$

Here $t = (8!)(6!)(4!)(1!)$, $t_1 = (8-6)(8-4)(8-1)$, $t_2 = (6-4)(6-1)$, $t_3 = (4-1)$, and $t_4 = 1$. Then $t/[(t_1)(t_2)(t_3)(t_4)] = 414{,}720 = (8)(6)(5)(3)(1)(6)(4)(3)(1)(4)(2)(1)(1)$.

3.85 Consider a partition of n as $a_1 + a_2 + \cdots + a_m = n$, where $a_1 \geq a_2 \geq \cdots \geq a_m$. Corresponding to this partition, we can arrange the first n positive integers in m rows and a_1 columns such that: (i) the first numbers in each row form the first column, the second numbers in each row form the second column, etc., and (ii) row i contains a_i numbers for each i, and (iii) the numbers in each row and in each column are increasing. Any array thus constructed is known as a **Young tableau** for the sequence a_i, and the ordered tuple $[a_1 \quad a_2 \quad \cdots \quad a_m]$ is the **shape** of the tableau. Each number in the tableau is called a **part**. A Young tableau is also known as a **standard tableau**. For example, the partition $9 = 4 + 3 + 2$ defines a tableau with 3 rows and 4 columns. We use the integers from 1 to 9 and construct a Young tableau as follows:

$$\begin{matrix} 1 & 4 & 6 & 9 \\ 2 & 3 & 7 & \\ 5 & 8 & & \end{matrix}$$

Show that the number of distinct Young tableaux corresponding to the partition $a_1 + a_2 + \cdots + a_m = n$ where the summands are nonincreasing is $(n!)/[\Pi\, h_{ij}]$, where the product is over all the h_{ij}], where the product is over all the h_{ij} and the h_{ij} are the numbers in the H-tableau of the partition as defined in Problem 3.81. [This result is known as the **Frame–Robinson–Thrall theorem**. What we have here is a partition of n in 2 dimensions (plane partitions) with some restrictions. There are generalizations of plane partitions to higher dimensions, but a generalization of the results of Young tableaux to higher dimensions is an open problem.]

It is enough to prove that the number of distinct Young tableaux corresponding to the partition $n = a_1 + a_2 + \cdots + a_m$ is equal to $(n!)(t_1)(t_2)\cdots(t_m)/t$, where the numbers t and t_i are as in Problem 3.84. Suppose the number of Young tableaux is $f(a_1, a_2, \ldots, a_m)$. Let us write $x_i = h_{i1}$ for each i. Then we have to show that $f(a) = f(a_1, a_2, \ldots, a_m) = (n!)\,\Pi\,(x_i - x_j)/[(x_1!)(x_2!)\cdots(x_m!)]$, in which the product in the numerator is for all i and j, where $j > i \geq 1$.

The proof is by induction on n. It is obviously true for $n = 1$ and $n = 2$. Suppose that the theorem is valid for partitions of integers less than n. We shall show it is valid for the partition $n = a_1 + a_2 + \cdots + a_m$ as well.

Now $$n - 1 = a_1 + a_2 + \cdots + a_{k-1} + (a_k - 1) + a_{k+1} + \cdots + a_m$$

Thus $$f(a_1, a_2, \ldots, a_{k-1}, (a_k - 1), a_{k+2}, \ldots, a_m) = \frac{(P)}{(Q)}$$

where $$P = \left[(n-1)! \prod (x_i - x_j) \prod (x_i - x_k + 1) \prod (x_k - 1 - x_j)\right]$$

and $$Q = x_1! x_2! \cdots x_{k-1}!(x_k - 1)! x_{k+1}! \cdots x_m!$$

The first product in P is for all $j > i$ where i is not equal to k, the second product is for $k > i$, and the third product is for $j > k$. Then

$$f(a_1, a_2, \ldots, a_{k-1}, (a_k - 1), a_{k+1}, \ldots, a_m) = \left[\frac{(x_k) f(a)}{n}\right] \prod \left[\frac{x_k - 1 - x_j}{x_k - x_j}\right]$$

where the product is for all j and k such that $j \neq k$.

In a fixed Young tableau Y' with n cells associated with a given partition of n, the number n appears as the last number in row k for some k.

If the cell that contains n is deleted, we have the partition of

$$n - 1 = a_1 + a_2 + \cdots + a_{k-1} + (a_k - 1) + a_{k+1} + \cdots + a_m$$

which is a partition of $n - 1$. The number of distinct Young tableaux for this partition is

$$f(a_1, a_2, \ldots, a_{k-1}, (a_k - 1), a_{k+1}, \ldots, a_m)$$

by the induction hypothesis. Now k can vary from 1 to m. So the total number of all Young tableaux for the given partition of n is

$$\sum f(a_1, a_2, \ldots, a_{k-1}, a_k - 1, a_{k+1}, \ldots, a_m) = \left[\frac{f(a)}{n}\right] \sum \left\{x_k \prod \frac{x_k - x_j - 1}{x_k - x_j}\right\}$$

where k varies from 1 to m and in the product k and j are unequal. This equals

$$\left[\frac{f(a)}{n}\right] \sum \left\{x_k \prod \left[1 - \frac{1}{x_k - x_j}\right]\right\}$$

Now $$\sum x_k = \sum a_k + \sum (m - k) = n + (m-1) + (m-2) + \cdots + 1 = n + \frac{m(m-1)}{2}$$

The double series $\sum \sum x_k/(x_k - x_j)$ has $m(m-1)$ terms, and terms in pairs like $x_k/(x_k - x_j)$ and $x_j/(x_j - x_k)$ add up to 1. Therefore, the sum of this double series is $m(m-1)/2$. The number of Young tableaux is $[f(a)/n]\{n + m(m-1)/2 - m(m-1)/2 + 0\} = f(a)$ since the other terms in the series cancel out. Thus the theorem is also valid for n, and the proof by induction is complete.

3.86 Find the number of distinct Young tableaux associated with the partition $18 = 6 + 4 + 4 + 3 + 1$.

The number of summands is 5, and the largest summand is 6. So the H-tableau and Young tableaux corresponding to this partition will have 5 rows and 6 columns. The H-tableau for this partition is

10	8	7	5	2	1
7	5	4	2		
6	4	3	1		
4	2	1			
1					

The product of the h_{ij} in the H-tableau can be determined by using a row-by-row calculation: $(5600)(280)(72)(8)(1) = 903{,}168{,}000$. This number can also be calculated by using the first column of the H-tableau as $[(10!)(7!)(6!)(4!)]/[(3)(4)(6)(9)(1)(3)(6)(2)(5)(3)]$. Thus the number of distinct Young tableaux is $(18!)/903{,}168{,}000$, which is equal to 7,088,796,000.

3.87 (*The Generalized Balloting Problem*) In an election involving m candidates A_i $(i = 1, 2, \ldots, m)$, candidate A_i received a_i votes, $a_i \geq a_{i+1}$ for each i and $a_m > 0$. If the ballots are cast one at a time, find the probability that A_{i+1} is never ahead of A_i for each i.

If the total number of votes is $n = a_1 + a_2 + \cdots + a_m$, then any Young tableau corresponding to this partition of n is a possible voting pattern and conversely. Here is an illustrative example involving 3 candidates A, B, and C, who got 5, 3, and 2 votes, respectively. The partition here is $10 = 5 + 3 + 2$. Now consider the following Young tableau associated with this partition.

$$\begin{array}{llllll} \text{A:} & 1 & 3 & 6 & 7 & 9 \\ \text{B:} & 2 & 4 & 10 & & \\ \text{C:} & 5 & 8 & & & \end{array}$$

This tableau indicates that the first ballot went to A, the second ballot went to B, the third ballot went to A, the fourth ballot went to B, the fifth ballot went to C, and so on. Let h be the product of the h_{ij} of the H-tableau of the partition. Then the number of Young tableaux is $n!/h$. The total number of permutations is $n!/[(a_1!)(a_2!)\cdots(a_m!)]$. Thus the probability is $[(a_1!)(a_2!)\cdots(a_m!)]/h$. In the above example h is computed as 8064, and the probability that A trails B and B trails C at no time is $(5!)(3!)(2!)/8064 = 1440/8064 = 0.1786$.

Supplementary Problems

3.88 Reconcile (*iii*) of Example 3 with the result of Problem 2.22. (*Hint*: Use Problem 1.46 and Theorem 2.2.)

3.89 Show that the function $g(x)$ determined in Example 10 actually generates the sequence $\langle a_n \rangle$ of Example 9. [*Hint*: $(1-x)^{-4} = \Sigma\,[(n+3)(n+2)(n+1)/6]x^n$.]

3.90 Show that the ordinary generating function, $g(x)$, and the exponential generating function, $G(x)$, of a given sequence are formally related via

$$g(x) = \int_0^\infty e^{-u} G(xu)\, du$$

[*Hint*: Recall the integral definition of the gamma function and the fact $\Gamma(n+1) = n!$.]

3.91 Check Problem 3.10 by showing that there are 34 experiments such that the total is 18 and the green die shows an odd number. The coefficient of x^{18} in the generating function $(x + x^2 + \cdots + x^6)^3(x + x^3 + x^5)$ is $298 - 264 = 34$.

3.92 By reinterpreting Problem 3.12 as a pigeonhole problem, obtain a new proof of Theorem 1.3. The ordinary generating function is the product of n polynomials $f_i(x)$ with unit coefficients where $i = 1, 2, \ldots, n$. The degree of $f_i(x)$ is p_i. While computing the coefficient of x^t where $t = p_1 + p_2 + \cdots + p_n - n + 1$ there is at least one j such that the highest degree term in $f_j(x)$ is considered in the counting process.

3.93 Obtain Euler's theorem (Problem 3.29) by specializing Problem 2.24. (*Hint*: Put $r = 2$ in Problem 2.24.)

3.94 From Problems 3.32 and 3.24(*b*), obtain the ordinary generating function of $\langle q^{\#}(r, n)\rangle_{r\geq 0}$.
Ans. $x^{C(n+1,\,2)}/(1-x)(1-x^2)(1-x^3)\cdots(1-x^n)$

3.95 With reference to Problem 3.50, find the ordinary generating function of $\langle s_m(n)\rangle_{m\geq 0}$ (*a*) directly; (*b*) from the exponential generating function via Problem 3.83.

Ans. $1+\dfrac{1}{1-x}+\dfrac{1}{1-2x}+\cdots+\dfrac{1}{1-(n-1)x}$

3.96 Solve the following linear recurrence relations with constant coefficients:

(*a*) $$a_n = 2a_{n-1} + 3a_{n-2} \qquad \text{with } a_0 = 1,\ a_1 = 2$$

(*b*) $$a_n = 6a_{n-1} - 9a_{n-2} \qquad \text{with } a_0 = 1,\ a_1 = 4$$

(*c*) $$a_{n+3} = 3a_{n+2} + 4a_{n+1} - 12a_n \qquad \text{with } a_0 = 0,\ a_1 = -11,\ a_2 = -15$$

(*d*) $$a_n = a_{n-1} + 2a_{n-2} + 4(3^n) \qquad \text{with } a_0 = 11,\ a_1 = 28$$

(*e*) $$a_n = 4(a_{n-1} - a_{n-2}) + 2^n$$

(*f*) $$a_n = 3a_{n-1} - 4n + 3(2^n)$$

[*Hint*: Let $a_n = p_n + q_n$, where $p_n = 3p_{n-1} - 4n$ and $q_n = 3q_{n-1} + 3(2^n)$.]

Ans.
(*a*) $a_n = 3^n + (-1)^n$ (*d*) $a_n = (-1)^n + 2^n + 3^{n+2}$
(*b*) $a_n = 3^{n-1}(3+n)$ (*e*) $a_n = (C_0 + C_1 n + \frac{1}{2}n^2)2^n$
(*c*) $a_n = 2(-2)^n + 2^n - 3^{n+1}$ (*f*) $a_n = C3^n + 2n + 3(1 - 2^{n+1})$

3.97 Let $f(n)$ be the number of n-letter sequences that can be formed using the letters P, Q, R, S, and T, if each sequence must involve an odd number of P's. Find a recurrence relation for $\langle f(n)\rangle$. (*Hint*: Categorize the sequences as starting with P or not starting with P.) *Ans.* $f(n) = 3f(n-1) + 5^{n-1}$, with $f(1) = 1$

3.98 Find the ordinary generating function of the sequence defined by

$$a_{n+2} - 3a_{n+1} - 4a_n = 0 \qquad \text{with } f(1) = 1,\ f(2) = 3$$

Ans. $x/(1-3x-4x^2)$

3.99 Laplace's equation in one dimension is $\phi''(x) = 0$, with general solution $\phi(x) = A + Bx$. (*a*) Approximate Laplace's equation by a finite-difference (recurrence) relation on the integer points. (*b*) Solve the relation of (*a*).
Ans. (*a*) $\phi(n+1 - 2\phi(n) + \phi(n-1) = 0$; (*b*) $\phi(n) = C_0 + C_1 n$

3.100 Let $f(r;n)$ denote the number of *self-conjugate* partitions of r into exactly n parts; set $f(0;0) \equiv 1$. (*a*) Establish the (2-index) recurrence relation

$$f(r+2;\ n+1) = f(r;\ n) + f(r-2n+1;\ n)$$

(*b*) Show that $f(r;n) = f((n+1)^2 - r - 2;\ n)$. [*Hints*: (*a*) First prove that

$$f(r;\ n) = \sum_{i=0}^{n-1} f(r-2n+1;\ i)$$

by conditioning on the size of the second-largest elbow in the star diagram. (*b*) Consider the complement of the star diagram in a suitable square of side $n+1$.]

3.101 Find the ordinary generating function which will give the number of solutions in nonnegative integers of $a+b+c+d=10$.
Ans. The number of solutions is the coefficient of x^{10} in $g(x)=(1+x+x^2+\cdots+x^{10})^4$.

3.102 Find the ordinary generating function of the numbers of solutions in integers of $x+y+z=r$ ($r=0,1,2,\ldots$), if (*a*) each variable is nonnegative and at most 3; (*b*) each variable is at least 2 and at most 5; and (*c*) $0\le x\le 6$, $2\le y\le 7$, $5\le z\le 8$, x is even and y is odd.
Ans. (*a*) $g_1(x)=(1+x+x^2+x^3)^3$; (*b*) $g_2(x)=(x^2+x^3+x^4+x^5)^3=x^6g_1(x)$;
(*c*) $g_3(x)=(1+x^2+x^4+x^6)(x^3+x^5+x^7)(x^5+x^6+x^7+x^8)$

3.103 Give the ordinary generating functions of the sequences (*a*) $\langle 2n^2\rangle$, and (*b*) $\langle (n+1)n(n-1)\rangle$.
Ans. (*a*) $2x(1+x)(1-x)^{-3}$; (*b*) $6x^2(1-x)^{-4}$

3.104 Find the ordinary generating function associated with the problem of counting the ways of collecting \$20 from a group of 10 people if 9 people pay at least \$1 and at most \$3 and 1 person pays either \$1 or \$5 or \$10.
Ans. The number of ways is the coefficient of x^{20} in $g(x)=(x^1+x^2+x^3)^9(x^1+x^5+x^{10})$.

3.105 Find the ordinary generating function of the numbers of solutions in integers of $u_1+u_2+\cdots+u_{10}=n$, where each variable is at least -2 and at most 2.
Ans. For $i=1,2,\ldots,10$ let $u_i=-2+w_i$. Then $w_1+w_2+\cdots+w_{10}=n+20$, where each variable is nonnegative and at most 4. The generating function is $g(y)=(1+y+y^2+y^3+y^4)^{10}$ and the desired number is the coefficient of y^{n+20} in $g(y)$.

3.106 Find the generating function for the number of ways of changing \$1 using quarters, dimes, nickels, and pennies, if the number of pennies is at most 10.
Ans. Because the number of pennies is 0 or 5 or 10, the generating function is

$$(1+x^5+x^{10})(1+x^5+x^{10}+x^{15}+\cdots+x^{100})(1+x^{10}+x^{20}+\cdots+x^{100})(1+x^{25}+x^{50}+\cdots+x^{100})$$

3.107 Find the ordinary generating functions of the following sequences: (*a*) $\langle n(n-1)\rangle$; (*b*) $\langle n^2\rangle$; and (*c*) $\langle n^3\rangle$.
Ans. (*a*) $2x^2/(1-x)^3$; (*b*) $x(1+x)/(1-x)^3$; (*c*) $x(1+4x+x^2)/(1-x)^4$

3.108 Find the ordinary generating function of the sequence $\langle 1,2,3,4,5,5,5,\ldots\rangle$. [*Hint*: Consider the convolution of $\langle 1,1,1,1,1,0,0,0,\ldots\rangle$ and $\langle 1,1,1,\ldots\rangle$.] *Ans.* $(1-x^5)(1-x)^{-2}$

3.109 Prove that the $(n-1)$st Fibonacci number is given by the terminating series

$$f(n)=\frac{1}{2^n}[C(n+1,1)+5C(n+1,3)+5^2C(n+1,5)+\cdots$$

[*Hint*: Apply the binomial theorem to (*ii*) of Problem 3.60.]

3.110 Prove by induction or otherwise that the Fibonacci numbers satisfy

$$(a)\ \sum_{i=0}^{n} f(i)=f(n+2)-1 \qquad (b)\ \sum_{i=0}^{n}[f(i)]^2=f(n)f(n+1)$$

Ans. The results can be easily established by induction.

3.111 Infer from Problem 3.66 that

$$f(n+1)f(n-1) - [f(n)]^2 = (-1)^{n+1}$$

Ans. The left-hand side is the determinant of F_{n+1} and the right-hand side is the determinant of A^{n+1}.

3.112 Establish the Binet formula

$$L_n = \left(\frac{1+\sqrt{5}}{2}\right)^n + \left(\frac{1-\sqrt{5}}{2}\right)^n \qquad (n = 1, 2, 3, \ldots)$$

for the Lucas numbers. *Ans.* Use the formula $L_n = f(n) + f(n-1)$.

3.113 The sequence $\langle n! \rangle$, $\langle 1, 1, 1, \ldots \rangle$, and $\langle D_n \rangle$ have the respective exponential generating functions $(1-x)^{-1}$, e^x, and $G(x)$. Infer at once Eq. (*) of Problem 3.73(*b*). (*Hint*: See Problem 2.25.)

3.114 Using Problem 3.74, derive the following integral representation of the central binomial coefficient:

$$C(2k, k) = \frac{2}{\pi}\int_0^{\pi/2} (2 \sin \theta)^{2k}\, d\theta \qquad (k = 0, 1, 2, \ldots)$$

On simplifying the definite integral by using the reduction formula established in Problem 3.74, we arrive at the binomial coefficient $C(2n, n)$.

3.115 Solve

$$f(n) = 5f\left(\frac{n}{2}\right) - 6f\left(\frac{n}{4}\right) + n \qquad \text{with } f(1) = 2,\ f(2) = 1$$

Ans. $f(2^r) = (1 - 2r)2^r + 3^r$

3.116 Consider the regular pentagon A_n in which the length of each side is $n-1$ units. Divide each side into segments of unit length and put a dot on each of the dividing points and on the vertices. Now fix one of the vertices of the pentagon (as an initial vertex) and take the two sides of A_n meeting at that vertex. Measure $n-2$ units along each of these sides starting from the fixed vertex and complete the smaller regular pentagon A_{n-1} which is nestled inside A_n sharing a common vertex. Divide the 3 new sides of A_{n-1} also into segments of unit length and put a dot on each dividing point on these sides. Continue this process until a polygon A_2 in which each side is of unit length is obtained. Define $P(n)$ = Total number of dots in A_n when A_2 is completed. Of course, $P(2) = 5$. Define $P(1) = 1$. The sequence $\langle P(n) \rangle = \langle 1, 5, 12, 22, \ldots \rangle$ is the sequence of **pentagonal numbers**. Obtain a formula for $P(n)$. *Ans.* $P(n) = n(3n-1)/2$.

3.117 (*Euler's identity*). Show that $(1-x)(1-x^2)(1-x^3)\ldots$ is equal to $1 + \sum_{n=1}^{\infty} Q(n)x^n$ where $Q(n) = (-1)^k$ if $n = (1/2)(3k^2 \pm k)$ and 0 otherwise.
[*Hint*: In Problem 3.37, $q^{\#}(n, E) - q^{\#}(n, O) = Q(n)$.]

3.118 (*Euler's Pentagonal Theorem*). Obtain the following recurrence relation for the partition $p(n)$:

$$p(n) = \sum_{k \geq 1} (-1)^{k-1}\left\{p\left(n - \frac{3k^2+k}{2}\right) + p\left(n - \frac{3k^2-k}{2}\right)\right\}$$

for every $n \geq 3$.
[*Hint*: Use Problem 3.117.]

Chapter 4

Group Theory in Combinatorics

4.1 THE BURNSIDE–FROBENIUS THEOREM

Definition: A nonempty set G with a binary operation $\circ$ defined on it constitutes a **group** $(G, \circ)$ if the following four properties hold. (*i*) For all x and y in G, $x \circ y$ is in G. (In **multiplicative notation** one writes xy instead of $x \circ y$.) (*ii*) There exists an **identity element** e in G such that $x \circ e = e \circ x = x$ for all x in G. (*iii*) Corresponding to each element x in G, there exists an **inverse element** x^{-1} in G such that $x \circ x^{-1} = x^{-1} \circ x = e$. (*iv*) For every x, y, and z in G the elements $x \circ (y \circ z)$ and $(x \circ y) \circ z$ are identical.

The associativity property (*iv*) allows us to write $x \circ y \circ z$ for the triple product. We usually write $a \circ b$ as ab and $(G, \circ)$ as G if there is no risk of ambiguity. A subset H of G is called a **subgroup** of $(G, \circ)$ if $(H, \circ)$ is a group. If G is a finite set with $|G| = n$, then $(G, \circ)$ is a **finite group of order** $\boldsymbol{n}$.

Example 1. The **symmetric difference** of sets A and B is defined by

$$A * B = (A \cup B) - (A \cap B)$$

i.e., $A * B$ is the set of elements that belong to A or to B but not to both. Let X be an n set and let 2^X be its power set (Problem 1.22). We shall verify that $(2^X, *)$ is a finite group of order 2^n.

(i) Obvious.

(ii) $e = \emptyset$, the null subset of X. For if $A \in 2^X$, the set of elements that belong to just one of A and $\emptyset$ is precisely A.

(iii) For every $A \in 2^X$, $A^{-1} = A$. Indeed, the set of elements that belong to just one of A and A is empty.

(iv) Write $P \equiv A * (B * C)$ and $Q \equiv (A * B) * C$. If $x \in P$, either (1) $x \in A$ and $x \in B$ and $x \in C$; or (2) $x \in A$ and $x \notin B$ and $x \notin C$; or (3) $x \notin A$ and $x \in B$ and $x \notin C$; or (4) $x \notin A$ and $x \notin B$ and $x \in C$. In case (1) $x \notin A * B$ and $x \in C$, so that $x \in Q$; in case (2) $x \in A * B$ and $x \notin C$, so that $x \in Q$; in case (3) $x \in A * B$ and $x \notin C$, so that $x \in Q$; in case (4) $x \notin A * B$ and $x \in C$, so that $x \in Q$. The conclusion is that $P \subseteq Q$. A similar analysis shows that $Q \subseteq P$. Hence, $P = Q$.

The **symmetric group** S_n—the group of the permutations of an n-set—was defined in Problems 1.65–1.68. The paramount importance of this group resides in the following fact: given any finite group G, there is a value of n such that S_n possesses a subgroup that is structurally identical with G (see Problem 4.54).

Definition: Suppose that G is a fixed subgroup of the symmetric group of a finite set X and that x is a given element of X. Let

$$Gx \equiv \{g(x) : g \in G\}$$

$$G_x \equiv \{g \in G : g(x) = x\}$$

$$F(g) \equiv \{z \in X : g(z) = z\}$$

In words, Gx (the **orbit of x with respect to G**) is the set of all images of the given element x under the permutations in G; G_x (the **stabilizer of x in G**) is the set of all permutations in G that have x as a fixed point; $F(g)$ (the **permutation character of g** in X) is the set of all fixed points of a given permutation $g \in G$. [Note that G_x is never empty—*why*?—but that $F(g)$ may be so.]

Theorem 4.1. Suppose that a finite set X possesses exactly k *distinct* orbits with respect to a group G of permutations of X. Then:

(i) For every $x \in X$, $|Gx|\,|G_x| = |G|$.
(ii) $\Sigma_{x\in X} |G_x| = k\,|G|$.
(iii) $\Sigma_{x\in X} |G_x| = \Sigma_{g\in G} |F(g)|$.

Proofs. (*i*) See Problem 4.22; (*ii*) see Problem 4.24; and (*iii*) see Problem 4.25.
Parts (*ii*) and (*iii*) of Theorem 4.1 imply

Theorem 4.2 (Burnside–Frobenius Theorem):

$$\sum_{g\in G} |F(g)| = k\,|G|$$

Notational Warning: The finite set X whose orbits are the "subject" of Theorems 4.1 and 4.2 is, in coloring problems, replaced by the finite set C of all color distributions. In those problems X has a different significance (it stands for the set of objects being colored).

4.2 PERMUTATION GROUPS AND THEIR CYCLE INDICES

Problem 1.158 introduced the concept of the cycle representation of a permutation f of $X = \{1, 2, \ldots, n\}$. The following algorithm produces this representation:

1. Choose an element i of X (usually $i = 1$). Find the image of i under the mapping f, then the image of the image, then ..., until the image j appears such that $f(j) = i$. Thus the cycle $(i \cdots j)$ has been generated.
2. Choose an element of X not found in any one of the cycles already generated, and use this element as element i in step 1, thereby generating a new cycle.
3. Repeat step 2 until X has been exhausted.

It is clear that the cycle representation of a permutation is unique up to the order of the cycles in the composition and up to the choice, within each cycle, of the leading element.

Example 2. Given $X = \{1, 2, \ldots, 8\}$ and

$$1\,2\,3\,4\,5\,6\,7\,8 \xrightarrow{f} 3\,2\,5\,1\,4\,8\,6\,7$$

Starting with 1: $f(1) = 3$, $f(3) = 5$, $f(5) = 4$, $f(4) = 1$. Thus we have a cycle of length 4 which may be denoted by (1 3 5 4) [or (3 5 4 1) or (5 4 1 3) or (4 1 3 5)].

Starting with 2: $f(2) = 2$. We have the cycle (2), of length 1.

Starting with 6: $f(6) = 8$, $f(8) = 7$, $f(7) = 6$. Now we have a cycle of length 3 which may be denoted by (6 8 7) [or (8 7 6) or (7 6 8)].

The sum of the lengths has reached $4 + 1 + 3 = 8 = |X|$, which means we are finished: the cycle representation of f is (1 3 5 4) (2) (6 8 7) (or $\cdots$).

Definition: Let the cycle representation of f, a permutation of an n set, consist of a_1 cycles of length 1, a_2 cycles of length $2, \ldots, a_i$ cycles of length $i, \ldots$. Then the **type** of f is the vector $[a_1 \;\; a_2 \;\; \cdots \;\; a_n]$, and the **weight** of the type is the positive integer $W = 1^{a_1} 2^{a_2} \cdots n^{a_n}$.

Example 3. The permutation of Example 2 has 1 cycle of length 1, 1 cycle of length 3, and 1 cycle of length 4. The type of this permutation is $[1 \;\; 0 \;\; 1 \;\; 1 \;\; 0 \;\; 0 \;\; 0 \;\; 0]$. The weight of this type is $1^1 3^1 4^1 = 12$.

Definition: Let G denote a group, of order m, of permutations of an n-set and let $g \in G$ be of type $[a_1 \;\; a_2 \;\; \cdots \;\; a_n]$. The **cycle index of g** is the monic multinomial

$$Z(g; x_1, x_2, \ldots, x_n) \equiv x_1^{a_1} x_2^{a_2} \cdots x_n^{a_n}$$

and the **cycle index of G** is the multinomial

$$Z(G; x_1, x_2, \cdots x_n) \equiv \frac{1}{m} \sum_{g \in G} Z(g; x_1, x_2, \ldots, x_n)$$

(The letter Z stands for 'Zyklenzeiger' used by Pólya.)

Example 4. Suppose the 4 vertices of a square are labeled 1, 2, 3, and 4, clockwise. A clockwise rotation through an angle of 0°, 90°, 180° or 270° takes the square into itself. Thus there are 4 **circular** or **cyclic symmetries.** In addition, there are 4 **dihedral symmetries** that are obtained by reflection of the square in the 2 diagonals and in the 2 lines bisecting opposite sides. Consequently, the symmetries of the square compose a subgroup G of order 8 of S_4; the elements of G are as follows:

(i) The permutation induced by rotating the square clockwise through 0° is $g_1 = e = (1)(2)(3)(4)$, with cycle index x_1^4.

(ii) The permutation induced by rotating the square clockwise through 90° is $g_2 = (1\,2\,3\,4)$ with cycle index x_4^1.

(iii) The permutation induced by rotating the square clockwise through 180° is $g_3 = (1\,3)(2\,4)$, with cycle index x_2^2.

(iv) The permutation induced by rotating the square clockwise through 270° is $g_4 = (1\,4\,3\,2)$, with cycle index x_4^1.

(v) The permutation induced by reflection in the line joining the midpoints of $\overline{1\,2}$ and $\overline{3\,4}$ is $g_5 = (1\,2)(3\,4)$, with cycle index x_2^2.

(vi) The permutation induced by reflection in the line joining the midpoints of $\overline{1\,4}$ and $\overline{2\,3}$ is $g_6 = (1\,4)(2\,3)$, with cycle index x_2^2.

(vii) The permutation induced by reflection in the diagonal joining corners 2 and 4 is $g_7 = (2)(4)(1\,3)$, with cycle index $x_1^2 x_2^1$.

(viii) The permutation induced by reflection in the diagonal joining corners 1 and 3 is $g_8 = (1)(3)(2\,4)$, with cycle index $x_1^2 x_2^1$.

The cycle index of G is therefore

$$Z(G; x_1, x_2, x_3, x_4) = \tfrac{1}{8}(x_1^4 + 2x_1^2 x_2 + 3x_2^2 + 2x_4)$$

4.3 PÓLYA'S ENUMERATION THEOREMS

Definition: A function f from a finite set X to a finite set of **colors** Y is called a **coloring** of X. Two colorings f and ϕ in the set C of all colorings of X are said to be **equivalent (indistinguishable) with respect to a group G of permutations of X** if there exists a permutation π in G such that $f(x) = \phi(\pi(x))$ for all x in X. In other words, if we attach names to the elements of X—so that G may be considered a group of "renamings"—then we do not distinguish between 2 colorings of X that become identical under some renaming in G. Clearly, the relation of indistinguishability is reflexive, symmetric, and transitive; i.e., an equivalence relation.

Definition: The equivalence classes into which C is partitioned by the indistinguishability relation are called the **patterns in C** (with respect to the group G).

A basic question: For specified X, Y and G, how many patterns will there be?

Theorem 4.3 (***Pólya's First Enumeration Theorem***). Let C be the set of all functions (colorings) from an n-set X to an r set Y ($n \geq 2$). Let G be a group of permutations of X, with cycle index $Z(G; x_1, x_2, \ldots, x_n)$. Then the number of patterns in C with respect to G is $Z(G; r, r, \ldots, r)$.

Proof. See Problem 4.61.

Example 5. If, in Theorem 4.3, $G = \{e\}$, then any 2 colorings are distinguishable, so that the number of patterns is the number of colorings. Because

$$Z(G; x_1, x_2, \ldots, x_n) = Z(e; x_1, x_2, \ldots, x_n) = x_1^n$$

The theorem gives this number as r^n (in agreement with the product rule of Chapter 1).

Example 6. Consider the group G of symmetries of the square, as developed in Example 4. (*i*) Count the distinguishable (with respect to G) colorings of the 4 vertices, if each vertex is to be either red or blue. (*ii*) Exhibit in a diagram the patterns, enumerated in (*i*).

(i) From Example 4,

$$Z(G; 2, 2, 2, 2) = \tfrac{1}{8}(2^4 + 2 \cdot 2^3 + 3 \cdot 2^2 + 2 \cdot 2) = 6$$

(ii) Figure 4-1 shows *how* the $2^4 = 16$ colorings fall into 6 patterns. Within a pattern, reflection of a coloring in the indicated symmetry axis of the square gives rise to the (equivalent) coloring immediately to the right. (Note that 2 colorings may be equivalent under more than 1 symmetry. For example, the second coloring in pattern IV is also a 90° rotation of the first.)

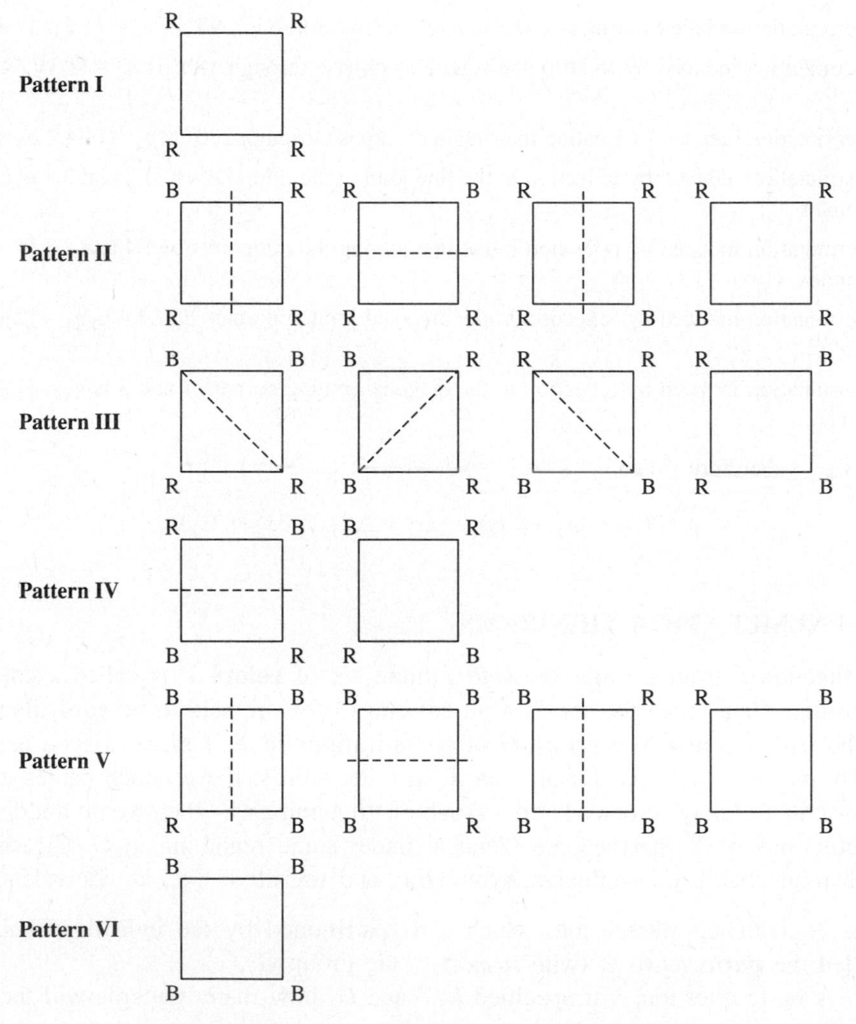

Fig. 4-1

It is important to recognize how the partitioning of the colorings of the square into patterns follows the modes of division of the vertex set X into a subset X_1 with all elements red and a subset X_2 with all elements blue. ("Division," not "partitioning," because X_1 or X_2 may be empty.) Thus, pattern I comprehends all divisions for which $|X_1| = 4$ and $|X_2| = 0$; II, $|X_1| = 3$ and $|X_2| = 1$; III *and* IV, $|X_1| = 2$ and $|X_2| = 2$; V, $|X_1| = 1$ and $|X_2| = 3$; and VI, $|X_1| = 0$ and $|X_2| = 4$.

Is it possible that the cycle index might be "milked" to yield not just the number of distinct patterns [as in Example 6(*i*)] but a complete *inventory* of these patterns [as in Example 6(*ii*)]? An affirmative answer is given by Pólya's remarkable second theorem, which derives from the cycle index a multinomial generating function for the numbers of patterns corresponding to the various divisions of X into color classes.

Before stating Pólya's second theorem (its proof is developed in Problem 4.66), we must slightly complicate the notation. As before, X will denote an n-set on which a group G of permutations is defined. But now Y is an abstract r-set, on which a **weight function** $w(y)$ is defined. Almost always w will be color-valued—putting us back where we were before—but occassionally (see Example 8 and Problem 4.71) we shall want another sort of weight. In any case, functions from X to Y will still be called "colorings," and C will still denote the set of all colorings.

Definition: Let the weight function w map Y into a set of r colors, $\{w(y_1), w(y_2), \ldots, w(y_r)\}$. The **pattern inventory** (of C) **with respect to** G is the multinomial

$$\mathrm{PI}(G; w(y_1), w(y_2), \ldots, w(y_r)) \equiv \sum_{\substack{n_1+n_2+\cdots+n_r=n \\ n_i \geq 0}} \tau(n_1, n_2, \ldots, n_r)[w(y_1)]^{n_1}[w(y_2)]^{n_2} \cdots [w(y_r)]^{n_r}$$

The coefficient $\tau(n_1, n_2, \ldots, n_r)$ gives the number of distinguishable (with respect to G) colorings (= number of patterns) that assign color $w(y_1)$ to n_1 elements of X; color $w(y_2)$ to n_2 elements; . . . ; color $w(y_r)$ to n_r elements. The summation is over the sizes of the color classes into which X is divided; by Problem 1.139 the sum consists of $C(n+r-1, r-1)$ terms.

Theorem 4.4 ***(Pólya's Second Enumeration Theorem).*** The pattern inventory, $\mathrm{PI}(G; w(y_1), w(y_2), \ldots, w(y_r))$, is the value of the cycle index, $Z(G; x_1, x_2, \ldots, x_n)$, at

$$x_i = [w(y_1)]^i + [w(y_2)]^i + \cdots + [w(y_r)]^i \qquad (i = 1, 2, \ldots, n)$$

Proof. See Problem 4.66.

Example 7. Rework Example 6 by means of the pattern inventory (PI).

Here $n = 4$ and $r = 2$; let $w(y_1) = R$ (red) and $w(y_2) = \mathrm{B}$ (blue). From Example 4,

$$Z(G; x_1, x_2, x_3, x_4) = \tfrac{1}{8}(x_1^4 + 2x_1^2x_2 + 3x_2^2 + 2x_4)$$

whence, by Theorem 4.4,

$$\mathrm{PI}(G; \mathrm{R}, \mathrm{B}) = \tfrac{1}{8}[(\mathrm{R}+\mathrm{B})^4 + 2(\mathrm{R}+\mathrm{B})^2(\mathrm{R}^2+\mathrm{B}^2) + 3(\mathrm{R}^2+\mathrm{B}^2)^2 + 2(\mathrm{R}^4+\mathrm{B}^4)]$$

$$= \mathrm{R}^4 + \mathrm{R}^3\mathrm{B} + 2\mathrm{R}^2\mathrm{B}^2 + \mathrm{R}\mathrm{B}^3 + \mathrm{B}^4$$

From the pattern inventory one reads that there exists 1 pattern in which 4 vertices are colored red (and 0 vertices are colored blue); 1 pattern in which 3 are red; 2 patterns in which 2 are red; 1 pattern in which 1 is red; and 1 pattern in which 0 are red (and 4 are blue)—for a total of 6 patterns. Incorporating as it does practically the entire content of Fig. 4-1, the pattern inventory represents a complete solution to the problem.

Example 8. Derive Theorem 4.3 from Theorem 4.4.

Use the special weight function $w(y) \equiv 1$ in the definition of the pattern inventory to get

$$\mathrm{PI}(G; 1, 1, \ldots, 1) = \sum_{\substack{n_1+n_2+\cdots+n_r=n \\ n_i \geq 0}} \tau(n_1, n_2, \ldots, n_r)$$

$$= \text{total number of patterns}$$

But, by Theorem 4.4 with the same weight function,

$$\mathrm{PI}(G; 1, 1, \ldots, 1) = Z(G; r, r, \ldots, r)$$

Solved Problems

THE BURNSIDE–FROBENIUS THEOREM

4.1 Prove that, in any group, (*a*) the identity element is unique; and (*b*) the inverse of any element is unique.

(*a*) Suppose there existed two identities, e and f. Then, since e is a right-identity and f is a left-identity, $f = f \circ e = e$.

(*b*) If element x had 2 inverses, y and z, the associative law would give

$$(y \circ x) \circ z = y \circ (x \circ z) \quad \text{or} \quad e \circ z = y \circ e \quad \text{or} \quad z = y$$

4.2 Groups $(G, \circ)$ and $(G', \circ')$ are **isomorphic** (identical in structure) if there exists a one-to-one correspondence f between G and G' such that $f(x \circ y) = f(x) \circ' f(y)$, for all x and y in G. (*Shorter:* Isomorphic groups have the same multiplication table.) Show that, under an isomorphism, (*a*) the identity elements of G and G' correspond; (*b*) if u and v are inverses in G, then $f(u)$ and $f(v)$ are inverses in G'. and (c) give an example of 2 isomorphic groups of order n.

(*a*) In G, $x \circ e = e \circ x = x$; whence $f(x \circ e) = f(e \circ x) = f(x)$, or

$$f(x) \circ' f(e) = f(e) \circ' f(x) = f(x)$$

which shows that *the* identity [Problem 4.1(*a*)] in G' is $e' = f(e)$.

(*b*) From $u \circ v = v \circ u = e$ and part (*a*),

$$f(u) \circ' f(v) = f(v) \circ' f(v) = e'$$

(*c*) One such pair is composed of the group of rotational symmetries of a regular n-gon and the group $(\{0, 1, 2, \ldots, n-1\}, +)$, where the operation $+$ is addition modulo n.

4.3 If x is an element in a group $(G, \circ)$, we write $x \circ x$ as x^2, $x \circ x^2 = x^2 \circ x$ as x^3, and so on. Similarly, $x^{-1} \circ x^{-1}$ is written as x^{-2}, $x^{-1} \circ x^{-2}$ as x^{-3}, etc. Thus the ***k*th power**, x^k, of the element x is well defined when k is any nonzero integer; we make the natural definition $x^0 \equiv e$. The group G is said to be a **cyclic group** if it contains an element x such that every element of G is a power of x. In this case we say that G is **generated by** x, and we write $G = \langle x \rangle$. If x generates G and if the powers of x are all distinct, G is an **infinite cyclic group**. Show that the set of all integers under the binary operation of addition is an infinite cyclic group.

If Z is the set of all integers, $(Z, +)$ is a group because all 4 group axioms are satisfied by the structure. Let $G = \langle a \rangle$ be an infinite cyclic group. The mapping $f: Z \to G$ defined by $f(z) = a^z$ is obviously a bijection, and we have

$$f(z + w) = a^{z+w} = a^z \circ a^w = f(z) \circ f(w)$$

Therefore the 2 groups are isomorphic, which means that $(Z, +)$ is the infinite cyclic group $\langle 1 \rangle$.

4.4 A group $(G, \circ)$ is **abelian** if $x \circ y = y \circ x$ for every x and y in G. Show that every cyclic group is abelian.

Let x and y be 2 elements in a cyclic group G generated by g. Because $x = g^m$ and $y = g^n$ for some integers m and n,

$$x \circ y = g^m \circ g^n = g^{m+n} = g^{n+m} = g^n \circ g^m = y \circ x$$

4.5 If x is an element of $(G, \circ)$ and if there exists a positive integer m such that x^m is the identity element e in G, then x is said to be **of finite order.** If x is of finite order, the smallest positive integer k such that $x^k = e$ is the **order** of x in G. Show that if the order of x is k, then $C = \{e, x, x^2, \ldots, x^{k-1}\}$ is a (cyclic) subgroup of G, of order k.

Because $e \in C$, it is only necessary to show that (*i*) C is closed under $\circ$; and (*ii*) $u \in C$ implies $u^{-1} \in C$.

(i) If $x^m, x^n \in C$ and $m + n \equiv r \pmod{k}$,

$$x^m \circ x^n = x^{m+n} = x^{qk+r} = x^{qk} \circ x^r = e^q \circ x^r = x^r$$

and $0 \leq r \leq k - 1$.

(ii) If $u = x^m$ is in C, $v = x^{k-m}$ also is in C, and

$$u \circ v = v \circ u = x^k = e$$

4.6 Show that if G is a finite group of order n, then all elements of G are of finite order and no order exceeds n.

If $x \in G$, the elements in $\{x^k : k = 0, 1, \ldots\}$ cannot be all distinct, since G is finite. Hence there must exist integers p and q, where $p > q \geq 0$, such that $x^p = x^q$, which implies $x^{p-q} = e$. So x is of finite order—say, k. Because x generates a *sub*group of order k (Problem 4.5), $k \leq n$.

4.7 If G is a cyclic group of order n, find the number of distinct generators of G.

Suppose $G = \langle x \rangle = \{e, x^1, x^2, \ldots, x^{n-1}\}$. Let m be any positive integer less than, and relatively prime to, n; consider the cyclic group $G' = \langle x^m \rangle = \{e, x^{1m}, x^{2m}, \ldots, x^{(n-1)m}\}$. To establish that $G' = G$ it suffices to show that the elements of G' are distinct. Suppose, on the contrary, that for some $0 \leq b < a \leq n - 1$ we had $x^{am} = x^{bm}$. Then necessarily

$$(a - b)m = cn$$

for some integer c. But, m and n being relatively prime, this equation would require that n divide $a - b$, an impossibility since $0 < a - b < n$.

If, on the other hand, $m = rp$ and $n = sp (r < s < n)$, then

$$(x^m)^s = (x^n)^r = e$$

so that the order of x^m—and consequently the order of $\langle x^m \rangle$—is smaller than n. The conclusion is that G has just as many generators as there are positive integers less than, and relatively prime to, n; i.e., $\phi(n)$ generators (Problem 2.45).

4.8 Let G denote the set of all 2×2 matrices such that the first row is $[1 \quad m]$, where m is an integer, and the second row is $[0 \quad 1]$. Show that G is an infinite cyclic group under matrix multiplication. Find the generator of this group.

Because

$$\begin{bmatrix} 1 & r \\ 0 & 1 \end{bmatrix} \begin{bmatrix} 1 & s \\ 0 & 1 \end{bmatrix} = \begin{bmatrix} 1 & r+s \\ 0 & 1 \end{bmatrix}$$

the one-to-one correspondence

$$\begin{bmatrix} 1 & m \\ 0 & 1 \end{bmatrix} \leftrightarrow m$$

establishes that G is isomorphic (Problem 4.2) to the infinite cyclic group of the integers under addition. Since $(Z, +)$ is generated by 1, G is generated by

$$\begin{bmatrix} 1 & 1 \\ 0 & 1 \end{bmatrix}$$

4.9 Prove that a subgroup of a cyclic group is cyclic.

Let G' be a subgroup of $G = \langle x \rangle$. Every element in G' is of the form x^k, let m be the smallest positive k for which x^k is in G'. Now, for any integer k, the division algorithm gives $k = qm + r$, where $0 \le r < m$. Hence,

$$x^k = x^{qm+r} = (x^m)^q x^r = ux^r$$

in which $u \in G'$ because $x^m \in G'$ and G' is closed under multiplication. It follows that u^{-1} also belongs to G'; so that, if $x^k \in G$,

$$x^r = u^{-1} x^k \in G'$$

If r were positive, this would violate the minimality of m. Therefore, $r = 0$ and each x^k in G' may be written as $(x^m)^q$; i.e., $G' = \langle x^m \rangle$.

4.10 If G and G' are 2 groups, the **direct product** of G and G' is the set of all ordered pairs,

$$G \times G' = \{(g, g') : g \in G, g' \in G'\}$$

endowed with the binary operation defined by

$$(g_1, g_1')(g_2, g_2') = (g_1 g_2, g_1' g_2')$$

Show that $G \times G'$ is a group.

By definition, the product of 2 elements in $G \times G'$ is in $G \times G'$. Further,

$$(g, g')(e, e') = (ge, g'e') = (g, g') = (eg, e'g') = (e, e')(g, g')$$

Thus (e, e') is the identity in the direct product. Also,

$$(g, g')(g^{-1}, g'^{-1}) = (g^{-1}, g'^{-1})(g, g') = (e, e')$$

so each element in the direct product has an inverse element. The associativity rule is obviously satisfied in the product structure. Thus $G \times G'$ is a group.

4.11 A finite cyclic group of order m is denoted by C_m. Show that if m and n are relatively prime, the direct product $C_m \times C_n$ is a cyclic group.

As the direct product is a finite group of order mn, it suffices to prove that it contains an element of order mn. Let C_m be generated by x and C_n be generated by y, and let k be the order of the element $z = (x, y)$ of the direct product. Then $z^k = (e, e') = (x^k, y^k)$, which implies that k is a multiple of m and a multiple of n. Since k is the *smallest* positive integer p such that $z^p = (e, e')$, it follows that k is the least common multiple of m and n. But, if m and n are relatively prime, their least common multiple is their product. Thus the order of z is mn.

4.12 Show that a subset H of a group is a subgroup if and only if xy^{-1} is in H whenever x and y are in H.

The necessity of the condition is obvious. Assume, then, that xy^{-1} is in H whenever x and y are in H. (*i*) If x is any element in H, then $xx^{-1} = e$ is in H. (*ii*) Given e and x in H, $ex^{-1} = x^{-1}$ is in H. (*iii*) If x and y are in H, then y^{-1} is in H, and so $x(y^{-1})^{-1} = xy$ is in H.

4.13 Show that a subset H of a *finite* group is a subgroup if and only if H is closed with respect to multiplication.

Again only sufficiency need be proved. Let x and y be 2 elements in H, and let the order of y be m. Then $y^m = e$ implies $y^{m-1} = y^{-1}$; and, by hypothesis, y^{m-1} is in H. Thus x and y^{-1} are in H, and the conclusion follows from Problem 4.12.

4.14 If H is a subgroup of G and x is an element of G, the set $xH \equiv \{xh : h \in H\}$ is called the **left coset** of H with respect to x. (The **right coset** of H with respect to x is $Hx \equiv \{hx : h \in H\}$.) Show that $xH = yH$ if and only if $y^{-1}x$ is in H.

Suppose that $xH = yH$. Since e is in H, $xe = x$ is in xH, and therefore x is in yH. So there exists an h in H such that $x = yh$. That is, H contains $h = y^{-1}x$.

On the other hand, suppose that x and y are 2 elements of G such that $y^{-1}x$ is in H. Then there exists an element h in H such that $x = yh$. Let z be in xH; there exists an h' in H such that

$$z = xh' = (yh)h' = y(hh') = yh''$$

This implies that z is in yH. Likewise, every element in yH is in xH. So $xH = yH$.

4.15 Show that if H is a subgroup of G and if x and y are in G, then either $xH \cap yH$ is empty or $xH = yH$.

If $xH \cap yH$ is not empty, there exists an element z which is in xH and also in yH. Hence there exist h and h' in H such that $z = xh = yh'$, which in turn implies that $y^{-1}x = h'h^{-1}$ is in H. Problem 4.14 now gives $xH = yH$.

4.16 Show that the class of distinct left cosets of a subgroup H of a group G constitutes a partition of the group. (The same is true of the class of distinct right cosets of H.)

Let x be any element of G. Then x is an element of xH, since $x = xe$ and e is in H. Thus every element of G is in at least 1 left coset of H. Two distinct left cosets have no elements in common, as established in Problem 4.15. Thus the left cosets of H make up a partition of G.

4.17 Show that if H ($|H| = k$) is a finite subgroup of G, then every left (right) coset of H has cardinality k.

Let $H = \{h_1, h_2, h_3, \ldots, h_k\}$ and let x be any element of G. Then $xH = \{xh_1, xh_2, \ldots, xh_k\}$. The elements of xH must be distinct, for $xh_i = xh_j$ would imply $h_i = h_j$. Hence, $|xH| = k$.

4.18 Prove *Lagrange's theorem:* The order of a finite group is divisible by the order of any subgroup.

Let the group be of order n and let a given subgroup, of order s, have r distinct left cosets. Then, by Problems 4.16 and 4.17, $rs = n$.

4.19 Infer from Lagrange's theorem that (*a*) the order of any element of a finite group G divides the order of G; and (*b*) if the order of G is a prime number p, then G is isomorphic to C_p (Problem 4.11).

(*a*) An element of order s generates a subgroup C_s, so that s must divide the order of G.

(*b*) By (*a*), any element of G besides the identity is of order p. Hence the subgroup G_p generated by that element must exhaust G. [Note the agreement with Problem 4.7: C_p has $p - 1 = \phi(p)$ distinct generators.]

4.20 (*Characterization Theorem for Cyclic Groups*) If G is a group of order $n \geq 2$, the following are equivalent: (*i*) G is a cyclic group. (*ii*) For each divisor d of n, the cardinality of $\{x \in G : x^d = e\}$ is d. (*iii*) For each divisor d of n, the cardinality of $\{x \in G : \text{the order of } x \text{ is } d\}$ is $\phi(d)$.

(*i*) ⇒ (*ii*) Suppose G is generated by x. Let $n = dk$ and consider the collection

$$Y \equiv \{x^0, x^k, x^{2k}, x^{3k}, \ldots, x^{(d-1)k}\}$$

The d elements in this collection are distinct (because x is of order n). The typical element x^{ik} of Y satisfies

$$(x^{ik})^d = (x^{dk})^i = (x^n)^i = e^i = e$$

Thus Y is a subset of $\{x \in G : x^d = e\}$.

Conversely, let y be any element of G such that $y^d = e$. Since x is a generator of G, there exists an integer $0 \leq m \leq n-1$ such that $y = x^m$; therefore, $x^{md} = e$. But x is of order n; so that, for some integer r,

$$md = rn = rdk \qquad \text{or} \qquad m = rk$$

Thus $y = x^{rk}$, with $0 \leq r \leq d-1$ (because $r/d = m/n$), which means that $y \in Y$. Consequently, $\{x \in G : x^d = e\}$ and Y are identical sets, so that

$$|\{x \in G : x^d = e\}| = |Y| = d$$

(ii)* ⇒ *(iii) Let y be an element of G, of order c. Then $y^d = e$ if and only if $c \mid d$ (c divides d). Consequently, $\{x \in G : x^d = e\}$ may be partitioned in such manner that the ith cell consists of all elements of G whose order equals the ith divisor of d. Define $f(c)$ to be the number of elements of order c, and specialize d to a divisor of n. Then, by *(ii)*,

$$\sum_{c \mid d} f(c) = d$$

Inversion of (*) by the Möbius formula (Problem 2.57) yields

$$f(d) = \sum_{c \mid d} \mu(c)(d \mid c) = \phi(d)$$

The second equality following from Problem 2.51.

(iii)* ⇒ *(i) By *(iii)*, with $d = n$, there exist $\phi(n) \geq 1$ elements of order n in G. Hence $G = C_n$.

4.21 Given a finite set X and a group G of permutations of X, prove that the distinct orbits with respect to G constitute a partition of X.

On X define a binary relation $\mathcal{R}$ by

$$x \mathcal{R} y \Leftrightarrow y = g(x) \text{ for some } g \in G$$

It is trivial to show that $\mathcal{R}$ is an equivalence relation, the equivalance classes of which constitute a partition of X. But these equivalance classes are just the distinct orbits of X.

4.22 Prove Theorem 4.1(*i*).

Problem 4.90 shows G_x to be a subgroup of G; our plan is to produce a bijection between Gx and the set L of distinct left cosets of G_x, whereupon the theorem follows from the proof of Lagrange's theorem (Problem 4.18).

Let $u \in Gx$; i.e., $u = g(x)$ for some $g \in G$. Consider the mapping $u \to gG_x$ from Gx to L.

(i) The mapping is onto. In fact, if $lG_x \in L$, we have—l being a permutation of X—$l(x) = y$ ($y \in X$). This means that $y \in Gx$ and $y \to lG_x$.

(ii) The mapping is one-one. Let u and v belong to Gx: $u = g(x)$ and $v = h(x)$, for $g, h \in G$. Suppose that in L, $gG_x = hG_x$. Then, by Problem 4.14, $h^{-1}g \in G_x$, which implies

$$h^{-1}(g(x)) = x \qquad \text{or} \qquad g(x) = h(x) \qquad \text{or} \qquad u = v$$

Thus our mapping is the desired bijection.

4.23 With reference to Problems 4.21 and 4.22, show that if x and y are in the same orbit,

$$|G_x| = |G_y|$$

By Problem 4.21, $y = h(x)$ for some $h \in G$. Thus the set

$$G(x, y) \equiv \{g \in G : g(x) = y\}$$

is known to be nonempty (it contains the element h). Now,

$$g \in G(x, y) \Rightarrow g(x) = h(x) \Rightarrow h^{-1}g \in G_x \Rightarrow g \in hG_x \tag{*}$$

Thus $G(x, y) \subset hG_x$. But the implications in (*) can be reversed, whence $hG_x \subset G(x, y)$. Consequently, $G(x, y) = hG_x$; Problem 4.17 gives

$$|G(x, y)| = |G_x|$$

Similarly it is established that $G(x, y)$ is a right coset of G_y, so that

$$|G(x, y)| = |G_y|$$

Hence, $|G_x| = |G_y|$.

4.24 Prove Theorem 4.1(*ii*).

By Problem 4.21, there exist elements $x_1, x_2, \ldots, x_k$ such that $\{Gx_1, Gx_2, \ldots, Gx_k\}$ is a partition of X. This lets us write

$$\sum_{x \in X} |G_x| = \sum_{i=1}^{k} \sum_{x \in Gx_i} |G_x|$$

But, by Problem 4.23, $|G_x|$ has the constant value $|G_{x_i}|$ over G_{x_i} (since $x_i \in Gx_i$). Hence,

$$\sum_{x \in X} |G_x| = \sum_{i=1}^{k} |Gx_i|\,|G_{x_i}| = \sum_{i=1}^{k} |G| = k\,|G|$$

where the second equality comes from Theorem 4.1(*i*).

4.25 Prove Theorem 4.1(*iii*).

In the sum $\Sigma_{g \in G} |F(g)|$ the *count* (Section 2.3) of any $x \in X$ is $|G_x|$; therefore,

$$\sum_{g \in G} |F(g)| = \sum_{x \in X} |G_x|$$

4.26 Show that the Burnside–Frobenius theorem holds for $X = \{a, b, c, d\}$ and $G = \{g_1, g_2, g_3, g_4\}$, where g_1 maps each element into itself; g_2 maps a and b into each other and c and d into each other; g_3 maps a and c into each other and b and d into each other; g_4 maps a and d into each other and b and c into each other,

First we verify that G is a group. Of course, g_1 is the identity. We find:

$$g_2g_3 = g_4 = g_3g_2 \qquad g_2g_4 = g_3 = g_4g_2 \qquad g_3g_4 = g_2 = g_4g_3$$

Also, each element in G is its own inverse. Thus G is a group; $|G| = 4$.

Because $Ga = Gb = Gc = Gd = X$, there is only 1 orbit ($k = 1$). Further, $F(g_1) = X$ and $F(g_i)$ is empty when $i = 2, 3, 4$. Hence,

$$\sum_{g \in G} |F(g)| = |X| = 4 = (1)(4) = k\,|G|$$

4.27 (*Example 6 Revisited*) Consider a square the corners of which may be independently colored either red or blue; the set C of colorings will have $2^4 = 16$ elements. Verify that the Burnside–Frobenius theorem holds if indistinguishability of colorings is defined by the symmetry group of the square (Example 4).

First, reread the notational warning that follows Theorem 4.2. Next, identify the 16 elements of C by reading Fig. 4-1 by rows:

f_1: No corner is blue.
f_2: Only top = left corner is blue.
f_3: Only top = right corner is blue.
f_4: Only bottom = right corner is blue.
f_5: Only bottom = left corner is blue.
f_6: Only top corners are blue.
f_7: Only left-hand corners are blue.
f_8: Only bottom corners are blue.
f_9: Only right-hand corners are blue.
f_{10}: Only top right and bottom = left corners are blue.
f_{11}: Only top left and bottom = right corners are blue.
f_{12}: Only bottom left corner is red.
f_{13}: Only bottom right corner is red.
f_{14}: Only top right corner is red.
f_{15}: Only top left corner is red.
f_{16}: No corner is red.

To apply the Burnside–Frobenius theorem to the set C, it is necessary to translate the group G of permutations *of the corners* into a group G' of permutations *of the colorings*. The formal correspondence $g \to g'$ is developed in Problem 4.59. [Informally, the idea is this: you can hold the color pattern fixed and rotate the square "under it," or you can hold the square fixed and rotate the color pattern "over it" (in the opposite direction); it comes to the same thing. Likewise for reflections.] For now, it is enough to point out that, under the correspondence, the fixed-point set $F(g')$ consists of those colorings f that are taken into themselves by the corner permutation g. Hence, with the elements of G labeled as in Example 4, we have:

$$F(g'_1) = C \qquad F(g'_5) = \{f_1, f_6, f_8, f_{16}\}$$
$$F(g'_2) = \{f_1, f_{16}\} \qquad F(g'_6) = \{f_1, f_7, f_9, f_{16}\}$$
$$F(g'_3) = \{f_1, f_{10}, f_{11}, f_{16}\} \qquad F(g'_7) = \{f_1, f_3, f_5, f_{10}, f_{11}, f_{12}, f_{14}, f_{16}\}$$
$$F(g'_4) = \{f_1, f_{16}\} \qquad F(g'_8) = \{f_1, f_2, f_4, f_{10}, f_{11}, f_{13}, f_{15}, f_{16}\}$$

The distinct orbits of C are precisely the "patterns" of Fig. 4-1. Thus, $k = 6$, and

$$\sum_{g' \in G'} |F(g')| = 16 + 2 + 4 + 2 + 4 + 4 + 8 + 8 = 48 = (6)(8) = k\,|G'|$$

4.28 Use the Burnside–Frobenius theorem to find the number of distinguishable colorings—with respect to the symmetry group of the square—of a 3×3 chessboard, if 2 cells must be colored black and the others white.

The set C of colorings has $C(9, 2) = 36$ elements; the permutation groups G and G' are those of Problem 4.27. As g'_1 is the identity, $F(g'_1) = C$ and $|F(g'_1)| = 36$. There is no coloring which will come back to itself after rotating the square through 90° or 270°; hence, $|F(g_2')| = |F(g_4') = 0$. There are 4 colorings that are preserved under a 180° rotation. [See Fig. 4-2. When any of the cell pairs (2, 8), (4, 6), (1, 9), (3, 7) is colored black, the whole pattern is invariant.] Thus, $|F(g_3')| = 4$. There are 6 colorings which are invariant under g_5': those with black pairs (1, 3), (4, 6), (7, 9), (2, 5), (2, 8), (5, 8), so $|F(g_5')| = 6$. Similarly,

$$|F(g_6')| = |F(g_7')| = |F(g_8')| = 6$$

1	2	3
4	5	6
7	8	9

Fig. 4-2

The theorem then gives the number of distinguishable colorings (distinct orbits of C) as

$$k = \frac{36 + 4 + 4(6)}{8} = 8$$

4.29 From the Burnside–Frobenius theorem, obtain the number of ways of seating n people around a circular table. (Compare Problem 1.34.)

For purposes of analysis let the people instead be seated at the vertices of a regular n-gon. We then have a coloring problem: in how many ways can n colors (people) be assigned to the n vertices if indistinguishability is with respect to the group of rotational symmetries of the n-gon? The set C of colorings is of cardinality $n!$, and the groups G and G' of vertex and coloring permutations are of order n. The only element of G' with a nonempty fixed-point set is the identity, for which $|F(e')| = n!$. Hence the theorem gives the number of distinct orbits in C (the number of distinct seating arrangements) as

$$k = \frac{n!}{n} = (n-1)!$$

4.30 Find the number of ways of coloring the corners of a regular pentagon using 3 colors if indistinguishability is with respect to the subgroup of rotational symmetries.

There are $3^5 = 243$ ways of coloring the 5 corners if rotational symmetries are ignored; thus we have a set C of 243 elements. The group G' of rotational symmetries has 5 elements. Let g_1' be the identity; then $F(g_1')$ has 243 elements. The other rotations will preserve a color configuration if and only if it involves a single color. This means that $F(g_2')$, $F(g_3')$, $F(g_4')$, $F(g_5')$ have 3 elements each. Thus the number of colorings is $(\frac{1}{5})(243 + 12) = 51$.

4.31 (*The Burnside–Frobenius Theorem with Weights*) Suppose that $X_1, X_2, \ldots, X_k$ are the distinct orbits in the set $X = \{x_1, x_2, \ldots, x_n\}$, with respect to the permutation group $G = \{g_1, g_2, \ldots, g_m\}$. On X define a weight function $\omega(x)$—weights may be numbers or algebraic symbols—with the property that whenever x_r and x_s are in the same orbit, $\omega(x_r) = \omega(x_s)$. Use the following recipe to induce a weight function on G:

$$W(g_i) = \sum_{x \in F(g_i)} \omega(x) \qquad (i = 1, 2, \ldots, m)$$

that is, the weight of a permutation in G is the total weight of its fixed points in X. Prove that

$$\sum_{i=1}^{m} W(g_i) = \left(\sum_{p=1}^{k} \omega_p \right) m \qquad (*)$$

in which $\omega_p (p = 1, 2, \ldots, k)$ is the unique value assumed by $\omega(x)$ over X_p.

Let t be an element of X_p, so that $X_p = Gt$ and $\omega(t) = \omega_p$. By definition of the stabilizer, t contributes its weight to exactly $|G_t|$ summands on the left side of (*); its contribution is thus $|G_t|\, \omega_p$. Because $|G_x|$ is constant over X_p (Problem 4.23), any other element of X_p makes the same contribution. Consequently, the net contribution of X_p to the left side is

$$|Gt| \times |G_t|\, \omega_p = m\omega_p$$

wherein Theorem 4.1(i) was used. As this weight is precisely reflected in the right side of (*), the equation is proved.

PERMUTATION GROUPS AND THEIR CYCLE INDICES

4.32 (*Cauchy's Formula*) Show that the number of permutations of $X = \{1, 2, \ldots, n\}$ that are of type $[a_1 \quad a_2 \quad \cdots \quad a_n]$ is

$$\frac{n!}{Wa_1!\, a_2! \cdots a_n!}$$

where $W = 1^{a_1} 2^{a_2} \cdots n^{a_n}$ is the weight of the type.

The number of ways of partitioning X into a_1 cells of cardinality 1, a_2 cells of cardinality 2, $\ldots$, a_n cells of cardinality n, is given by Theorem 2.3 as

$$N = \frac{n!}{[a_1!\,(1!)^{a_1}][a_2!\,(2!)^{a_2}] \cdots [a_n!\,(n!)^{a_n}]}$$

But a *cell* is not the same as a *cycle*. In fact, a cell with q elements gives rise to $(q-1)!$ distinct cycles of length q—1 for each circular permutation of the elements (see Problem 1.34). Hence the desired number is

$$N[(1-1)!]^{a_1}[(2-1)!]^{a_2} \cdots [(n-1)!]^{a_n} = \frac{n!}{[a_1!\,1^{a_1}][a_2!\,2^{a_2}] \cdots [a_n!\,n^{a_n}]}$$

$$= \frac{n!}{Wa_1!\, a_2! \cdots a_n!}$$

4.33 Into how many types does S_n fall?

$[a_1 \quad a_2 \quad \cdots \quad a_n]$ is a type-vector if and only if it is a solution-vector of

$$1a_1 + 2a_2 + \cdots + na_n = n \qquad (a_i \text{ is a nonnegative integer})$$

By Section 3.2 (see page 105) there are exactly $p(n)$ types.

4.34 Two permutations f and g of X are said to be **conjugate** if there exists a permutation h of X such that $hf = gh$. Show that 2 permutations are conjugate if and only if they are of the same type.

If f and g are conjugate permutations, there exists a permutation h such that $g = hfh^{-1}$. Suppose $C = (x_1 x_2 \cdots x_r)$ is a cycle of f, of length r. Then

$$f(x_1) = x_2 \qquad f(x_2) = x_3 \qquad \cdots \qquad f(x_r) = x_1$$

Let $h(x_i) = y_i$, for each i. Then

$$g(y_i) = h(f(h^{-1}(y_i)) = h(f(x_i) = h(x_{i+1}) = y_{i+1}$$

in which the subscript is to be evaluated modulo r. Thus every cycle of length r corresponds to a cycle of g of length r, and vice versa. So f and g are of the same type.

On the other hand, assume that f and g are of the same type, and let $C = (x_1 x_2 \cdots x_r)$ be a cycle of f. Then g has a cycle of the form $C' = (y_1 y_2 \cdots y_r)$. Define $h(x_i) = y_i$ over C and similarly over every other cycle of f; this makes h a bijection from X to X, or a permutation of X. We have:

$$h(f(x_i)) = h(x_{i+1}) = y_{i+1} = g(y_i) = g(h(x_i))$$

So f and g are conjugate.

4.35 Show that if f and g are permutations, fg and gf are of the same type.

This follows from Problem 4.34 and

$$fg = fgff^{-1} = f(gf)f^{-1}$$

4.36 Prove that conjugate permutations have the same number of fixed points.

A fixed point is a cycle of length 1. Since conjugate permutations are of the same type (Problem 4.34), they have, in particular, the same number a_1 of cycles of length 1.

4.37 Consider a regular n-gon with its vertices marked $1, 2, \ldots, n$ in clockwise sequence. A clockwise rotation in the plane through an angle of $360°/n$ takes the figure into itself. This rotation f maps 1 into 2, 2 into 3, ..., n into 1; thus $f = (1\ 2 \cdots n)$ is a permutation of the vertex set. The cyclic group $\{e, f, f^2, \ldots, f^{n-1}\}$ generated by f is the group of rotational symmetries of the n-gon. List the 6 elements of the group of rotational symmetries of a regular hexagon and their types.

The zero rotation $e = (1)(2)(3)(4)(5)(6)$ is of type $[6\ \ 0\ \ 0\ \ 0\ \ 0\ \ 0]$. The type of $f = (1\,2\,3\,4\,5\,6)$ is

$$[0\ \ 0\ \ 0\ \ 0\ \ 0\ \ 1]$$

Now, $f(f(1)) = 3$; $f(f(2)) = 4$; $f(f(3)) = 6$; $f(f(4)) = 6$; $f(f(5)) = 1$ $f(f(6)) = 2$. Thus,

$$f^2 = (1\,3\,5)(2\,4\,6) \quad \text{of type} \quad [0\ \ 0\ \ 2\ \ 0\ \ 0\ \ 0]$$

$$f^3 = (1\,4)(2\,5)(3\,6) \quad \text{of type} \quad [0\ \ 3\ \ 0\ \ 0\ \ 0\ \ 0]$$

$$f^4 = (1\,5\,3)(2\,6\,4) \quad \text{of type} \quad [0\ \ 0\ \ 2\ \ 0\ \ 0\ \ 0]$$

$$f^5 = (1\,6\,5\,4\,3\,2) \quad \text{of type} \quad [0\ \ 0\ \ 0\ \ 0\ \ 0\ \ 1]$$

4.38 Display the complete group of symmetries of a regular $2m$-gon.

The group of rotational symmetries, as found in Problem 4.37, must be enlarged to include the dihedral symmetries resulting from (*i*) reflection in the line joining the midpoints of a pair of opposite edges, and (*ii*) reflection in the line joining a pair of opposite vertices. Let g denote the vertex permutation corresponding to reflection in the line joining the midpoints of the edges $\overline{m\ m+1}$ and $\overline{2m\ 1}$:

$$g = (1\ \overline{2m})(2\ \overline{2m-1})(3\ \overline{2m-2}) \cdots (m\ \overline{m+1})$$

(*Hint: Draw a picture.*) Then we have:

$$\begin{aligned}
gf(1) &= g(2) &&= 2m-1\\
gf(2) &= g(3) &&= 2m-2\\
&\cdots\cdots\cdots\cdots\\
gf(m-1) &= g(m) &&= m+1\\
gf(m) &= g(m+1) &&= m\\
gf(m+1) &= g(m+2) &&= m-1\\
&\cdots\cdots\cdots\cdots\\
gf(2m-1) &= g(2m) &&= 1\\
gf(2m) &= g(1) &&= 2m
\end{aligned}$$

or
$$gf = (\overline{2m})(\overline{1\ 2m-1})(\overline{2\ 2m-2})\cdots(\overline{m-1}\ \overline{m+1})(m)$$

It is seen that gf is the permutation induced by reflection in the line joining vertices m and $2m$.

Continuing in this fashion, one finds that gf^2 represents reflection in the line joining the midpoints of edges $\overline{m-1\ m}$ and $\overline{2m-1\ 2m}$, and that, in general, each postmultiplication by f rotates the axis of reflection counterclockwise through $360°/4m$. Hence the *set* of dihedral symmetries is

$$\{g, gf, gf^2, \ldots, gf^{2m-1}\}$$

and the complete *group* of symmetries is

$$\{e, f, f^2, \ldots, f^{2m-1}, g, gf, gf^2, \ldots, gf^{2m-1}\}$$

Note that the dihedral symmetries make up a left coset of the subgroup of rotational symmetries. The order of the complete group is $4m$, or twice the number of vertices.

4.39 Display the complete group of symmetries of a regular $(2m+1)$-gon.

The subgroup of rotational symmetries is, as always, $\langle f\rangle = \{e, f, f^2, \ldots, f^{2m}\}$. This time there is only one kind of dihedral symmetry axis: a line joining a vertex to the midpoint of the opposite edge. Let the vertex permutation h correspond to reflection in the axis from vertex 1 to the midpoint of edge $\overline{m+1}\ \overline{m+2}$; i.e.,

$$h = (1)(\overline{2\ 2m+1})(\overline{3\ 2m})\cdots(\overline{m+1}\ \overline{m+2})$$

(*Hint: Draw a picture.*) By a calculation like that in Problem 4.38, one shows that the dihedral symmetries compose the left coset $h\langle f\rangle$. The complete group of symmetries is therefore

$$\{e, f, f^2, \ldots, f^{2m}, h, hf, hf^2, \ldots, hf^{2m}\}$$

Comparing with Problem 4.38, one sees that the symmetry group of a regular n-gon has the same structure whether n is even or odd, and that the order of the group is always $2n$. This group is called the **dihedral group**, with symbol H_{2n}.

4.40 If $X = \{1, 2, 3, 4\}$ and $G = \{g_1, g_2, g_3, g_4\}$ is a group of permutations of X, where

$$g_1 = (1)(2)(3)(4) \qquad g_3 = (1)(2)(3\ 4)$$
$$g_2 = (1\ 2)(3)(4) \qquad g_4 = (1\ 2)(3\ 4)$$

find the cycle index of G.

Since X has 4 elements, the index has 4 variables x_i, where $i = 1, 2, 3, 4$.

The element g_1 has 4 cycles of length 1; so its contribution is x_1^4. Both g_2 and g_3 have 2 cycles of length 1 and 1 cycle of length 2; their contribution is $2x_1^2x_2$. The contribution of g_4 is x_2^2. Thus the cycle index of the group is ($|G| = 4$)

$$Z(G; x_1, x_2, x_3, x_4) = \tfrac{1}{4}(x_1^4 + 2x_1^2x_2 + x_2^2)$$

4.41 If $A = \{\alpha, \beta, \gamma, \delta\}$ and $H = \{h_1, h_2, h_3, h_4\}$ is a group of permutations of A, where

$$h_1 = (\alpha)(\beta)(\gamma)(\delta) \qquad h_3 = (\alpha\,\gamma)(\beta\,\delta)$$
$$h_2 = (\alpha\,\beta)(\gamma\,\delta) \qquad h_4 = (\alpha\,\delta)(\beta\,\gamma)$$

find the cycle index of H.

The contributions of the 4 elements of H are x_1^4, x_2^2, x_2^2, and x_2^2. Hence the cycle index is

$$Z(H; x_1, x_2, x_3, x_4) = \tfrac{1}{4}(x_1^4 + 3x_2^2)$$

4.42 *True or False:* Isomorphic groups have identical cycle indices.

False. As is shown in Problem 4.97(*b*), groups G of Problem 4.40 and H of Problem 4.41 are isomorphic.

4.43 If $f = (1\,2 \cdots n)$ is a permutation of $X = \{1, 2, \ldots, n\}$, then the cyclic group (of permutations) $G = \langle f \rangle$ is of order n. Prove that in the cycle representation of any element of G all cycles are of same length. (In other words, the type of any element is a vector with n components of which exactly 1 is nonzero).

The type of $f^0 = e$ is $[n \quad 0 \quad 0 \quad 0 \quad \cdots \quad 0]$. Let $m = m(i)$ be the length of the shortest cycle in the cycle representation of $f^i (1 \le i \le n-1)$ and let x be an element in some cycle of f^i of length m. Then $f^{im}(x) = (f^i)^m(x) = x$.

Now, if $y \in X$, both x and y belong to the same cycle of the permutation $f = (1\,2 \cdots n)$. This implies that there exists r such that $f^r(x) = y$. Consequently,

$$(f^i)^m(y) = f^{im}f^r(x) = f^rf^{im}(x) = f^r(x) = y$$

So the element y belongs to a cycle in f^i whose length divides m. But m is the length of the *shortest* cycle in f^i Thus, every cycle in f^i is of length $m(i)$, which common length must be a divisor of n.

4.44 For the group of Problem 4.43 show that

$$Z(G; x_1, x_2, \ldots, x_n) = \frac{1}{n}\sum_{d|n} \phi(d)x_d^{n/d}$$

Let $g \in G$ be of order d. It is easy to see from Problem 4.43 that the cycle representation of g is composed of n/d cycles of length d. (If the common length were not d, it would have to be a divisor of d. But then the order of g would be smaller than d.) Hence,

$$Z(g; x_1, x_2, \ldots, x_n) = x_d^{n/d}$$

But, by Problem 4.20(*iii*), for each divisor d of n, G contains exactly $\phi(d)$ elements of order d; the result follows at once.

4.45. Verify Problem 4.44 for $G = \langle f \rangle$, where $f = (1\,2\,3\,4)$.

The elements of G and their cycle indices are as follows:

Element g	$Z(g; x_1, x_2, x_3, x_4)$
$f^0 = (1)(2)(3)(4)$	x_1^4
$f^1 = (1\,2\,3\,4)$	x_4^1
$f^2 = (1\,3)(2\,4)$	x_2^2
$f^3 = (1\,4\,3\,2)$	x_4^1

Thus
$$Z(G; x_1, x_2, x_3, x_4) = \frac{1}{|G|} \sum_{g \in G} Z(g; x_1, x_2, x_3, x_4) = \tfrac{1}{4}(x_1^4 + x_2^2 + 2x_4)$$

On the other hand, the divisors of $n = 4$ are $d = 1, 2, 4$. From Problem 2.45,

$$\phi(1) = 1 \qquad \phi(2) = 2(1 - \tfrac{1}{2}) = 1 \qquad \phi(4) = 4(1 - \tfrac{1}{2}) = 2$$

and we have

$$\tfrac{1}{4} \sum_{d \mid 4} \phi(d) x_d^{4/d} = \tfrac{1}{4}(1x_1^4 + 1x_2^2 + 2x_4^1)$$

4.46 Obtain the cycle index of the dihedral group H_{2n} (Problem 4.39).

The subgroup of rotational symmetries is precisely the group G of Problems 4.43 and 4.44. Hence the net contribution of the subgroup is

$$U = \sum_{d \mid n} \phi(d) x_d^{n/d}$$

Two cases must be considered in treating the dihedral symmetries: *(i)* n is even, and *(ii)* n is odd.

(i) By Problem 4.38, there are $n/2$ reflections of type

$$[0 \quad n/2 \quad 0 \quad 0 \quad \cdots \quad 0]$$

(no fixed points) and $n/2$ reflections of type

$$[2 \quad (n-2)/2 \quad 0 \quad 0 \quad \cdots \quad 0]$$

(2 fixed points), for a net contribution of

$$V = \frac{n}{2}[x_2^{n/2} + x_1^2 x_2^{(n-2)/2}]$$

(ii) By Problem 4.39, there are n reflections of type

$$[1 \quad (n-1)/2 \quad 0 \quad 0 \quad \cdots \quad 0]$$

for a net contribution of

$$V' = n x_1 x_2^{(n-1)/2}$$

we have found

$$Z(H_{2n}; x_1, x_2, \ldots, x_n) = \begin{cases} \dfrac{1}{2n}(U + V) & n \text{ even} \\ \dfrac{1}{2n}(U + V') & n \text{ odd} \end{cases}$$

4.47 The length of a stick is n feet. The individual feet are marked consecutively $1, 2, 3, \ldots, n$. The only symmetries are rotations about the center through 0° and 180°. Obtain the cycle index of this permutation group.

Here $G = \{e, g\}$. (*i*) If $n = 2k$,

$$g = (1\,\overline{2k})(2\,\overline{2k-1})(3\,\overline{2k-2})\cdots(k\,\overline{k+1})$$

and so the cycle index is

$$\tfrac{1}{2}(x_1^{2k} + x_2^k)$$

(*ii*) If $n = 2k + 1$,

$$g = (1\,\overline{2k+1})(2\,\overline{2k})\cdots(k\,\overline{k+2})(k+1)$$

and so the cycle index is

$$\tfrac{1}{2}(x_1^{2k+1} + x_1 x_2^k)$$

4.48 Find the cycle index of the group of vertex permutations induced by the rotational symmetries of the cube.

Looking down on the cube, label the top vertices 1, 2, 3, 4 in clockwise order; label the bottom vertices so that 5 is below 1, 6 is below 2, 7 is below 3, and 8 is below 4. The permutations composing the symmetry group G fall into the following classes:

(i) The identity $e = (1)(2)(3)(4)(5)(6)(7)(8)$, with index x_1^8.

(ii) The permutation $(1\,3)(2\,4)(5\,7)(6\,8)$, corresponding to the rotation through 180° (either clockwise or counterclockwise) about the line joining the midpoints of the top and bottom faces, plus 2 analogous permutations. The net cycle index is $3x_2^4$.

(iii) The permutation $(1\,2\,3\,4)(5\,6\,7\,8)$, corresponding to a clockwise rotation through 90° about the line joining the midpoints of the top and bottom faces, plus 2 analogous permutations. The net cycle index is $3x_4^2$.

(iv) As in (iii) but counterclockwise; $3x_4^2$.

(v) (*Hint: Draw a picture.*) The permutation $(1\,2)(3\,5)(4\,6)(7\,8)$, corresponding to a (clockwise or counterclockwise) rotation of 180° about the line joining the midpoints of the opposite sides $\overline{1\,2}$ and $\overline{7\,8}$, plus 5 analogous permutations. The net cycle index is $6x_2^4$.

(vi) The permutation $(1)(2\,4\,5)(3\,8\,6)(7)$, corresponding to a clockwise rotation through 120° about the diagonal $\overline{1\,7}$, plus 3 analogous permutations. The net cycle index is $4x_1^2x_3^2$.

(vii) As in (*vi*) but counterclockwise; $4x_1^2x_3^2$. The order of G is $1 + 3 + 3 + 3 + 6 + 4 + 4 = 24$, and

$$Z(G; x_1, x_2, \ldots, x_8) = \tfrac{1}{24}(x_1^8 + 9x_2^4 + 8x_1^2x_3^2 + 6x_4^2)$$

4.49 Find the cycle index of the group of face permutations induced by the rotational symmetries of the cube.

Looking down on the cube, label the top face 1, the bottom face 2, and the lateral faces 3, 4, 5, 6 (clockwise). Following the classification of Problem 4.48, one has:

(i) $e = (1)(2)(3)(4)(5)(6)$; index, x_1^6

(ii) Three permutations like $(1)(2)(3\,5)(4\,6)$; net index, $3x_1^2x_2^2$

(iii) Three permutations like $(1)(2)(3\,4\,5\,6)$; net index, $3x_1^2x_4$

(iv) Net index, $3x_1^2x_4$

(v) Six permutations like $(1\,5)(2\,3)(4\,6)$; net index, $6x_2^3$

(vi) Four permutations like $(1\,5\,4)(2\,3\,6)$; net index, $4x_3^2$

(vii) Net index, $4x_3^2$

Thus
$$Z(G; x_1, x_2, \ldots, x_6) = \tfrac{1}{24}(x_1^6 + 3x_1^2x_2^2 + 6x_2^3 + 6x_1^2x_4 + 8x_3^2)$$

4.50 If 2 polyhedra are **geometric duals**, the vertices of the one are in unique correspondence with the faces of the other. One pair of duals are the **cube** and the **regular octahedron**. The line segments determined by the centers of adjacent faces of a cube are the edges of an inscribed octahedron. Reciprocally (see Fig. 4-3), each vertex of the cube is centered over a face of the octahedron, which implies that a cube can be inscribed in an octahedron. Find the cycle indices of *(a)* the group of vertex permutations, and *(b)* the group of face permutations, induced by the rotational symmetries of the octahedron.

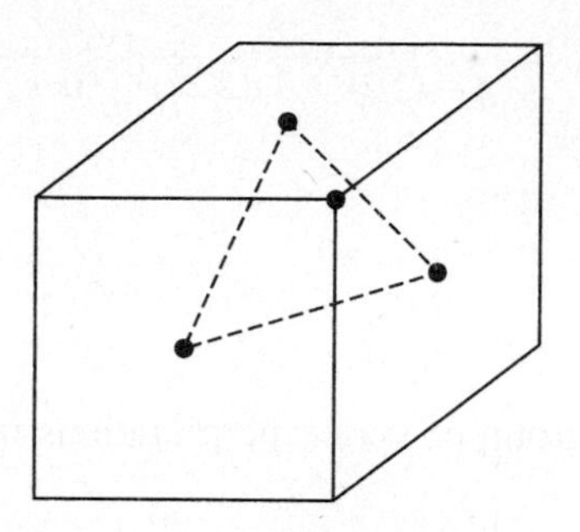

Fig. 4-3

A cube and an octahedron obviously have the same rotational symmetries. Thus, by the vertex-face duality: *(a)* See Problem 4.49. *(b)* See Problem 4.48.

4.51 A **regular tetrahedron** has 4 vertices, 4 faces (congruent equilateral triangles), and 6 edges. Find the cycle index of the group of permutations of the 4 vertices (or 4 faces) induced by the rotational symmetries of the regular tetrahedron.

If we join the centers of pairs of adjacent faces of a regular tetrahedron, we get another regular tetrahedron. In other words, the dual of a regular tetrahedron is a regular tetrahedron. Thus the group of permutations is the same whether we consider faces or vertices. Let the vertices be marked 1, 2, 3, and 4. The rotational symmetries are as follows (*a cardboard model helps*):

(*i*) The identity $e = (1)(2)(3)(4)$.

(ii) Join a vertex (say, vertex 1) to the center of the opposite face, creating an axis L. Rotate the solid about L through 120° clockwise or 120° counterclockwise, generating the permutations $(1)(2\,3\,4)$ and $(1)(2\,4\,3)$. Now do the same thing for the other 3 vertices, obtaining 8 permutations in all.

(iii) Join the midpoints of 2 nonintersecting edges (say, $\overline{1\,2}$ and $\overline{3\,4}$), creating an axis L'. A 180° rotation (either sense) about L' generates the permutation $(1\,2)(3\,4)$. There are 2 more permutations of this type.

Thus G has 12 elements, and

$$Z(G; x_1, x_2, x_3, x_4) = \tfrac{1}{12}(x_1^4 + 8x_1x_3 + 3x_2^2)$$

4.52 Obtain the cycle index of the group of permutations of the 6 edges induced by the rotational symmetries of the regular tetrahedron.

The tetrahedron is $ABCD$ (vertices), with edges AB, AC, AD, BC, BD, and CD marked as 1, 2, 3, 4, 5, and 6, respectively. The 12 rotational symmetries listed in Problem 4.51 generate the following edge permutations:

(i) $e = (1)(2)(3)(4)(5)(6)$.

(ii) $(1\,2\,3)(4\,6\,5)$ and $(1\,3\,2)(4\,5\,6)$ about vertex A, and a similar pair for each of the other 3 vertices.

(iii) $(1)(6)(2\,5)(3\,4)$ and 2 similar permutations.

Hence $$Z(G; x_1, x_2, \ldots, x_6) = \tfrac{1}{12}(x_1^6 + 3x_1^2x_2^2 + 8x_3^2)$$

4.53 Given a group G of permutations of $X = \{x_1, x_2, \ldots, x_n\}$ and a group H of permutations of $Y = \{y_1, y_2, \ldots, y_m\}$, where X and Y are disjoint. (*a*) Prove that the definition

$$(g, h)(v) = \begin{cases} g(v) & \text{if } v \in X \\ h(v) & \text{if } v \in Y \end{cases} \qquad (*)$$

for every $(g, h) \in G \times H$ (Problem 4.10), makes $G \times H$ a group *of permutations of* $X \cup Y$. (*b*) Determine the cycle index of $G \times H$.

(*a*) That $G \times H$ is a group is guaranteed by Problem 4.10; we are only required to show that $(g, h)(\ \ )$ is a bijection from $X \cup Y$ to itself. But this is obvious from (*): X is mapped one-one onto X (because g is a permutation of X), and Y is mapped one-one onto Y (because h is a permutation of Y), and so the disjoint union is mapped one-one onto itself.

(*b*) Temporarily assume that $n \geq m$. Let $g \in G$ and $h \in H$ be of respective types

$$[a_1 \quad a_2 \quad \cdots \quad a_n] \qquad \text{and} \qquad [b_1 \quad b_2 \quad \cdots \quad b_m]$$

Then, since $(g, h) \in G \times H$ has the cycle representation

$$\underbrace{(\quad)(\quad)\cdots(\quad)}_{g}\underbrace{(\quad)(\quad)\cdots(\quad)}_{h}$$

it is of type

$$[a_1 + b_1 \quad a_2 + b_2 \quad \cdots \quad a_m + b_m \quad a_{m+1} \quad a_{m+2} \quad \cdots \quad \overbrace{0 \quad 0 \quad \cdots \quad 0}^{m \text{ entries}}]$$

Hence we have

$$\begin{aligned} Z(G \times H); u_1, u_2, \ldots, u_{n+m}) &= \frac{1}{|G|\,|H|} \sum_{\substack{g \in G \\ h \in H}} u_1^{a_1+b_1} u_2^{a_2+b_2} \cdots u_m^{a_m+b_m} u_{m+1}^{a_{m+1}+0} u_{m+2}^{a_{m+2}+0} \cdots u_n^{a_n+0} \\ &= \left(\frac{1}{|G|} \sum_{g \in G} u_1^{a_1} u_2^{a_2} \cdots u_n^{a_n}\right)\left(\frac{1}{|H|} \sum_{h \in H} u_1^{b_1} u_2^{b_2} \cdots u_m^{b_m}\right) \\ &= Z(G; u_1, u_2, \ldots, u_n)\, Z(H; u_1, u_2, \ldots, u_m) \end{aligned}$$

Because our result is symmetric in m and n, it is also valid when $m \geq n$.

4.54 (*Cayley's Theorem*) Show that every finite group is isomorphic to a group of permutations.

The idea behind the proof is very simple: because each element of a group G has its inverse, the rows of the multiplication table for G must be distinct permutations of G. Thus, given the finite group $(G, \circ)$, where $G = \{g_1, g_2, \ldots, g_m\}$, define m distinct permutations of G by

$$\pi_1(g) = g_1 \circ g \qquad \pi_2(g) = g_2 \circ g \qquad \cdots \qquad \pi_m(g) = g_m \circ g$$

Consider the group $(G', \circ')$, where $G' = \{\pi_1, \pi_2, \ldots, \pi_m\}$ and where $\circ'$ denotes multiplication of permutations as defined in Problem 1.65. The mapping $f: G \to G'$ defined by

$$f(g_i) = \pi_i \qquad (i = 1, 2, \ldots, m)$$

is obviously a bijection. Indeed, it is an isomorphism; for, if $g_i \circ g_j = g_k$, then for each $g \in G$,

$$\begin{aligned} \pi_k(g) &= g_k \circ g = (g_i \circ g_j) \circ g = g_i \circ (g_j \circ g) \\ &= g_i \circ \pi_j(g) = \pi_i(\pi_j(g)) = (\pi_i \circ' \pi_j)(g) \end{aligned}$$

i.e., $\pi_i \circ' \pi_j = \pi_k$ (f preserves group multiplication).

4.55 (*a*) How may one define a cycle index for an arbitrary finite group? (*b*) Illustrate the procedure for the group of subsets of $X = \{a, b\}$ under the symmetric difference (see Example 1).

(*a*) In the notation of Problem 4.54, take the cycle index of the permutation group G' to be the cycle index of the abstract group G.

(*b*) The multiplication table for $G = \{\emptyset, \{a\}, \{b\}, X\}$ is

*	$\emptyset$	$\{a\}$	$\{b\}$	X			
$\emptyset$	$\emptyset$	$\{a\}$	$\{b\}$	X	$\rightarrow$	π_1:	$(*)(\{a\})(\{b\})(X)$
$\{a\}$	$\{a\}$	$\emptyset$	X	$\{b\}$	$\rightarrow$	π_2:	$(\emptyset\,\{a\})(\{b\}\,X)$
$\{b\}$	$\{b\}$	X	$\emptyset$	$\{a\}$	$\rightarrow$	π_3:	$(\emptyset\,\{b\})(\{a\}\,X)$
X	X	$\{b\}$	$\{a\}$	$\emptyset$	$\rightarrow$	π_4:	$(\emptyset\,X)(\{a\}\,\{b\})$

To the right of each row is shown the cycle representation of the permutation in G' generated by that row. These 4 permutations have the respective cycle indices $x_1^4, x_2^2, x_2^2, x_2^2$; hence,

$$Z(G; x_1, x_2, x_3, x_4) = Z(G'; x_1, x_2, x_3, x_4) = \tfrac{1}{4}(x_1^4 + 3x_2^2)$$

4.56 In Problem 4.55 it is seen that $\emptyset$ (the identity element of G) is of order 1 and that each cycle of the image permutation π_1 is of length 1. Likewise, the other 3 elements are of order 2, and their image permutations have all cycles of length 2. Prove that in general (following the notation of Problem 4.54 but replacing $\circ$ by juxtaposition), if $g_i \in G$ is of order k, then every cycle in the representation of $\pi_i \in G'$ is of length k.

Obtain the cycle representation of π_i, where by definition $\pi_i(g) = g_i g$, by the recipe of Section 4.2. Thus, starting with g_1 (say), generate the cycle

$$(g_1 \; g_i g_1 \; g_i^2 g_1 \cdots g_i^{k-1} g_1)$$

which is of length k. Choosing an element g_α that is not present in the cycle, generate the new cycle

$$(g_\alpha \; g_i g_\alpha \; g_i^2 g_\alpha \cdots g_i^{k-1} g_\alpha)$$

also of length k. Continue until G is exhausted, obtaining finally m/k cycles of length k.

4.57 Let G be the set of all 3×3 matrices A that have $[1 \quad a \quad b]$ as the first row, $[0 \quad 1 \quad c]$ as the second row, and $[0 \quad 0 \quad 1]$ as the third row; the numbers a, b, and c are elements of the set $\{0, 1, 2\}$. If scalar addition and multiplication are modulo 3, show that G is a group under ordinary matrix multiplication. Determine the cycle index of G.

(i) G is closed under multiplication:

$$\begin{bmatrix} 1 & a & b \\ 0 & 1 & c \\ 0 & 0 & 1 \end{bmatrix}\begin{bmatrix} 1 & a' & b' \\ 0 & 1 & c' \\ 0 & 0 & 1 \end{bmatrix} = \begin{bmatrix} 1 & a' + a & b' + ac' + b \\ 0 & 1 & c' + c \\ 0 & 0 & 1 \end{bmatrix}$$

(ii) The 3×3 identity matrix belongs to $G (a = b = c = 0)$.

(iii)
$$\begin{bmatrix} 1 & a & b \\ 0 & 1 & c \\ 0 & 0 & 1 \end{bmatrix}^{-1} = \begin{bmatrix} 1 & -a & ac - b \\ 0 & 1 & -c \\ 0 & 0 & 1 \end{bmatrix}$$

(iv) Matrix multiplication is associative.

Thus G is a multiplicative group, of order 27.

By direct computation,

$$\begin{bmatrix} 1 & a & b \\ 0 & 1 & c \\ 0 & 0 & 1 \end{bmatrix}^2 = \begin{bmatrix} 1 & 2a & 2b+ac \\ 0 & 1 & 2c \\ 0 & 0 & 1 \end{bmatrix}$$

$$\begin{bmatrix} 1 & a & b \\ 0 & 1 & c \\ 0 & 0 & 1 \end{bmatrix}^3 = \begin{bmatrix} 1 & 0 & 0 \\ 0 & 1 & 0 \\ 0 & 0 & 1 \end{bmatrix}$$

Thus one element of G (the identity) is of order 1 and the other 26 elements are of order 3. It follows from Problems 4.55(*a*) and 4.56 that

$$Z(G; x_1, x_2, \ldots, x_{27}) = \tfrac{1}{27}(x_1^{27} + 26x_3^9)$$

4.58 A regular polytope (a solid in which all faces are congruent polygons and each vertex is incident with the same number of faces) with 12 vertices, 20 faces (congruent equilateral triangles), and 30 edges is called a **regular icosahedron.** Identify the rotational symmetries of this solid, and obtain the cycle indices of the groups of (*a*) vertex permutations, and (*b*) face permutations.

The rotational symmetries are as follows:

(i) The zero rotation.

(ii) Exactly 5 faces meet at each vertex. Therefore, about an axis joining opposite vertices, clockwise rotations of the solid through 72° (*not* 60°), 144°, 216°, and 288° bring it into itself. There are $(12/2)(4) = 24$ symmetries of this type.

(iii) About an axis joining the centers of opposite faces, clockwise rotations of 120° and 240° take the solid into itself. There are $(20/2)(2) = 20$ symmetries of this type.

(iv) About an axis joining the midpoints of opposite edges, a clockwise rotation of 180° takes the solid into itself. There are $(30/2)(1) = 15$ symmetries of this type.

The rotation group is of order $1 + 24 + 20 + 15 = 60$.

(*a*) The induced vertex permutations are of the following types:

(i) [12 0 0 ··· 0].

(ii) Here, 2 vertices stay fixed, and the others move in 2 disjoint cycles of length 5; type [2 0 0 0 2 0 0 ··· 0].

(iii) The 12 vertices move in disjoint cycles of length 3; [0 0 4 0 0 ··· 0].

(iv) The 12 vertices move in disjoint cycles of length 2; [0 6 0 0 ··· 0].

Summing the cycle indices of these permutations and dividing by the order of the group, one finds

$$Z(G; x_1, x_2, \ldots, x_{12}) = \tfrac{1}{60}(x_1^{12} + 15x_2^6 + 20x_3^4 + 24x_1^2x_5^2)$$

(*b*) The induced face permutations are of the following types:

(i) [20 0 0 ··· 0].

(ii) The 20 faces move in disjoint cycles of length 5; [0 0 0 0 4 0 0 ··· 0].

(iii) Two faces stay fixed, and the remaining 18 move in disjoint cycles of length 3 (*a model helps here*); [2 0 6 0 0 ··· 0].

(iv) The 20 faces move in disjoint cycles of length 2; [0 10 0 0 ··· 0].

Hence
$$Z(G; x_1, x_2, \ldots, x_{20}) = \tfrac{1}{60}(x_1^{20} + 15x_2^{10} + 20x_1^2x_3^6 + 24x_5^4)$$

PÓLYA'S ENUMERATION THEOREMS

4.59 Let C be the (finite) set of all functions f from a finite set X to a finite set Y, and let G be a group of permutations of X. For each π in G, define a mapping π' from C to C by

$$\pi'(f(x)) = f(\pi(x)) \qquad \text{(for each } x \in X \text{ and each } f \in C)$$

Prove that (*a*) π' is a permutation of C, and (*b*) $G' = \{\pi' : \pi \in G\}$ is a group.

(*a*) If $\pi'(f_1) = \pi'(f_2)$, then $f_1(\pi(x)) = f_2(\pi(x))$ for every $x \in X$, which implies that $f_1(t) = f_2(t)$ for every $t \in X$. So $f_1 = f_2$ (π' is injective). Also, π' is surjective; in fact, for any $f \in C$,

$$f(x) = f(\pi(\pi^{-1}(x))) = \pi'(f(\pi^{-1}(x))) \equiv \pi'((f\pi^{-1})(x))$$

Hence, as a bijection, π' is a permutation of C.

(*b*) By Problem 4.13 it suffices to show that G' is closed with respect to multiplication (composition). Let π_1 and π_2 in G respectively determine π_1', and π_2' in G'. Our assertion is that $\pi_1\pi_2$ in G determines $\pi_1'\pi_2'$ in G; i.e., $(\pi_1\pi_2)' = \pi_1'\pi_2'$.
Proof:

$$\begin{aligned}(\pi_1\pi_2)'(f(x)) &= f((\pi_1\pi_2)(x)) = f(\pi_1(\pi_2(x))) \\ &= \pi_1'(f(\pi_2(x))) = \pi_1'(\pi_2'(f(x))) = (\pi_1'\pi_2')(f(x))\end{aligned}$$

Observe that the above proof *almost* establishes that G and G' are isomorphic groups. The rest of the job is done in Problem 4.60.

4.60 With reference to Problem 4.59, prove that if Y contains at least 2 elements, the mapping from G to G' defined by

$$\pi \to \pi'$$

is an isomorphism.

Because the mapping is a surjection that preserves group multiplication [Problem 4.59(*b*)], all that remains is to show that the mapping is an injection. This is easy: if $\pi_1 \neq \pi_2$, there exists a $t \in X$ such that $\pi_1(t) \neq \pi_2(t)$. Because C contains *all* functions from X to Y, it must contain a function ϕ that maps the distinct X points $\pi_1(t)$ and $\pi_2(t)$ into 2 distinct Y points (which exist by hypothesis); that is, $\phi(\pi_1(t) \neq \phi(\pi_2(t))$, which implies

$$\pi_1'(\phi(t)) \neq \pi_2'(\phi(t))$$

But if π_1' and π_2' differ for a single value of the function $\phi \in C$, they must represent distinct permutations of C.

4.61 Prove Theorem 4.3.

The patterns in C with respect to G (a permutation group on X) are the distinct orbits in C with respect to G, and these in turn—by the isomorphism of Problem 4.60—are the distinct orbits in C' with respect to G' (a permutation group on C). Their number is given by the Burnside–Frobenius theorem (Theorem 4.2) as

$$k = \frac{1}{|G'|} \sum_{\pi' \in G'} |F(\pi')| \qquad (*')$$

where $F(\pi') = \{f \in C : \pi'(f) = f\}$. Now, because $\pi'(f) = f$ if and only if $f(\pi(x)) = f(x)$ for all $x \in X$ and because $|G'| = |G|$, one can convert $(*')$ back to X and G:

$$k = \frac{1}{|G|} \sum_{\pi \in G} |\{f \in C : f(\pi(x)) = f(x) \text{ for all } x \in X\}| \qquad (*)$$

Now, if $f(\pi(x)) = f(x)$ and if $(x_1\, x_2 \cdots x_j)$ is a cycle of π,

$$f(x_1) = f(x_2) = \cdots = f(x_j)$$

that is, f is constant over each cycle of π. Conversely, if f is constant over each cycle of π and if $(x\, x_t \cdots x_u)$ is the cycle involving the arbitrary element $x \in X$,

$$f(\pi(x)) = f(x_t) = f(x)$$

It follows that the summand in the right-hand side of (*) is just the number of ways of coloring X with $r \geq 2$ colors so that elements in the same cycle of the permutation π are given the same color. If π is of type $[a_1 \quad a_2 \quad \cdots \quad a_n]$, this number of ways is $r^{a_1+a_2+\cdots+a_n}$; Eq. (*) becomes

$$k = \frac{1}{|G|} \sum_{\pi \in G} r^{a_1+a_2+\cdots+a_n} \equiv \frac{1}{|G|} \sum_{\pi \in G} Z(\pi; r, r, \ldots, r) \equiv Z(G; r, r, \ldots, r)$$

4.62 Infer Fermat's Little Theorem (Problem 1.185) from Theorem 4.3.

By Problem 4.92 the cycle index of the cyclic group of prime order p (the group of rotational symmetries of the regular p-gon is

$$Z(C_p; x_1, x_2, \ldots, x_p) = \frac{1}{p}[x_1^p + (p-1)x_p]$$

Then Theorem 4.3 gives for the number of colorings in r colors of the p-gon that are distinguishable with respect to C_p:

$$Z(C_p; r, r, \ldots, r) = \frac{r^p - r}{p} + r$$

Thus
$$\frac{r^p - r}{p} = Z(C_p; r, r, \ldots, r) - r = \text{an integer}$$

4.63 In a military mess the food trays are rectangular and divided into 4 equal rectangular compartments. Find the number of distinguishable ways of filling a tray with 4 foods if the long dimension must be parallel to the table edge.

Label the corners of the tray 1, 2, 3, and 4 (clockwise), where $\overline{1\,2}$ is a longer side. The symmetry group G of the rectangle is composed of the following permutations:

(1)(2)(3)(4)	[the zero rotation]
(1 3)(2 4)	[180° rotation]
(1 2)(3 4)	[reflection in perpendicular axis]
(1 4)(2 3)	[reflection in parallel axis]

Hence, $Z(G; x_1, x_2, x_3, x_4) = \frac{1}{4}(x_1^4 + 3x_2^2)$ and

$$Z(G; 4, 4, 4, 4) = \tfrac{1}{4}(4^4 + 3 \cdot 4^2) = 76 \text{ ways}$$

4.64 As usual, let G be a group of permutations of $X = \{x_1, x_2, \ldots, x_n\}$, and let C be the set of all functions from X to $Y = \{y_1, y_2, \ldots, y_r\}$. If $w(y)$ is a given weight function on Y, we induce a weight function $\omega(f)$ on C by the formula

$$\omega(f) = [w(f(x_1))][w(f(x_2))] \cdots [w(f(x_n))]$$

(*a*) If f and ϕ in C are equivalent with respect to G (see the definition in Section 4.3), prove that $\omega(f)=\omega(\phi)$.

(*b*) Denote by $C_1, C_2, \ldots, C_k$ the distinct patterns in C; let $\omega(C_i)$ $(i=1,2,\ldots,k)$ stand for the constant value of ω over C_i [see (*a*)]. Show that the pattern inventory of C (see the definition following Example 6) can be expressed as

$$\mathrm{PI}(G; w(y_1), w(y_2), \ldots, w(y_r)) = \sum_{i=1}^{k} \omega(C_i)$$

(*a*) Since f and ϕ are equivalent, there exists a permutation π of X such that $f(x)=\phi(\pi(x))$ for all x in X. Therefore,

$$\begin{aligned}\omega(f) &= [w(f(x_1))][w(f(x_2))]\cdots[w(f(x_n))] \\ &= [w(\phi(\pi(x_1)))][w(\phi(\pi(x_2)))]\cdots[w(\phi(\pi(x_n)))] \\ &= [w(\phi(x'_1))][w(\phi(x'_2))]\cdots[w(\phi(x'_n))] = \omega(\phi)\end{aligned}$$

(*b*) In a permutation of colored objects, the *numbers* of red objects, of green objects, etc., clearly do not change. It follows that all colorings f making up a given pattern, C_i, in C are characterized by the same "assignment vector" $(n_1, n_2, \ldots, n_r)$. This means that any $f \in C_i$ maps n_1 elements of X into y_1, n_2 elements into $y_2, \ldots, n_r$ elements into y_r; so that the weight of C_i is

$$\omega(C_i) = \omega(f) = [w(y_1)]^{n_1}[w(y_2)]^{n_2}\cdots[w(y_r)]^{n_r}$$

Now, in the definition of the pattern inventory, the coefficient $\tau(n_1, n_2, \ldots, n_r)$ is defined to be the number of patterns answering to the vector $(n_1, n_2, \ldots, n_r)$. Hence we can write:

$$\begin{aligned}\sum_{i=1}^{k} \omega(C_i) &\equiv \text{total weight of the } k \text{ patterns} \\ &= \sum_{(n_1,n_2,\ldots,n_r)} [\text{total weight of patterns answering to } (n_1, n_2, \ldots, n_r)] \\ &= \sum_{(n_1,n_2,\ldots,n_r)} \tau(n_1, n_2, \ldots, n_r)[w(y_1)]^{n_1}[w(y_2)]^{n_2}\cdots[w(y_r)]^{n_r} \\ &\equiv \mathrm{PI}(G; w(y_1), w(y_2), \ldots, w(y_r))\end{aligned}$$

4.65 Let $X=\{1,2,3,4\}$, $Y=\{y_1, y_2\}$; $w(y_1)=\mathrm{R}$, $w(y_2)=\mathrm{B}$; and

$$G=\{(1)(2)(3)(4), (1\,2)(3\,4), (1\,3)(2\,4), (1\,4)(2\,3)\}$$

Use Problem 4.64(*b*) to find the pattern inventory for the set C of all functions from X to Y.

The cycle index is $\frac{1}{4}(x_1^4+3x_2^2)$ (this is the group of Problems 4.41 and 4.63). By Pólya's first theorem, with $r=|Y|=2$, the number of patterns in C is

$$k=\tfrac{1}{4}(2^4+3\cdot 2^2)=7$$

To visualize these 7 patterns, it is helpful to have a concrete model of X and G. Fortunately, we have several available. Let us use the one provided by Example 4: if X is identified with the vertex set of the square, then G will be the subgroup $\{g_1, g_3, g_5, g_6\}$ of the full symmetry group D_8. Now, there are 5 possible values of the assignment vector (n_1, n_2):

$$(0,4) \qquad (1,3) \qquad (2,2) \qquad (3,1) \qquad (4,0)$$

Obviously, $(0,4)$ and $(4,0)$ each determine a single pattern (there's only one way to paint all vertices the same color); the respective weights of these patterns are $\omega(C_1)=\mathrm{B}^4$ and $\omega(C_2)=\mathrm{R}^4$. Similarly, $(1,3)$ and $(3,1)$

generate 1 pattern apiece (there is a reflection or rotation in G that will give the odd-colored vertex any desired location); $\omega(C_3) = \mathrm{RB}^3$, $\omega(C_4) = \mathrm{R}^3\mathrm{B}$. By elimination (2, 2) must give rise to $7 - 4 = 3$ patterns, with

$$\omega(C_5) = \omega(C_6) = \omega(C_7) = \mathrm{R}^2\mathrm{B}^2$$

And indeed it does, as is shown in Fig. 4-4. Thus,

$$\mathrm{PI}(G; \mathrm{R}, \mathrm{B}) = \sum_{i=1}^{7} \omega(C_i) = \mathrm{R}^4 + \mathrm{R}^3\mathrm{B} + 3\mathrm{R}^2\mathrm{B}^2 + \mathrm{RB}^3 + \mathrm{B}^4$$

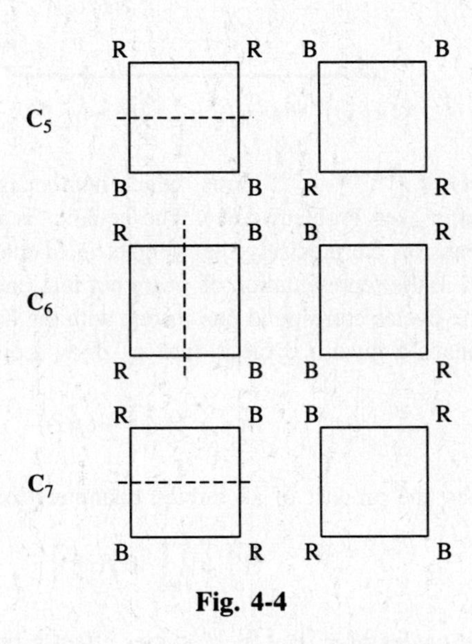

Fig. 4-4

4.66 Prove Theorem 4.4.

The derivation of Pólya's second theorem from the weighted Burnside–Frobenius theorem (Problem 4.31) runs parallel to that of his first theorem from Theorem 4.2. First of all, note that, by Problem 4.64(*a*), the weights function $\omega(f)$ has the required constancy property needed for an application of the weighted Burnside–Frobenius theorem to the orbits in C with respect to the permutation group G' of Problems 4.59 and 4.60. (Again heed the notational warning.) Making the application and recalling the results of Problem 4.64(*b*), we get—in complete analogy with (*′) of Problem 4.61—

$$\mathrm{PI}(G; w(y_1), w(y_2), \ldots, w(y_r)) = \sum \omega(C_i) = \frac{1}{|G'|} \sum_{\pi' \in G'} W(\pi') \qquad (i')$$

where

$$W(\pi') \equiv \sum_{f \in F(\pi')} \omega(f)$$

Convert back to X and G, exactly as in Problem 4.61:

$$\mathrm{PI} = \frac{1}{|G|} \sum_{\pi \in G} \left\{ \sum_{\substack{f \in C \,:\, f(\pi(x)) = f(x) \\ (\text{all } x)}} [w(f(x_1))][w(f(x_2))] \cdots [w(f(x_n))] \right\} \qquad (i)$$

According to Problem 4.61, the inner summation in (i) may be taken over all functions $f(x)$ that are constant over each cycle of π. Let π be of type $[a_1 \quad a_2 \quad \cdots \quad a_n]$ and define a horrendous multinomial in the $w(y_i)$ as follows:

$$\Omega \equiv \overbrace{[w(y_1)+w(y_2)+\cdots+w(y_r)]\cdots[w(y_1)+w(y_2)+\cdots+w(y_r)]}^{a_1 \text{ factors}}$$
$$\times\overbrace{[w(y_1)^2+w(y_2)^2+\cdots+w(y_r)^2]\cdots[w(y_1)^2+w(y_2)^2+\cdots+w(y_r)^2]}^{a_2 \text{ factors}}$$
$$\times\overbrace{[w(y_1)^3+w(y_2)^3+\cdots+w(y_r)^3]\cdots[w(y_1)^3+w(y_2)^3+\cdots+w(y_r)^3]}^{a_3 \text{ factors}}$$
$$\times \quad \cdots\cdots\cdots\cdots\cdots\cdots\cdots\cdots\cdots\cdots\cdots\cdots$$
$$\times\overbrace{[w(y_1)^n+w(y_2)^n+\cdots+w(y_r)^n]\cdots[w(y_1)^n+w(y_2)^n+\cdots+w(y_r)^n]}^{a_n \text{ factors}}$$

The expansion of Ω consists of $r^{a_1+a_2+\cdots+a_n}$ terms, which number is also the number of functions $f(x)$ that are constant over each cycle of π (see Problem 4.61). The equality is no accident; we now demonstrate that the individual terms in the expansion are precisely the weights ω of the individual functions $f(x)$.

Suppose that the cycles in the representation of π are put into one-to-one correspondence with the factors of Ω in the natural way: the one-cycles correspond one-to-one with the first a_1 factors; the 2 cycles, with the next a_2 factors; and so on. If $f(x)$ maps a given j cycle T into y_ν, draw a circle around the quantity

$$w(y_\nu)^j = \prod_{x\in T} w(f(x))$$

The expansion term given by the product of all circled quantities (one in each factor of Ω) will equal

$$\prod_U \Big[\prod_{x\in U} w(f(x))\Big]$$

in which U runs through all cycles of π. But these cycles effect a partition of X; so our expansion term is just

$$\prod_{x\in X} w(f(x)) = \omega(f)$$

We have just proved that the inner sum in (i) has the value Ω. But, by construction,

$$\Omega = Z(\pi; x_1, x_2, \ldots, x_n)\Big|_{\substack{x_j = w(y_1)^j + w(y_2)^j + \cdots + w(y_r)^j \\ (j=1,2,\ldots n)}}$$

whence Theorem 4.4 follows at once.

4.67 Rework Problem 4.65 by means of Theorem 4.4.

The pattern inventory is obtained by substituting $x_1 = \mathrm{R}+\mathrm{B}$ and $x_2 = \mathrm{R}^2+\mathrm{B}^2$ in the cycle index $\frac{1}{4}(x_1^4 + 3x_2^2)$:

$$\begin{aligned}\mathrm{PI}(G; \mathrm{R}, \mathrm{B}) &= \tfrac{1}{4}[(\mathrm{R}+\mathrm{B})^4 + 3(\mathrm{R}^2+\mathrm{B}^2)^2] \\ &= \mathrm{R}^4 + \mathrm{R}^3\mathrm{B} + 3\mathrm{R}^2\mathrm{B}^2 + \mathrm{R}\mathrm{B}^3 + \mathrm{B}^4\end{aligned}$$

4.68 Find the number of distinguishable necklaces consisting of 7 stones, of which 2 stones are red, 3 stones are blue, and 2 stones are green, when *(a)* only rotational symmetries (of a regular polygon with 7 vertices) are considered; and *(b)* both rotational and reflectional symmetries are considered.

(a) The group here is cyclic and of prime order; from Problem 4.92,

$$Z(C_7; x_1, x_2, \ldots, x_7) = \tfrac{1}{7}(x_1^7 + 6x_7)$$

By Theorem 4.4,

$$\mathrm{PI}(C_7; \mathrm{R}, \mathrm{B}, \mathrm{G}) = \tfrac{1}{7}[(\mathrm{R}+\mathrm{B}+\mathrm{G})^7 + 6(\mathrm{R}^7\mathrm{B}^7 + \mathrm{G}^7)]$$

The number we seek, $\tau(2,3,2)$, will be $\frac{1}{7}$ times the coefficient of $\mathrm{R}^2\mathrm{B}^3\mathrm{G}^2$ in $(\mathrm{R}+\mathrm{B}+\mathrm{G})^7$. The multinomial theorem (Problem 2.1) gives

$$\tau(2,3,2) = \frac{1}{7}\,\frac{7!}{2!\,3!\,2!} = 30$$

(*b*) The group is the dihedral group H_{14}. From Problem 4.94(*b*) and Theorem 4.4,

$$\mathrm{PI}(H_{14}; \mathrm{R}, \mathrm{B}, \mathrm{G}) = \tfrac{1}{2}\mathrm{PI}(C_7; \mathrm{R}, \mathrm{B}, \mathrm{G}) + \tfrac{1}{2}(\mathrm{R}+\mathrm{B}+\mathrm{G})(\mathrm{R}^2+\mathrm{B}^2+\mathrm{G}^2)^3$$

Using the result of (*a*), we have

$$\tau(2,3,2) = \frac{1}{2}(30) + \frac{1}{2}\left(0 + \frac{3!}{1!\,1!\,1!} + 0\right) = 18$$

4.69 Repeat Problem 4.68 for 9 stones, of which 2 are red, 3 are blue, and 4 are green.

(*a*) By Problem 4.44,

$$Z(C_9; x_1, x_2, \ldots, x_9) = \tfrac{1}{9}[\phi(1)x_1^9 + \phi(3)x_3^3 + \phi(9)x_9] = \tfrac{1}{9}(x_1^9 + 2x_3^3 + 6x_9)$$

so that the pattern inventory is

$$\mathrm{PI}(C_9; \mathrm{R}, \mathrm{B}, \mathrm{G}) = \tfrac{1}{9}\,[(\mathrm{R}+\mathrm{B}+\mathrm{G})^9 + 2(\mathrm{R}^3+\mathrm{B}^3+\mathrm{G}^3)^3 + 6(\mathrm{R}^9+\mathrm{B}^9+\mathrm{G}^9)]$$

and

$$\tau(2,3,4) = \frac{1}{9}\left[\frac{9!}{2!\,3!\,4!} + 0 + 0\right] = 140$$

(*b*) By Problem 4.46,

$$Z(H_{18}; x_1, x_2, \ldots, x_9) = \tfrac{1}{18}(x_1^9 2x_3^3 + 6x_9 + 9x_1x_2^4)$$

whence

$$\mathrm{PI}(H_{18}; \mathrm{R}, \mathrm{B}, \mathrm{G}) = \tfrac{1}{2}\mathrm{PI}(C_9; \mathrm{R}, \mathrm{B}, \mathrm{G}) + \tfrac{1}{2}(\mathrm{R}+\mathrm{B}+\mathrm{G})(\mathrm{R}^2+\mathrm{B}^2+\mathrm{G}^2)^4$$

and

$$\tau(2,3,4) = \frac{1}{2}(140) + \frac{1}{2}\left(0 + \frac{4!}{1!\,1!\,2!} + 0\right) = 76$$

4.70 With respect to the rotational symmetries of a cube, in how many ways can the faces be painted red, blue, or green, if each color must be used at least once?

Here it is best to proceed indirectly. From Problem 4.49,

$$Z(G; x_1, x_2, \ldots, x_6) = \tfrac{1}{24}(x_1^6 + 3x_1^2x_2^2 + 6x_2^3 + 6x_1^2x_4 + 8x_3^2)$$

By Theorem 4.3 there are

$$Z(G; 3, 3, \ldots, 3) = 57 \text{ colorings in R-B-G, R-B, R-G, B-G, R, B, or G}$$

$$Z(G; 2, 2, \ldots, 2) = \begin{cases} 10 \text{ colorings in R-B, R, or B} \\ 10 \text{ colorings in R-G, R, or G} \\ 10 \text{ colorings in B-G, B, or G} \end{cases}$$

Hence the required number (of R-B-G's) is $57 - 10 - 10 - 10 + 3 = 30$.

4.71 Of the 57 patterns of Problem 4.70, how many involve 0 red faces? 1 red face? $\cdots$ 6 red faces?

On the set of colors $Y = \{R, B, G\}$, define a weight function by

$$w(R) = t \qquad w(B) = w(G) = 1$$

For this weighting, Theorem 4.4 gives the pattern inventory as

$$\begin{aligned}PI(G; t, 1, 1) &= \tfrac{1}{24}[(t+2)^6 + 3(t+2)^2(t^2+2)^2 + 6(t^2+2)^3 + 6(t+2)^2(t^4+2) + 8(t^3+2)^2] \\ &= t^6 + 2t^5 + 6t^4 + 10t^3 + 16t^2 + 12t + 10\end{aligned}$$

The coefficient of t^m $(m = 0, 1, \ldots, 6)$ in the right-hand side gives the number of patterns involving m red faces.

4.72 In Problem 4.47 each 1-foot segment can be painted one of r colors. (*a*) How many patterns are possible? (*b*) If $n = 8$ and $r = 3$ (R, B, and G), in how many patterns are 2 segments R, 4 segments B, and 2 segments G?

(*a*)
$$\text{Number of patterns} = \tfrac{1}{2}[r^n + r^{\lfloor (n+1)/2 \rfloor}]$$

(*b*) The cycle index is $\frac{1}{2}(x_1^8 + x_2^4)$, and so the pattern inventory is

$$\tfrac{1}{2}[(R+B+G)^8 + (R^2+B^2+G^2)^4]$$

The required coefficient of $R^2B^4G^2$ in the pattern inventory is

$$\frac{1}{2}\left(\frac{8!}{2!\,4!\,2!} + \frac{4!}{1!\,2!\,1!}\right) = 216$$

4.73 Find the number of ways, under the rotational group, of coloring the vertices and faces of a regular octahedron so that 4 vertices are red, 2 vertices are blue, 4 faces are green, and 4 faces are yellow.

Because the vertex coloring and the face coloring are independent, we may treat them separately and then use the product rule. By Problem 4.50, the cycle index of the group of vertex permutations is

$$\tfrac{1}{24}(x_1^6 + 3x_1^2x_2^2 + 6x_2^3 + 6x_1^2x_4 + 8x_3^2)$$

Therefore, the pattern inventory for red (R) and blue (B) is

$$\tfrac{1}{24}[(R+B)^6 + 3(R+B)^2(R^2+B^2)^2 + 6(R^2+B^2)^3 + 6(R+B)^2(R^4+B^4) + 8(R^3+B^3)^2]$$

The coefficient of R^4B^2 in the pattern inventory is

$$\frac{1}{24}\left[\frac{6!}{4!\,2!} + 3(3) + 6\left(\frac{3!}{2!\,1!}\right) + 0\right] = 2$$

The cycle index of the group of face permutations is

$$\tfrac{1}{24}(x_1^8 + 9x_2^4 + 8x_1^2x_3^2 + 6x_4^2)$$

The pattern inventory for green (G) and yellow (Y) is

$$\tfrac{1}{24}[(G+Y)^8 + 9(G^2+Y^2)^4 + 8(G+Y)^2(G^3+Y^3)^2 + 6(G^4+Y^4)]$$

The coefficient of G^4Y^4 in the pattern inventory is

$$\frac{1}{24}\left[\frac{8!}{4!\,4!}+9\left(\frac{4!}{2!\,2!}\right)+8(4)+6(2)\right]=7$$

Thus there are $(2)(7) = 14$ ways of coloring.

4.74 The graphic representation of the hydrocarbon benzene (C_6H_6) as a regular hexagon with a CH group at each vertex is called a **benzene ring**. If 2 kinds of radicals can attach to each vertex, find the number of distinguishable compounds that can be formed from the benzene ring.

Here the group of symmetries is the dihedral group H_{12}. Setting $x_1 = x_2 = \cdots = x_6 = 2$ in the cycle index [Problem 4.94(*a*)], we obtain 13 patterns (compounds).

4.75 Rework Problem 4.28 by means of the pattern inventory. (In effect, this problem is a check of Problem 4.66.)

With reference to Fig. 4-2 and to Example 4, the permutations of the cells induced by the symmetries of the square are:

$$\begin{aligned}
\pi_1 &= (1)(2)(3)(4)(5)(6)(7)(8)(9) &&\rightarrow && x_1^9\\
\pi_2 &= (5)(1\,3\,9\,7)(2\,6\,8\,4) &&\rightarrow && x_1x_4^2\\
\pi_3 &= (5)(1\,9)(3\,7)(2\,8)(4\,6) &&\rightarrow && x_1x_2^4\\
\pi_4 &= (5)(1\,7\,9\,3)(2\,4\,8\,6) &&\rightarrow && x_1x_4^2\\
\pi_5 &= (2)(5)(8)(1\,3)(4\,6)(7\,9) &&\rightarrow && x_1^3x_2^3\\
\pi_6 &= (4)(5)(6)(1\,7)(2\,8)(3\,9) &&\rightarrow && x_1^3x_2^3\\
\pi_7 &= (3)(5)(7)(2\,6)(1\,9)(4\,8) &&\rightarrow && x_1^3x_2^3\\
\pi_8 &= (1)(5)(9)(2\,4)(3\,7)(6\,8) &&\rightarrow && x_1^3x_2^3
\end{aligned}$$

Hence the cycle index is

$$\tfrac{1}{8}(x_1^9 + 4x_1^3x_2^3 + x_1x_2^4 + 2x_1x_4^2)$$

and the pattern inventory is

$$\tfrac{1}{8}[(\mathrm{B}+\mathrm{W})^9 + 4(\mathrm{B}+\mathrm{W})^3(\mathrm{B}^2+\mathrm{W}^2)^3 + (\mathrm{B}+\mathrm{W})(\mathrm{B}^2+\mathrm{W}^2)^4 + 2(\mathrm{B}+\mathrm{W})(\mathrm{B}^4+\mathrm{W}^4)^2]$$

We want the coefficient of $\mathrm{B}^2\mathrm{W}^7$ in the pattern inventory; it is

$$\frac{1}{8}\left[\frac{9!}{2!\,7!}+4(6)+4+0\right]=8$$

in agreement with Problem 4.28.

4.76. Show that there are precisely 17,824 distinguishable (under rotations) vertex colorings of the regular dodecahedron, using 1 or 2 colors.

The regular icosahedron (Problem 4.58) will have as its geometric dual a solid with 20 vertices and 12 faces, each of which is a regular pentagon; this is the regular dodecahedron. Therefore, Problem 4.58(*b*) yields the cycle index as

$$Z(G; x_1, x_2, \ldots, x_{20}) = \tfrac{1}{60}(x_1^{20} + 15x_2^{10} + 20x_1^2x_3^6 + 24x_5^4)$$

The number of vertex colorings is then

$$Z(G; 2, 2, \ldots, 2) = \tfrac{1}{60}(1{,}069{,}440) = 17{,}824$$

Supplementary Problems

4.77 If $G = \langle x \rangle$ is a cyclic group of order 12, list the orders of x^k for $k = 0, 1, 2, \ldots, 11$.
Ans. 1, 12, 6, 4, 3, 12, 2, 12, 3, 4, 6, 12

4.78 Find the number of inequivalent ways of seating 4 men and 2 women at a rectangular dining table if the seats are situated as in Fig. 4-5.

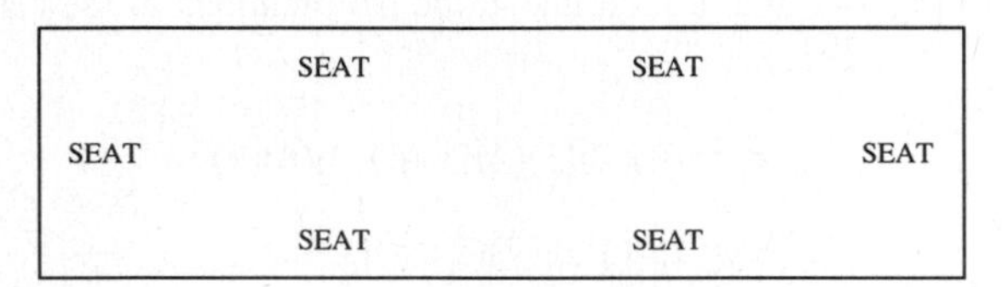

Fig. 4-5

Ans. $(64 + 16 + 8 + 8)/4 = 24$

4.79 Find the number of inequivalent ways of seating 2 men, 2 women, and 1 child at a round dining table.
Ans. $(3^5 + 3 + 3 + 3 + 3)/5 = 21$

4.80 Use the Burnside–Frobenius theorem to find the number of distinguishable ways of coloring the sides of a square using 2 colors. *Ans.* $(16 + 2 + 4 + 2 + 8 + 4 + 8 + 4)/8 = 6$

4.81 If f is the permutation that maps 1, 2, 3, 4, 5, 6, 7, 8, and 9 into 9, 8, 5, 4, 1, 6, 3, 2, and 7, write the disjoint-cycle representation of f. *Ans.* $(1\,9\,7\,3\,5)(2\,8)(4)(6)$.

4.82 If $x = (a\,b\,c\,d)$ and $y = (b\,d)$, express x^2, x^3, x^4, xy, and x^2y as products of disjoint cycles.
Ans. $(a\,c)(b\,d)$, $(a\,d\,c\,b)$, $(a)(b)(c)(d)$, $(a\,d)(b\,c)$, $(a\,c)(b)(d)$

4.83 If $x = (1\,2\,3\,4\,5\,6\,7\,8)$, (*a*) find x^k for $k = 2, 3, 4, 5, 6, 7, 8$; and (*b*) calculate the cycle index of $\langle x \rangle$, and check it against the result of Problem 4.44.
Ans. $(1\,3\,5\,7)(2\,4\,6\,8)$, $(1\,4\,7\,2\,5\,8\,3\,6)$, $(1\,5)(2\,6)(3\,7)(4\,8)$, $(1\,6\,3\,8\,5\,2\,7\,4)$, $(1\,7\,5\,3)(2\,8\,6\,4)$, $(1\,8\,7\,6\,5\,4\,3\,2)$, $(1)(2)(3)(4)(5)(6)(7)(8)$

4.84 If $x = (1\,2\,3\,4\,5\,6\,7\,8)$, find the groups generated by x^2 and x^4. *Ans.* $\{x^2, x^4, x^6, x^8\}$, $\{x^4, x^8\}$

4.85 If $x = (1\,3\,5\,7)(2\,4\,6)$ and $y = (1\,2\,3\,4\,5)$, find the order of xy. *Ans.* lcm $(4, 3) = 12$

4.86 Find the number of permutations of type $[3\quad 1\quad 0\quad 0\quad 0]$. *Ans.* 10

4.87 Express the permutation $(1\,2\,3\,4)(5\,6\,7)(1\,6\,7\,2\,9)(3\,4)$ as a product of *disjoint* cycles.
Ans. $(1\,7\,3)(2\,9)(5\,6)$

4.88 Prove **Cauchy's identity**,

$$\sum_{\substack{1a_1+2a_2+\cdots+na_n=n \\ a_i \geq 0}} \frac{1}{1^{a_1}2^{a_2}\cdots n^{a_n}a_1!\,a_2!\cdots a_n!} = 1$$

Ans. When we multiply the left hand side by $n!$, we should get the total number of permutations because of Cauchy's formula established in Problem 4.32

4.89 A complex number θ is a **primitive *n*th root of unity** if $\theta^n = 1$, but $\theta^k \neq 1$ for $k = 1, 2, \ldots, n-1$. Count the primitive nth roots of unity. (*Hint*: Recall Problem 4.7.) *Ans.* $\phi(n)$

4.90 Show that the stabilizer G_x is a subgroup of G.
Ans. Let g and h be in G_x. Then $g(h(x)) = g(x) = x$ which implies gh is in G_x

4.91 What is the result of choosing $w(x) \equiv 1$ in the weighted Burnside–Frobenius theorem (Problem 4.31)?
Ans. Theorem 4.2

4.92 Specialize the result of Problem 4.44 to the case $n = p$, a prime.

Ans. $Z(G; x_1, x_2, \ldots, x_p) = \dfrac{1}{p}[x_1^p + (p-1)x_p]$

4.93 Prove that $r^8 + r^4 + 2r^2 + 4r$ is divisible by 8, for all positive integers r.

Ans. $Z(C_8; x_1, x_2, \ldots, x_8) = \frac{1}{8}[\phi(1)x_1^8 + \phi(2)x_2^4 + \phi(4)x_4^2 + \phi(8)x_8]$

4.94 Evaluate the cycle indices of the dihedral groups (*a*) H_{12}, and (*b*) H_{14}.

Ans. (*a*) $\frac{1}{12}(x_1^6 + 3x_1^2x_2^2 + 4x_2^3 + 2x_3^2 + 2x_6)$; (*b*) $\frac{1}{14}(x_1^7 + 7x_1x_2^3 + 6x_7)$

4.95 Making use of Problem 4.32, compute the cycle index of the symmetric group S_5.

Ans. $\dfrac{1}{5!}(x_1^5 + 10x_1^3x_2 + 20x_1^2x_3 + 15x_1x_2^2 + 30x_1x_4 + 20x_2x_3 + 24x_5)$

4.96 Find the cycle index of the group of edge permutations induced by the rotational symmetries of the cube.
Ans. $Z(G; x_1, x_2, \ldots, x_{12}) = \frac{1}{24}(x_1^{12} + 6x_1^2x_2^5 + 3x_2^6 + 8x_3^4 + 6x_4^3)$

4.97 (*Klein's Four-Group*) (*a*) Find the cycle index of the group K of vertex permutations induced by the symmetries of a rhombus. (*b*) Show that the groups of Problems 4.40, 4.41, and 4.55(*b*) are all isomorphic to K.
Ans. (*a*) $Z(K; x_1, x_2, x_3, x_4) = \frac{1}{4}(x_1^4 + 2x_1x_2 + x_2^2)$. (*b*) *Hint*: Consider the bijection $g_i \langle - - - \rangle h_i$ from Problems 4.40 and 4.41 and the multiplication rule defined in Problem 4.55(*b*).

4.98 Refer to Problem 4.58. Obtain the cycle index of the group of edge permutations induced by the rotational symmetries of the regular icosahedron.
Ans. $Z(G; x_1, x_2, \ldots, x_{30}) = \frac{1}{60}(x_1^{30} + 15x_1^2x_2^{14} + 20x_3^{10} + 24x_5^6)$

4.99 Given a pyramid with a square base and congruent triangular lateral faces, the rotational symmetries of the figure (which are those of the base) induce a group G of permutations of the 5 faces and a group $G^* = C_4$ of permutations of the 4 lateral faces. (*a*) Determine the cycle index of G. (*b*) Find the number of distinguishable colorings of the 5 faces using 4 colors. (*c*) Repeat (*b*) if the base must differ in color from any lateral face.
Ans. (*a*) $Z(G; x_1, x_2, \ldots, x_5) = x_1\, Z(C_4; x_1, x_2, x_3, x_4) = \frac{1}{4}(x_1^5 + x_1x_2^2 + 2x_1x_4)$; (*b*) 280; (*c*) $4Z(C_4; 3, 3, 3, 3) = 96$

4.100 (*a*) Repeat Problem 4.99(*b*), taking into account both rotational and reflectional symmetries. (*b*) Obtain the cycle index of the group of edge permutations induced by all the symmetries of the pyramid.
Ans. (*a*) cycle index $= x_1 Z(H_8; x_1, x_2, x_3, x_4)$, and so the number is 220. (*b*) $\frac{1}{8}(x_1^8 + 4x_1^2x_2^3 + x_2^4 + 2x_4^2)$

4.101 Find the number of (rotationally) distinct ways of painting the faces of a regular dodecahedron in 3 or fewer colors. [*Hint*: See Problem 4.58(a).] *Ans.* 9099

4.102. Find the number of (rotationally) distinct ways of painting the faces of a regular icosahedron so that 4 faces are red and the other faces are blue. *Ans.* 96

4.103 Find the number of (rotationally) distinct ways of painting the faces of a cube using 6 colors so that each face is of a different color. *Ans.* 30

4.104 Find the number of (rotationally) distinct ways of coloring the vertices of a cube using at most 3 colors. *Ans.* 333

4.105 Find the number of distinguishable necklaces with 10 stones of at most 2 colors. (The symmetry group is H_{20}.) *Ans.* 78

4.106 Find the number of distinguishable necklaces that can be made using 5 diamonds and 7 rubies. *Ans.* 38

4.107 If the vertices of a square are painted in 3 or fewer colors, in how many patterns will 2 vertices be of 1 color and 2 of another color. *Ans.* 6

4.108 Find the number of ways, under the rotational group, of coloring a regular tetrahedron so that 2 vertices are red, 2 vertices are blue; 2 faces are green, 2 faces are yellow; 3 edges are black; 3 edges are white. (*Hint*: By inspection, there is only one vertex pattern and one face pattern.) *Ans.* 4

4.109 Prove that, for every integer r, $r^2(r^2+11) \equiv 0 \pmod{12}$. (*Hint*: Count the vertex colorings of a regular tetrahedron.)

4.110 How many distinct 7-horse merry-to-rounds are there with 2 red horses, 3 white horses, and 2 blue horses? *Ans.* 30

4.111 In a family of organic compounds a carbon atom occupies the center of a regular tetrahedron, the vertices of which are bonding sites for atomic hydrogen, an ethyl radical, a methyl radical, or atomic chlorine. Give the numbers of distinguishable compounds with 0, 1, 2, 3, and 4 hydrogens. (*Hint*: Follow Problem 4.71.) *Ans.* 15, 11, 6, 3, 1

4.112 Find the number of distinguishable ways of coloring the cells of a 3×3 chessboard so that 2 cells are red, 4 cells are white, and 3 cells are blue. (*Hint*: See Problem 4.75) *Ans.* 174

4.113 With respect to the group of rotational symmetries of the cube, in how many ways can 6 edges be colored red and the remaining 6 blue? (*Hint*: Use Problem 4.96.) *Ans.* 48

4.114 A square is divided into 8 triangles as in Fig. 4-6. (*a*) How many banners can be made on this model, if a triangle must be black, orange, or green? (*b*) How many of the possible banners have 2 black triangles, 4 orange triangles, and 2 green triangles? *Ans.* (*a*) 873; (*b*) 60

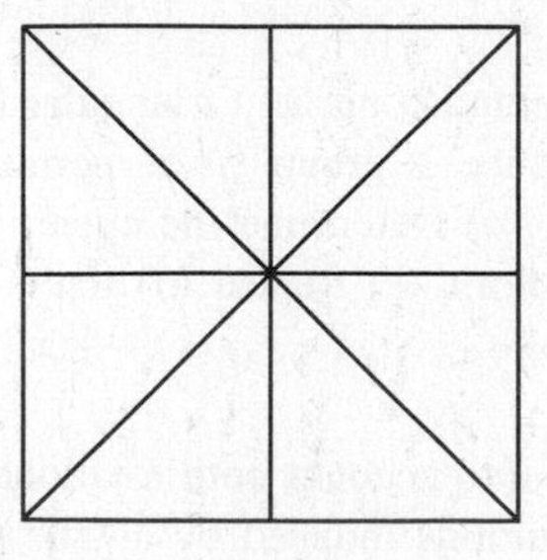

Fig. 4-6

Appendix

Graph Theory

A.1. INTRODUCTION

A graph in an informal sense can be considered as a collection of points, any pair of which may or may not be connected by a line. **Graph theory** is the study of such structures and their properties. In the past few decades, graph theory has developed into a systematic tool for analyzing problems in a variety of fields including combinatorics. Here we describe the terminology in graph theory that we have chosen to discuss in combinatorics. Note that terminology in graph theory has not yet been completely standardized. We also state a few important theorems in graph theory without proofs.

A.2. GRAPHS, MULTIGRAPHS, AND BIPARTITE GRAPHS

Formally, a **graph** $G = (V, E)$ consists of a finite nonempty set V of **vertices** (also known as **points** or **nodes**) and a set E of unordered pairs of distinct vertices. Such a pair $e = \{v_1, v_2\}$ is called an **edge** (also known as **line**, **link**, or **branch**) of the graph. In this case the edge **joins** the 2 vertices v_1 and v_2 which are **adjacent** to each other. Both the vertices are **incident** with edge e, and e is incident with both of them. The **order** of a graph is the number of vertices in it. Since V is finite, it is theoretically possible to represent any graph by a diagram. Sometimes the diagram of a graph is referred to as the graph itself.

The fact that E is a set imposes the restriction that when 2 distinct vertices are joined, there is at most 1 edge joining them. If we relax this restriction, the resulting structure $G = (V, E)$ is a **multigraph**, where E is a finite multiset. A graph $G = (V, E)$ is **simple** if E is not a multiset. A **bipartite graph** is a simple graph $G = (V, E)$ in which V can be partitioned into 2 subsets V_1 and V_2 such that any edge in E joins a vertex in V_1 and a vertex in V_2. In this case G is written symbolically as $G = (V_1, V_2, E)$. The **subgraph** $H = (W, F)$ of a graph $G = (V, E)$ is a graph in which W is a subset of V and F is a subset of E. A subgraph H of a graph G is a **spanning subgraph** if both G and H have the same set of vertices. The number of edges incident with a vertex v in a graph is called the **degree** of the vertex. A vertex is an **odd vertex** if its degree is odd. Otherwise it is an **even vertex**.

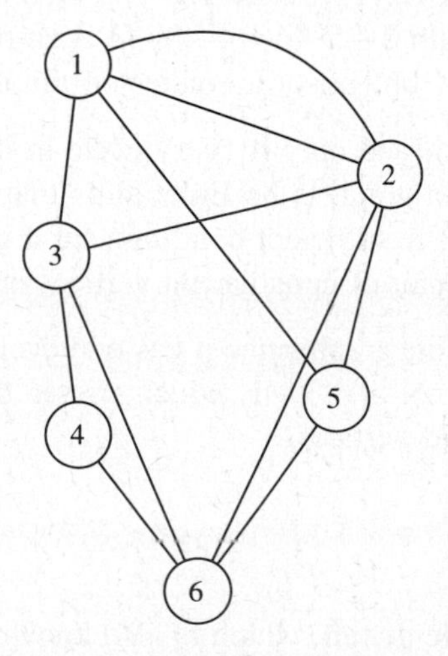

Fig. A-1

In Fig. A-1 we have the multigraph $G = (V, E)$ in which the set of vertices is $V = \{1, 2, 3, 4, 5, 6\}$. The edge $\{2, 3\}$ joins the vertices 2 and 3. So these 2 vertices are adjacent. Vertices 3 and 4 are incident to the edge {3, 4}, and the edge {3, 4} is incident to both these vertices. There are 2 edges joining the vertices 1 and 2. The vertex 3 is an even vertex since its degree is 4, whereas the vertex 2 is odd since its degree is 5. In Fig. A-2 we have the bipartite graph $G = (V_1, V_2, E)$, where $V_1 = \{a, p, q\}$ and $V_2 = \{b, r\}$.

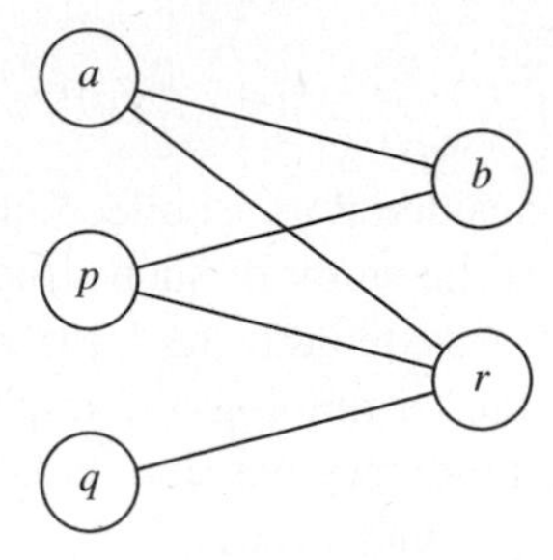

Fig. A-2

Theorem A.1. In any graph or multigraph, the number of odd vertices is even, and the sum of the degrees of all the vertices is twice the number of edges.

A.3. PATHS, CIRCUITS, AND CYCLES

A **path** between vertex u and vertex v in a graph is a finite alternating sequence of vertices and edges with terminal terms u and v such that any 2 consecutive terms are incident. If the graph is simple, the sequence can be written exclusively in terms of vertices. A path joining u and v is a **closed path** if u and v are the same. A closed path in which no 2 edges are the same is a **circuit**. A closed path in which no 2 vertices are the same is a **cycle**. Every cycle is a circuit. A path in which no 2 vertices are the same is a **simple path**. A graph is a **Eulerian graph** if it has a circuit that contains all the edges of the graph. A graph is a **Hamiltonian graph** if it has a cycle that contains all the vertices of the graph. An **even cycle** is a cycle in which the number of edges is even. A graph is said to be a **connected graph** if there is a path between every pair of vertices in the graph. A maximal connected subgraph of a graph G is called a **component** of G. If a connected graph $G = (V, E)$ becomes disconnected when an edge is deleted from E without deleting the 2 vertices incident to it from V, then that edge is a **bridge** in the graph.

In the graph G of Fig. A.1, the path 1, {1, 3}, 3, {3, 4}, 4, {4, 6}, 6 is a simple path between 1 and 6 which is written as 1—3—4—6. The closed path 2—5—6—4—3—2 is an odd cycle in G, whereas the closed path 3—4—6—5—2—6—3 is a circuit. It is obviously a connected graph in which no edge is a bridge.

Theorem A.2. (*i*) A graph is bipartite if and only if every cycle in it is an even cycle. (*ii*) A necessary and sufficient condition for a graph to be Eulerian is that it is connected and the degree of each vertex in it is even. (*iii*) A sufficient condition for a graph to be Hamiltonian is that the sum of the degrees of every pair of nonadjacent vertices is at least equal to the order of the graph.

The graph of Fig. A-1 is not a bipartite graph since it has odd cycles. It is obviously Hamiltonian because of the existence of the cycle 1—2—5—6—4—3—1, which passes through every vertex of the graph. The graph G is not Eulerian since it has odd vertices.

A.4. FORESTS AND TREES

A graph with no cycles is an **acyclic graph**, which is also known as a **forest**. Every forest is a bipartite graph. A **tree** is a connected acyclic graph. Each component of a forest is a tree. If a spanning subgraph of a graph is a tree, it is called a **spanning tree** in the graph. A graph is connected if and only if it has a spanning tree.

Theorem A.3. (*i*) A graph is a tree if and only if there is a unique path between every pair of vertices in it.

(*ii*) A graph is a tree if and only if it is connected and every edge in it is a bridge. (*iii*) A graph with n vertices is a tree if and only if it is connected and it has $n-1$ edges. (*iv*) A graph with n vertices is a tree if and only if it is acyclic and it has $n-1$ edges. (*v*) A graph $G=(V,E)$ is a tree if and only if it is acyclic and $G'=(V,E')$ has a unique cycle, where E' is obtained by adjoining one more edge to E.

In Fig. A-3 we have the graph $G=(V,E)$, which is a tree with 12 vertices and 11 edges. Notice that each edge in G is a bridge, and whenever 2 nonadjacent vertices are joined by an edge a unique cycle is created.

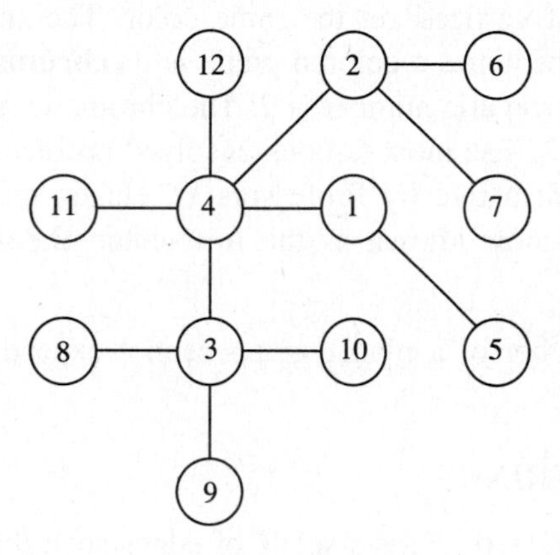

Fig. A-3

A.5. COMPLETE GRAPHS AND PLANAR GRAPHS

A graph is **complete** if there is exactly 1 edge joining every pair of vertices. A complete graph with n vertices is denoted by K_n. A bipartite graph $G=(V_1,V_2,E)$ is complete if there is an edge in E between every vertex in V_1 and every vertex in V_2. The complete bipartite graph in this case is denoted by $K_{m,n}$, where m and n are the cardinalities of V_1 and V_2, respectively. A graph is said to be **embedded** in a plane if it is possible to draw its diagram such that no 2 edges intersect except possibly at vertices. A graph is **planar** if it can be embedded in a plane. The regions defined by the edges of a planar graph are its **faces**. One of these faces is necessarily unbounded.

Theorem A.4. (*i*) The number of faces of a planar multigraph with n vertices and m edges is $m+2-n$. (*ii*) Both K_5 and $K_{3,3}$ are nonplanar.

In the planar graph of Fig. A-4, there are 5 vertices and 7 edges. The bounded faces are F_1, F_2, and F_3. The unbounded face is F_4. The total number of faces p is 4. Thus $n=5$, $m=7$, and $p=4$, which agrees with the assertion that $p=m+2-n$.

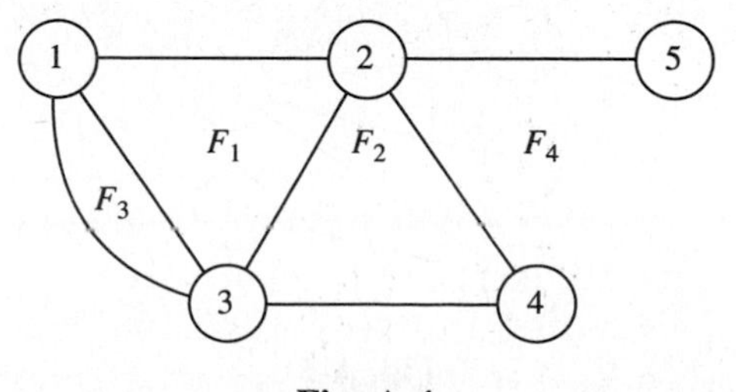

Fig. A-4

Two graphs are said to be **homeomorphic** (or identical to each other within vertices of degree 2) if they both can be obtained from the same graph G by introducing new vertices of degree 2 on its edges. Notice that insertion or deletion of such vertices on the edges of a graph does not affect considerations of planarity.

***Theorem A.5* (*Kuratowski's Theorem*).** A graph is planar if and only if it contains no subgraph that is homeomorphic to K_5 or $K_{3,3}$.

A.6. VERTEX COLORING OF GRAPHS

A graph G is said to be **colored** if it is possible to color the vertices of G such that (*i*) each vertex gets a unique color, and (*ii*) no 2 adjacent vertices get the same color. The smallest number of colors needed to color the vertices of G such that G becomes a colored graph is its **chromatic number**. A graph is bipartite if and only if it is connected and its chromatic number is 2. The chromatic number of the graph of Fig. A-1 is 4 and that of the graph of Fig. A-2 is 2. The most famous unsolved problem in mathematics was the four-color conjecture. It was settled in the affirmative by Professors W. Haken and K. I. Appel of the University of Illinois in 1976. This conjecture is now known as the **four-color theorem** and is given as the following assertion.

Theorem A.6. The chromatic number of a planar graph cannot exceed 4.

A.7. MATCHINGS AND COVERINGS

A **matching** in a simple graph $G = (V, E)$ is a set M of edges such that no 2 edges in M have a vertex in common. A **vertex-covering** is a set W of vertices in G such that every edge in E is incident to at least 1 vertex in W. A matching M in a bipartite graph (V_1, V_2, E) is a **complete matching** from V_1 to V_2 if every vertex in V_1 is incident to an edge in M. The cardinality of a matching in general cannot exceed the cardinality of a vertex-covering; and consequently, the largest size of a matching cannot exceed the smallest size of a vertex-covering in any graph. In the case of bipartite graphs, however, the following two famous theorems hold.

***Theorem A.7* (*König's Theorem*).** In a bipartite graph there exists a matching M and a vertex-covering W such that both M and W have the same cardinality. Equivalently, the size of a maximum matching in a bipartite graph is equal to the size of a minimum covering. (See Problem 2.109).

In the bipartite graph of Fig. A-5, the set $\{1, 5, 8\}$ is a minimum covering, and the set $\{\{1, 6\}, \{4, 8\}, \{5, 9\}\}$ is a maximum matching.

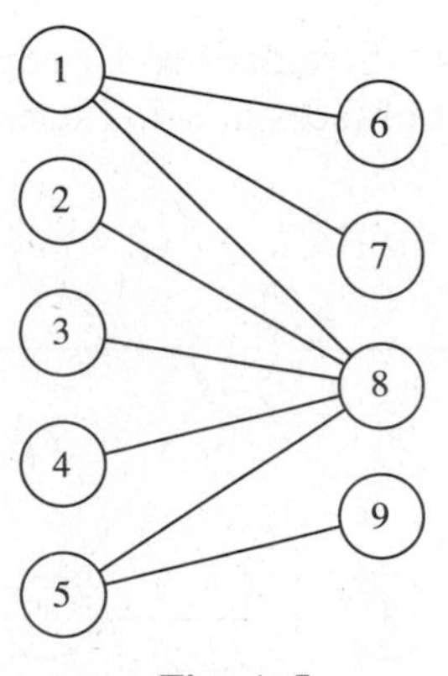

Fig. A-5

***Theorem A.8* (*Hall's Marriage theorem*).** In a bipartite graph (V_1, V_2, E), if A is a subset of V_1, let $f(A)$ be the set of vertices in V_2 that are joined to at least 1 vertex in A. Then a complete matching from V_1 to V_2 exists if and only if $|f(A)| \geq |A|$ for every subset A of V_1. [In a community of m unmarried women and m or more unmarried men, it is possible to match each woman with a man in the community she already knows if and only if the following condition (known as the **marriage condition**) is satisfied: every set of k women ($1 \leq k \leq m$) in the community collectively know at least k men in the community.] (See Theorem 2.7).

For example, consider the scenario consisting of 4 women w_1, w_2, w_3, and w_4, where w_1 knows 2 men m_1 and m_2, w_2 knows m_1 and m_3, w_3 knows m_2 and m_4, and w_4 knows m_3, m_4, and m_5. It is easy to verify that the marriage condition is satisfied for every subset of women in the group. One possible matching here is $\{w_1, m_2\}$, $\{w_2, m_3\}$, $\{w_3, m_4\}$, and $\{w_4, m_5\}$.

A.8. DIGRAPHS AND DILWORTH'S THEOREM

Recall that each edge in a graph is an unordered pair $\{v_1, v_2\}$ of vertices. In some situations, it is useful to give an orientation or a sense of direction for an edge. So instead of the unordered pair $\{v_1, v_2\}$, we consider the ordered pair (v_1, v_2) which represents the arc e from the vertex v_1 to the vertex v_2. In this case, e is incident from v_1 to v_2. The number of arcs incident to a vertex v is the **indegree** of v, and the number of arcs incident from v is its **outdegree**. In the figure of the graph, instead of drawing a line joining v_1 and v_2, an arrow is drawn from v_1 to v_2. Thus a **digraph** $G = (V, E)$ consists of a finite set V and a set E of **arcs** which are ordered pairs of distinct vertices. Any graph can be viewed as a digraph if each edge $\{v, w\}$ is replaced by 2 arcs (v, w) and (w, v). The **underlying graph** of a digraph is the graph obtained by converting each arc into an edge. A **dipath** (or **directed path**) from v_1 to v_k is a sequence of vertices $v_1, v_2, \ldots, v_k$ such that (v_i, v_{i+1}) is in E for $i = 1, 2, \ldots, k-1$. A dipath is a **simple dipath** if no 2 vertices in it are the same. A dipath from v to w is a **closed dipath** if $v = w$. A simple closed dipath is a **dicycle** (or **directed cycle**). A digraph is **acyclic** if it has no directed cycles. A digraph is **strongly connected** if there is a dipath in it from every vertex to every other vertex. A digraph is **weakly connected** if its underlying graph is a connected graph.

In Fig. A-6, we have a strongly connected digraph with 6 vertices in which 2—3—4—6—2 is a directed cycle. The underlying graph of this digraph is the graph shown in Fig. A-1.

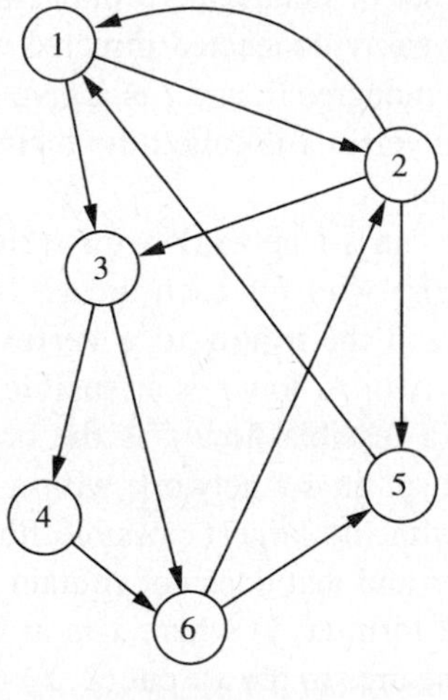

Fig. A-6

Theorem A.9 ***(Graph-Theoretic Version of Dilworth's Theorem).*** If E is the set of arcs in an acyclic digraph G, the maximum number of arcs in E with the property that no 2 of them belong to the same directed path is equal to the minimum number of arc-disjoint directed paths into which E can be partitioned. In other words, if A is a set of arcs in G, the minimum number of directed paths needed to cover the arcs in A is equal to the maximum number of arcs in A, no 2 of which belong to the same directed path. (See Problem 2.132.)

In the acyclic digraph of Fig. A-7, the set $\{(1,6), (2,6), (6,9), (2,4), (3,4)\}$ constitutes a set A of 5 arcs. The minimum number of directed paths needed to cover the arcs in the set A is 4. The maximum number of arcs in the set A, no two of which belong to the same directed path, is also 4.

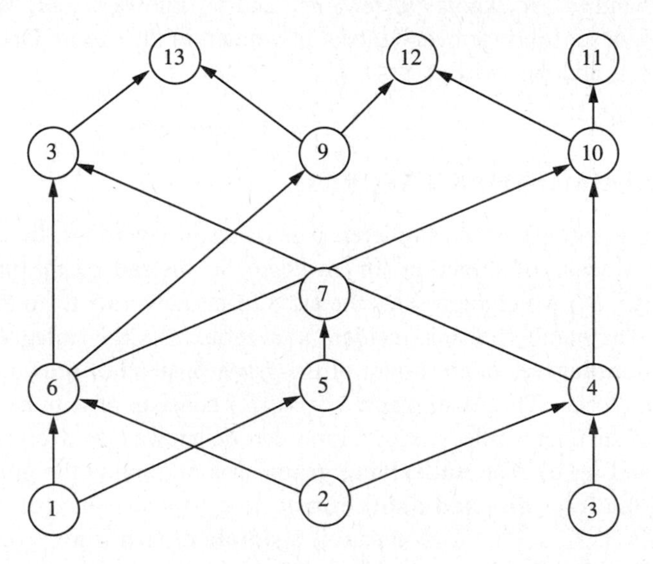

Fig. A-7

A.9. NETWORK FLOWS

A **network** (or **directed network**) is a graph (or a digraph) in which each edge e (arc e) is assigned a unique nonnegative integer $w(e)$ called its **weight**. Thus w is a function (the **weight function** of the network) from the set E of edges (or arcs) to the set of nonnegative integers. The weight $w(e)$ of the arc $e = (i, j)$ is written as $w(i, j)$ An **s-t network** is a weakly connected directed network $G = (V, E, s, t, w)$, where w is its weight function, s is a given vertex with indegree 0, and t is a given vertex with outdegree 0. The vertex s is called the **source** of the network, and the vertex t is called the **terminal.** A vertex is an **intermediate vertex** if it is neither the source nor the terminal.

A function f from the set of arcs in an s-t network with weight function w to the set of nonnegative integers is a **flow in the network** if $f(e) \leq w(e)$ for each arc e. The **outflow from a vertex v** is the sum $0(v) = \Sigma\{f(e) : e \text{ is of the form } (v, w)\}$, and the **inflow to a vertex v** is the sum $1(v) = \Sigma\{f(e) : e \text{ is of the form } (w, v)\}$. The **net flow at v** is $0(v) - 1(v)$. A flow f is a **feasible flow** if the net flow at every intermediate vertex is zero. The **flow value $v(f)$** of a feasible flow f is the outflow from s, which is also equal to the inflow to t. The maximum flow problem in an s-t network with a given weight function is the problem of finding a feasible flow in the network with the largest possible flow value.

If X is a set of vertices that contains s and that does not contain t and if Y is the relative complement of X in V, then the set (X, Y) of all arcs of the form (x, y) where x is in X and y is in Y is called an ***s-t cut*** of the network. The sum of the weights of all the arcs in the s-t cut (X, Y) is the **weight of the cut** and is denoted by $w(X, Y)$. An s-t cut (X, Y) for which $w(X, Y)$ is a minimum is called a **minimum cut** of the network. If f is any feasible flow and (X, Y) is any s-t cut, then $v(f) \leq C(X, Y)$. In particular, if (S, T) is a minimum cut, then $v(f) \leq C(S, T)$ for any feasible flow f.

Theorem A.10 **(*Max-Flow Min-Cut Theorem of Ford and Fulkerson*).** Let (V, E, s, t, w) be an s-t network, where w is a function from E to the set of nonnegative integers. Then there exists a feasible flow f and an s-t cut (S, T) such that: (*i*) $v(f) = C(S, T)$; (*ii*) if (i, j) is an arc of the network, where i is in S and j is in T, then $f(i, j) = w(i, j)$; and (*iii*) if (i, j) is an arc where i is in T and j is S, then $f(i, j) = 0$.

In the digraph of Fig. A-8, the source s is at vertex 1, and the terminal is at vertex 6. The capacity of each arc and the flow along that arc are written on the arc. The maximum amount that can be shipped from 1 to 6 without violating the capacity constraints is 7. If S is the set of vertices 1 and 3, and if T is the set of the remaining vertices, then the arcs in the cut (S, T) are $(1, 2)$ and $(3, 4)$, giving a cut value of $4 + 3 = 7$, which is the same as the maximum flow value. Furthermore, the flow in each arc from S to T is equal to its capacity.

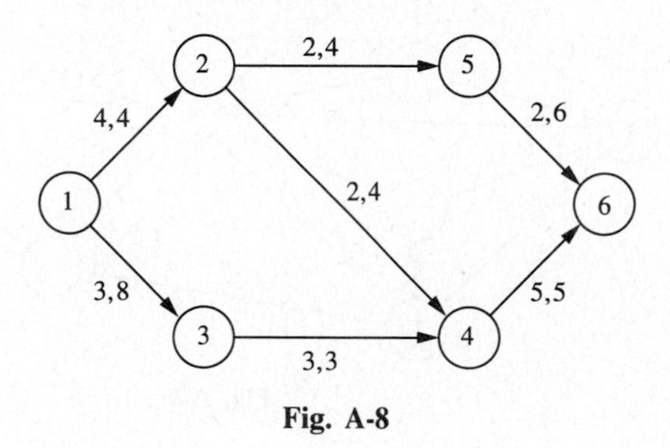

Fig. A-8

A.10. GENERALIZED s-t NETWORKS

An s-t network with a weight function w defined on its arcs is a **generalized s-t network** if there exists a weight function u from the set of vertices to the set of nonnegative integers. So if $V = \{1, 2, \ldots, n\}$ is the set of vertices, each arc (i, j) has a weight $w(i, j)$, and each vertex i has a weight $u(i)$.

A feasible flow f in the network is a **generalized feasible flow** if it also satisfies the following $n - 2$ vertex constraints: the outflow at every intermediate vertex i cannot exceed $u(i)$. A **generalized s-t cut** is a set X of arcs and vertices such that any dipath from the source to the terminal will use at least 1 element of X. The **weight of the generalized cut X** is the sum of the weights of all the elements in X.

Theorem A.11 ***(Generalized Max-Flow Min-Cut Theorem).*** The maximum value of a generalized s-t flow is equal to the weight of a minimum generalized s-t cut.

A.11. MENGER'S THEOREM AND CONNECTIVITIES IN DIGRAPHS AND GRAPHS

Two directed paths from u to v in a digraph are said to be: (*i*) **vertex-disjoint** if they have no vertices in common other than u and v and (*ii*) **arc-disjoint** if they have no arcs in common. The definitions of these concepts in the case of undirected graphs are analogous. We are now ready to state four different versions of Menger's theorem involving graph connectivity.

Theorem A.12 (Menger's Theorems).

(i) ***Vertex Form for Directed Graphs.*** The maximum number of vertex-disjoint directed paths from the vertex u to the vertex v in a digraph that has no arc from u to v is equal to the minimum number of vertices whose deletion destroys all directed paths from u to v in the digraph.

(ii) ***Arc Form for Directed Graphs.*** The maximum number of arc-disjoint directed paths from the vertex u to the vertex v in a digraph is equal to the minimum number of arcs whose deletion destroys all directed paths from u to v in the digraph.

(iii) ***Vertex Form for Undirected Graphs.*** The maximum number of vertex-disjoint paths between the vertex u and the vertex v in a graph that has no edge joining u and v is equal to the minimum number of vertices whose deletion destroys all paths between u and v in the graph.

(iv) ***Edge Form for Undirected Graphs.*** The maximum number of edge-disjoint paths between the vertex u and the vertex v in a graph is equal to the minimum number of edges whose deletion destroys all paths between u and v in the graph.

In Fig. A-9, the vertices 1 and 5 are not adjacent. The maximum number of vertex disjoint paths between 1 and 5 is 3. If the vertices 2, 3, and 4 are deleted, then there is no path between 1 and 5.

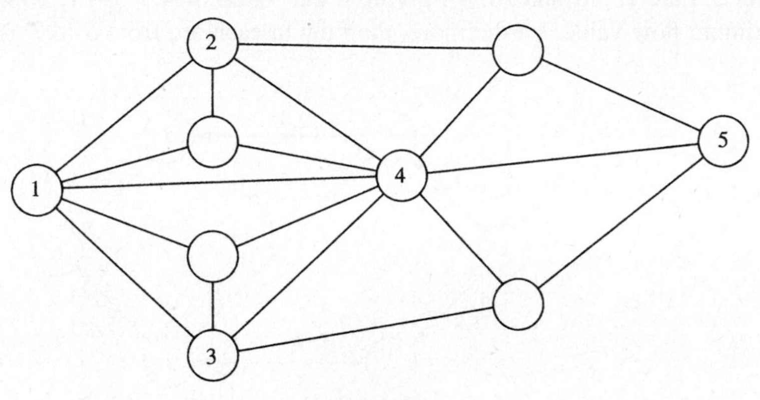

Fig. A-9

A.12. THE EQUIVALENCE THEOREM IN GRAPH THEORY AND COMBINATORICS

Theorem A.13. Dilworth's theorem, Ford–Fulkerson theorem, Hall's Marriage theorem, König's theorem, and Menger's theorem are equivalent. (See Problem 2.135 for a partial proof.)

Glossary

Antichain: (i) A collection F of subsets of a set X such that no set in F is properly contained in another set in F is an antichain in X. Also known as a clutter in X or a Sperner system in X. The cardinality of a maximum antichain in a finite set is obtained using Sperner's theorems. See Problems 2.116–2.118. (ii) A subset F of a partially ordered set P such that no two elements in F are comparable is an antichain in P. See Problem 2.128. (iii) A collection F of arcs in an acyclic digraph G such that no two arcs in F belong to the same directed path in G is an antichain in G. See Problem 2.132.

Bell's number B_n: The total number of partitions of a set with n elements. See Problem 3.48 (Dobinski's equality) in which B_n is expressed as the sum of an infinite series.

Bernoulli number: See Problem 3.45.

Bernoulli polynomial: See Problem 3.53.

Binet formula: (i) See Problem 3.60 for Fibonacci numbers. (ii) See Problem 3.112 for Lucas numbers.

Binomial coefficient $C(n, r)$: This number is $n!/[r!(n-r)!]$. Some interpretations of $C(n, r)$ are: (i) The number of ways to choose r elements from a set of n elements. (ii) The number of r-subsets of an n-set. (iii) The number of nondecreasing paths of length n from $(0, 0)$ to $(n-r, r)$ on the planar lattice of integral points. See Problem 1.58. (iv) The number of binary n-vectors with exactly r zeros.

Birkhoff–von Neumann Theorem: A square matrix A of nonnegative real numbers is doubly stochastic (i.e., each line sum is 1) if and only if it is a convex combination of permutation matrices. See Problem 2.179.

Burnside–Frobenius Theorem: See Theorem 4.2.

Catalan number C_n: This number is equal to $(1/n)C(2n-2, n-1)$. Some interpretations of C_n are: (i) The number of nondecreasing paths on the planar lattice of integral points from $(1, 0)$ to $(n, n-1)$ that lie entirely below the line $y = x$. See Problem 1.114. (ii) The number of monotonic increasing functions from the set $X = \{1, 2, \ldots, n-1\}$ to X. See Problem 1.117. (iii) The number of sequences of the form $\langle u_1, u_2, \ldots, u_{2n-2}\rangle$ such that (a) each u_i is either 1 or -1, (b) the sum of the $2n-2$ elements of the sequence is 0 and (c) $u_1 + u_2 + \cdots + u_k \geq 0$ for $1 \leq k \leq 2n-3$. See Problem 1.118. (iv) The number of sequences of the form $\langle u_1, u_2, \ldots, u_{2n-1}\rangle$ such that (a) each u_i is a nonnegative integer, (b) $u_1 = u_{2n-1} = 0$ and (c) $u_{i+1} - u_i$ is either -1 or 1 for each i. See Problem 1.119. (v) The number of ways in which parentheses can be inserted in a nonassociative product of n factors. See Problem 1.120. (vi) The number of diagonal triangulations of a convex polygon with $n+1$ vertices. See Problem 1.125. (vii) The number of binary vectors with n zeros and n ones such that every component after the first is preceded by more zeros than ones. See Problem 1.129. (viii) The number of ways of arranging $(2n-2)$ distinct real number as two decreasing subvectors $u = [u_1\ u_2 \ldots u_{n-1}]$ and $v = [v_1\ v_2 \ldots v_{n-1}]$ such that $u_i > v_i$ for each i. See Problem 1.130.

Cauchy's formula and Cauchy's identity: If f is a permutation of $X = \{1, 2, \ldots, n\}$ consisting of a_i cycles of length i ($i = 1, 2, \ldots, n$), then the type of f is the vector $[a_1\ a_2 \ldots a_n]$. Cauchy's formula gives the number of permutations of X for a given type. See Problem 4.32. For the connection between this formula and Cauchy's identity, see Problem 4.88.

Cayley's Theorem: Every finite group is isomorphic to a group of permutations on a finite set. See Problem 4.54.

Chromatic number: The minimum number of colors with which the vertices of a graph can be colored such that every two adjacent vertices have distinct colors.

Clutter in a set: See antichain.

Combination of n things taken r at a time: See binomial coefficient. For the definition of **generalized combinations,** see Section 2.1.

Convex polytope: A three-dimensional solid in which all faces are congruent polygons and each vertex is incident with the same number of faces. See Problem 4.58.

Convolution: See Theorem 3.1.

Convolution rule: This rule (also known as the Vandermonde identity) is $C(p+q,r)=\Sigma_{j=0}^{r} C(p,j)C(q,r-j)$. See Problem 1.43.
Cycle index: See Section 4.2 for the definition of a cycle index in a permutation group. For the extension to an arbitrary finite group see Problem 4.55.
Diagonal triangulation of a convex polygon: See Problem 1.123.
Derangement: A permutation P of $\{1, 2, \ldots, n\}$ such that $P(i)$ is not equal to i for each i. The number of permutations on a set of n elements is D_n. See Example 10 (Probleme des Recontres) in Chapter 3. See also Problems 2.25 and 3.73.
Dihedral group: The symmetry group H_{2n} of a regular n-gon. See Problem 4.39.
Dilworth's Theorem: (i) In a finite partially ordered set P the maximum cardinality of an antichain is equal to the minimum number of (disjoint) sets into which P can be partitioned. See Problem 2.129. (ii) In an acyclic digraph G the maximum number of arcs in a set A of arcs of G such that no two arcs in A belong to the same directed path is equal to the minimum number of arc-disjoint directed paths into which the set A can be partitioned. See Problem 2.132.
Dobinski's equality: See Bell number.
Dodecahedron: See Problems 4.76 and 4.101.
Duality principle of distribution: See Problem 2.116.
Dufree square: See Problem 3.34.
Euler's phi (totient) function: The number of positive integers not exceeding n and relatively prime to n is $\phi(n)$, where ϕ is Euler's phi function. See Problem 2.45.
Euler's identities connecting infinite products and infinite series: See Problems 3.35, 3.36, and 3.117.
Euler's pentagonal number: See Problem 3.116.
Euler's Pentagonal Theorem: See Problem 3.118.
Falling factorial polynomial: See Problem 1.132.
Fibonacci number: Numbers defined by $f(0)=f(1)=1$ and $f(n+2)=f(n+1)+f(n)$ for each $n \geq 0$. See Problems 3.60–3.63. The number of subsets (including the empty set) of $\{1, 2, \ldots, n\}$ such that no subset contains two consecutive integers is $f(n+1)$. See Problem 3.64.
Ford–Fulkerson Theorem: See Section A.9 in the Appendix.
Generating functions of a sequence: See Section 3.1 for the definitions of ordinary and exponential generating functions.
Forest: An undirected graph with no cycles. A tree is a connected forest. See A.4 in the Appendix.
Four Color Theorem: The chromatic number of a planar graph cannot exceed four. See A.6 in the Appendix.
Generalized permutation: See multinomial theorem.
Hall's Theorem: (Also known as Hall's Marriage Theorem.) Given a family of n sets, not necessarily distinct, it is possible to choose exactly one element from each set such that the chosen n elements are distinct if and only if the total number of elements in any subfamily of k of these sets is at least k for $k = 1, 2, \ldots, n$. The set of n elements thus chosen is called a system of distinct representatives (SDR). This necessary and sufficient condition which assures the existence of an SDR is called the marriage condition. See Theorem 2.7.
Hasse diagram of a poset: The representation of a partially ordered set as a directed graph. See Problem 2.131.
Hit polynomial: See Problem 2.92.
Icosahedron: See Problems 4.58 and 4.102.
Inclusion-exclusion principle: See Section 2.3 and the Sieve Formula (Theorem 2.5). For generalized inclusion-exclusion principle see Theorems 2.8 and 2.9 which are presented along with Problem 2.65.
Katona's Theorem: The graph-theoretic version of Sperner's theorems. See Problem 2.124.
König's Theorem: In a bipartite graph the cardinality of a maximum matching is equal to the cardinality of a minimum covering of the edges. See Problem 2.109.
König–Egervary Theorem: The term rank of a binary matrix (i.e., the largest number of ones that can be chosen from the matrix such that no two selected ones lie on the same row or column) is equal to the smallest number of lines (rows or columns) that can be deleted from the matrix so that the matrix becomes the zero matrix once these lines are deleted. See Problem 2.108.

Kuratowski's Theorem: A graph is planar if and only if it contains no graph that is homeomorphic to K_5 or $K_{3,3}$. See A.5 in the Appendix.
Lagrange's Theorem: The order of a finite group is divisible by the order of any subgroup. See Problem 4.18.
Latin square: If each row and each column of an $n \times n$ matrix A is a permutation of the first n positive integers, then the matrix A is a latin square of order n. Two latin squares of order n, $A = [a_{ij}]$ and $B = [b_{ij}]$ are orthogonal if the n^2 ordered pairs (a_{ij}, b_{ij}) are all distinct. See Problem 1.88. For the definition of a latin rectangle, see Problem 2.107.
Lucas number, L_n: These numbers are defined by $L_1 = 1$, $L_2 = 3$, and $L_{n+2} = L_{n+1} + L_n$ for $n \geq 1$. The number of subsets (including the empty set) of $\{1, 2, \ldots, n\}$ such that no subset contains two consecutive integers or 1 and n at the same time is L_n. (See Problem 3.65.)
LYM (Lubel–Yamamoto–Meschalkin) Inequality: See Problem 2.116.
Marriage condition: See Hall's Marriage Theorem.
Maximal [minimal]: Implies maximal [minimal] with respect to inclusion.
Maximum [minimum]: Implies maximal [minimal] with respect to cardinality.
Menage number: See Probleme des Menages.
Menger's Theorems: See Section A.11 in the Appendix.
Mirsky's Theorem: In a finite partially ordered set P the cardinality of a maximum chain is equal to the minimum number of disjoint antichains into which P can be partitioned. This theorem is known as the dual of Dilworth's Theorem. See Problem 2.130.
Möbius Function, $\mu(n)$: If n is 1 or a product of an even number of distinct primes $\mu(n) = 1$; if n is a product of an odd number of distinct primes $\mu(n) = -1$ and for $\mu(n) = 0$ for all other n. See Problem 2.51 and Problem 2.57 showing the relation between the Möbius function and Euler's phi function.
Multinomial number: If X is a collection of n objects that are not necessarily distinct, any arrangement (ordering) of these objects is a generalized permutation of X. If there are n_i objects of type i ($i = 1, 2, \ldots, k$), then the number of generalized permutations of X is denoted by $P(n; n_1, n_2, \ldots, n_k)$. If the elements in X are all distinct (i.e., X is a set), then the number of ordered partitions of X into k subsets $X_1, X_2, \ldots, X_k$ such that the cardinality of X_i is n_i ($i = 1, 2, \ldots, k$) is given by the multinomial number $C(n; n_1, n_2, \ldots, n_k)$ which is equal to $P(n; n_1, n_2, \ldots, n_k)$ as shown in Theorem 2.2.
Multinomial Theorem: See Problem 2.1.
Newton's identity: $C(n, r)C(r, k) = C(n, k)C\ (n - k, r - k)$. See Problem 1.44.
Number-theoretic function: A function whose domain is the set of positive integers. A number-theoretic function f is multiplicative if the range of f is closed under multiplication and $f(mn) = f(m)f(n)$ whenever m and n are relatively prime. See Problem 2.48.
Orbit: If G is a subgroup of the symmetric group of permutations on a finite set X, then the orbit of an element x in the set is the set $Gx = \{g(x) : g \in G\}$. The cardinality of Gx is a divisor of the order of G. See Problem 4.22.
Partially ordered set: Set Problem 2.128.
Partition of a set: (i) A family of pairwise disjoint nonempty subsets of a set X such that the union of all the sets in the family is X is a partition of X. The number of partitions of a set of cardinality n such that each partition has m sets is the Stirling number of the second kind denoted by $S(n, m)$. See Problem 1.146. The total number of partitions of a set of cardinality n is the Bell number B_n. See Problem 1.161. A partition of a nonempty subset of a set X is a partial partition of X. The number of partial partitions of a set of cardinality n is $B_{n+1} - 1$. See Problem 1.162. (ii) The number of ordered partitions of a set X into subsets $X_1, X_2, \ldots, X_k$ with cardinalities $n_1, n_2, \ldots, n_k$ respectively is the multinomial coefficient $C(n; n_1, n_2, \ldots, n_k)$, whereas the number of unordered partitions of X into p_1 subsets of cardinality n_1, p_2 subsets of cardinality $n_2, \ldots, p_k$ subsets of cardinality n_k is $[C(n;\ n_1, \ldots, n_1,\ n_2, \ldots, n_2, \ldots, n_k, \ldots, n_k)]/[p_1!\ p_2! \cdots p_k!]$. See Theorem 2.3.
Partition of a positive integer: (i) A partition of an integer r is a representation of r in the form $r = r_1 + r_2 + \cdots + r_k$; the integers r_i are the parts of the partition which satisfy the inequalities $r_1 \geq r_2 \geq \cdots \geq r_k \geq 1$. The number of partitions of r is denoted by $p(r)$. See Section 3.2. (ii) The number of partitions of r such that no part exceeds n is $p_n(r)$ and the number of partitions of r into at most n parts is $q_n(r)$. In Theorem 3.2, it is shown that $p_n(r) = q_n(r)$. (iii) The number of partitions of the positive integer N in which each part is

repeated fewer than r ($r > 1$) times is $f(N, r)$, and the number of partitions of N having no part divisible by r is $g(N, r)$. See Problem 2.24 proving $f(N, r) = g(N, r)$. (iv) The number of partitions of r with largest part equal to n is $p(r, n)$, and the number of partitions of r into exactly n parts is $q(n, r)$; in Problem 3.24 it is shown that $p(r, n) = q(r, n)$. (v) The number of partitions of r into unequal parts is $p^{\#}(r)$ and the number of partitions of r into (possibly repeated) odd parts is $p(r, \text{ODD})$. In Problem 3.29 it is proved that $p^{\#}(r) = P(R, \text{ODD})$. (vi) The number of partitions of r with distinct even parts is $p^{\#}(r, \text{EVEN})$, and the number of partitions of r into distinct odd parts is $p^{\#}(r, \text{ODD})$. See Problem 3.28 for their generating functions. (vii) The conjugate partition of a partition is obtained from its star diagram as in Theorem 3.2. A partition is self conjugate if it is equal to its conjugate. The number of self conjugate partitions of r is equal to $p^{\#}(r, \text{ODD})$. See Problem 3.33. (viii) The number of partitions of r into n distinct (unequal) parts is $q^{\#}(r, n)$. See Problem 3.32. (ix) The number of partitions of r into an even number of unequal parts is $q^{\#}(r, \text{E})$, and the number of partitions of r into an odd number of unequal parts is $q^{\#}(r, \text{O})$. See Problems 3.37, 3.117, and 3.118. (x) If K is a set of n distinct positive integers, $f_n(r)$ is the number of partitions of r selected (with replacement) from K. A generating function for $f_n(r)$ is established in Problem 3.8.

Pascal's identity: $C(n, r) = C(n-1, r) + C(n-1, r-1)$. See Problem 1.137.

Pattern inventory: See Section 4.3.

Permanent of a matrix: See Problem 2.76.

Permutation group: See Section 4.2.

Permutation matrix: A binary matrix P is a permutation matrix if PP^T is the identity matrix where P^T is the transpose of P. See Problem 2.115.

Pigeonhole principle: If n pigeonholes shelter $n + 1$ or more pigeons, at least one pigeonhole shelters at least two pigeons. See Section 1.3 and Problems 1.76–1.92.

Pólya's enumeration theorems: See Theorems 4.3 and 4.4.

Poset: See Problem 2.128.

Probleme des Menages: This is the problem of finding the number M_n of ways of seating $n \geq 3$ married couples at a circular table (with $2n$ numbered seats) so that sexes alternate and no husband and wife sit side by side. The number M_n is called the Menage number for n. See Problems 2.73 and 2.177.

Probleme des Recontres: See Derangement.

Rado's Theorem: See Problem 2.106.

Ramsey numbers: The Ramsey number $R(p, q)$ of two positive integers p and q is the smallest positive integer n with the property that each graph with n vertices contains either a set U of p vertices such that every vertex in U is adjacent to every other vertex in U or a set W of q vertices such that no two vertices in W are adjacent. See Problems 1.93–1.105.

Ranked poset: In a poset $Q = (X, <)$, the element x covers y if $y < x$ and $y \leq t \leq x$ imply $y = t$ or $t = x$. The poset Q is a ranked poset if there exists a function r defined on Q such that $r(x) = 0$ whenever x is a minimal element and $r(x) = r(y) + 1$ whenever x covers y. The set $Q_k = \{x : r(x) = k\}$ is the kth level set and the cardinality of Q_k is the kth Whitney number of Q. See Problems 2.137–2.142.

Recurrence relation: See Section 3.3.

Riemann zeta function: See Problem 2.62.

Rising factorial polynomial: The polynomial $[x]^n = x(x+1)(x+2) \cdots (x+n-1)$. See Problem 2.135. The number of ways of putting n distinct objects into m distinct boxes is $[m]^n$ if empty boxes are permitted and the (left to right) order of objects within a box is significant. See Problem 2.137. If empty boxes are not allowed and if m cannot exceed n, the number of ways is $(n!)C(n-1, m-1)$. See Problem 1.138.

Rook polynomial: See Problems 2.80–2.99.

Signless Stirling number: See Stirling number of the first kind.

Sieve formula: See Theorem 2.5.

Sperner's theorems: See antichain.

Stabilizer: If G is a fixed subgroup of the symmetric group of permutations of a finite set X, then the stabilizer of an element x is the set $\{g \in G : g(x) = x\}$ which is a subgroup of G. See Theorem 4.2 and Problem 4.90.

Star diagram: A graphical way of representing a partition of a positive integer using asterisks is a star diagram. See Section 3.2.

Stirling number of the first kind. The coefficient of x^r in the falling factorial polynomial $[x]_n = x(x-1)(x-2)\ldots(x-n+1)$ is $s(n,r)$, the Stirling number of the first kind. See Problem 1.132. The absolute value of $s(n,r)$ is called the Signless Stirling number and is denoted by $s'(n,r)$.
Stirling number of the second kind: The number of ways of partitioning a set of n elements into m (nonempty) subsets is $S(n,m)$, the Stirling number of the second kind (Problem 1.146) which is also equal to the number of ways of putting n distinct objects in m identical (distinguishable) boxes (Problem 1.147).
Symmetric chain: A sequence $\langle A_1, A_2, \ldots, A_k \rangle$ of subsets of a finite set X of cardinality n is a symmetric chain in X if (i) $A_i \subset A_{i+1}$ for $i = 1, 2, \ldots, k-1$, (ii) $|A_{i+1}| = |A_i| + 1$ for each i, and (iii) $|A_1| + |A_k| = n$. A symmetric chain decomposition (SCD) of the class $P(X)$ of all subsets of X is a partition of $P(X)$ into symmetric chains in X. See Problems 2.119–2.124.
Symmetric chain graph: See Problem 2.124.
Symmetric group, S_n: The group of permutations of a set with n elements.
Systems of distinct representatives (SDR): See Hall's Theorem.
Term rank of a binary matrix: See Problem 2.108.
Unimodal sequence: A sequence $\langle a_1, a_2, \ldots, a_n \rangle$ of real numbers is a unimodal sequence if there exists a positive integer j such that $a_1 < a_2 < \cdots < a_{j-1} \leq a_j > a_{j+1} > \cdots > a_n$. See Problem 1.74.
Vandermonde determinant: See Problem 3.71.
Whitney number: See Ranked poset.
Young's Tableau: See Problems 3.85–3.87.
Zeckendorf's Theorem: Every positive integer can be expressed as a sum of distinct pairwise nonconsecutive Fibonacci numbers. See Problems 3.67 and 3.68.
Zyklenzeiger: See Cycle index and Section 4.2.

Symbols Used (Notation)

Further Reading

Anderson, Ian, *Combinatorics of Finite Sets*, Oxford University Press, 1987.

Balakrishnan, V.K., *Network Optimization*, Chapman & Hall, 1995.

Biggs, Norman L., *Discrete Mathematics*, Oxford University Press, 1985.

Brualdi, Richard A. and Ryser, Herbert J., *Combinatorial Matrix Theory*, Cambridge University Press, 1991.

Comtet, Louis, *Advanced Combinatorics*, Reidel, 1974.

Graham, Ronald, Rothschild, Bruce L., and Spencer, Joel L., *Ramsey Theory (Second Edition)*, John Wiley, 1990.

Hall, Jr., Marshall, *Combinatorial Theory (Second Edition)*, Wiley-Interscience, 1986.

Krishnamurthy, V., *Combinatorics: Theory and Applications*, John Wiley, 1986.

van Lint, J.H. and Wilson, R.M., *A Course in Combinatorics*, Cambridge University Press, 1992.

Lovasz, Laszlo, *Combinatorial Problems and Exercises*, North-Holland, 1989.

Riordan, John, *An Introduction to Combinatorial Analysis*, John Wiley, 1958.

Roberts, Fred, *Applied Combinatorics*, Prentice Hall, 1984.

Stanley, Richard P., *Enumerative Combinatorics*, Wadsworth Publishing, 1986.

Tomescu, Ioan, *Problems in Combinatorics and Graph Theory*, John Wiley, 1985.

Trotter, William T., *Combinatorics and Partially Ordered Sets: Dimension Theory*, Johns Hopkins University Press, 1992.

Vilenkin, N., *Combinatorics*, Academic Press, 1971.

Wilf, Herbert, *Generatingfunctionology (Second Edition)*, Academic Press, 1994.

Index